I0064111

Quantum-like Networks

An Approach to Neural
Behavior through their
Mathematics and Logic

Other World Scientific Titles by the Author

Quanta, Logic and Spacetime: Variations on Finkelstein's Quantum Relativity
ISBN: 978-981-02-3255-9

Quanta, Logic and Spacetime
Second Edition
ISBN: 978-981-238-691-5

Quantum-like
Networks

An Approach to Neural
Behavior through their
Mathematics and Logic

| Stephen A Selesnick
University of Missouri-St Louis, USA

World Scientific

EW JERSEY · LONDON · SINGAPORE · BEIJING · SHANGHAI · HONG KONG · TAIPEI · CHENNAI · TOKYO

Published by

World Scientific Publishing Co. Pte. Ltd.

5 Toh Tuck Link, Singapore 596224

USA office: 27 Warren Street, Suite 401-402, Hackensack, NJ 07601

UK office: 57 Shelton Street, Covent Garden, London WC2H 9HE

Library of Congress Cataloging-in-Publication Data

Names: Selesnick, S. A. (Stephen Allan), author.

Title: Quantum-like networks : an approach to neural behavior through their
    mathematics and logic / Stephen A Selesnick.

Description: 1st. | New Jersey : World Scientific, [2023] |
    Includes bibliographical references and index.

Identifiers: LCCN 2022017522 | ISBN 9789811260698 (hardcover) |
    ISBN 9789811260704 (ebook for institutions) | ISBN 9789811260711 (ebook for individuals)

Subjects: LCSH: Neural networks (Neurobiology)--Mathematical models. |
    Quantum systems. | Quantum computing.

Classification: LCC QP363.3 .S35 2023 | DDC 612.8--dc23/eng/20220720

LC record available at https://lccn.loc.gov/2022017522

British Library Cataloguing-in-Publication Data

A catalogue record for this book is available from the British Library.

Copyright © 2023 by World Scientific Publishing Co. Pte. Ltd.

All rights reserved. This book, or parts thereof, may not be reproduced in any form or by any means, electronic or mechanical, including photocopying, recording or any information storage and retrieval system now known or to be invented, without written permission from the publisher.

For photocopying of material in this volume, please pay a copying fee through the Copyright Clearance Center, Inc., 222 Rosewood Drive, Danvers, MA 01923, USA. In this case permission to photocopy is not required from the publisher.

For any available supplementary material, please visit
https://www.worldscientific.com/worldscibooks/10.1142/12975#t=suppl

Desk Editor: Shaun Tan Yi Jie

Typeset by Stallion Press
Email: enquiries@stallionpress.com

NETWORK: Any thing reticulated or decussated, at equal distances, with interstices between the intersections.

Samuel Johnson (1709–1784)
A Dictionary of the English Language (1755)

Pay attention to the chatter we use to convince someone of the truth of a mathematical proposition. It sheds light on the function of these beliefs. I mean the chatter with which the intuition is awakened. By which, that is to say, the machine of a computing technique is set in motion.

Ludwig Wittgenstein (1889–1951)
Remarks on the Foundations of Mathematics
(From III, §27, 1942, translated by the author.)

Even if a minefield or the abyss should lie before me, I will march straight ahead without looking back.

Zhu Rongji (1928–★★★★)

Une simple largeur apparaît comme profondeur pour ceux qui sont allongés.
(Mere width appears as depth to those who are supine.)

Simplicien de Forcenex du Resnell-Trémoille
(1758–1810)
(Translated by Haifa Nouaime.)

For

Robert A. Taylor

doctor mirabilis and savior of networks

and

Joe Z. Tsien

seeker and finder indomitable

Introduction

This work was inspired essentially by the old question of whether or not brains compute. And if they do, what is it they compute and how do they do it? On the way to this goal, whether or not it has been reached — please see Chapter 4 in this connection — many other issues arose.

It is aimed at: mathematicians, physicists, computer scientists, and philosophers with an interest in the topics discussed; mathematicians with an interest in extensions of the range of applicability of mathematical resources apparently not previously so exploited; and neuroscientists and psychologists with strong mathematical leanings.

This having been said, let me assure the potential reader that almost no background in quantum physics, logic, network theory, computation or neuroscience is required for a reading of this work. A slight exposure to the basics of quantum mechanics and logic would be an advantage but, apart from requiring a knowledge of linear algebra, an effort has been made to render the work essentially self-contained. Where it is not, full references to the literature are supplied. Other than a knowledge of linear algebra, all that is required is an appreciation of the power of formalisms such as those to be found here.

We start from first principles, namely a logical take on very complex systems such as those found in living organisms. Our particular approach to the logic of such systems, while not entirely without precedent in its general philosophy, seems in detail to be new. The logic that results is close to the familiar Boolean logic of the classical neural net model of McCulloch and Pitts, differing from it in one respect only, namely in that disjunction does not distribute over conjunction (and, necessarily, vice versa). This has the consequence that disjunction in the logic is not truth functional: the proposition p OR q may be valid without either p or q being valid. Here OR is the disjunction in the logic at hand. It is this property that is characteristic of the logic of the quantum world and the *sole* reason we have used the term *quantum-like* to describe our networks. To this property may be ascribed most of the discomfort experienced by classical thinkers when confronting actual quantum theory. Disjunction manifests in that theory as quantum superposition, and is responsible for all the apparent puzzles associated with Schrödinger's cat, the entanglement of quantum states, etc. In quantum theory proper, these apparent conundrums followed from Heisenberg's formulation of "matrix mechanics" in 1925 and led to a revolutionary new ontology that is still provoking head scratching among sensitive physicists. In stark contrast to actual quantum theory, the quantum-like properties of our model do not require any new ontologies: rather they manifest in mundane ways, such as parallelism of computation, and constraints upon probable outcomes. (It is the aspect of extreme parallelism that lies behind the quest for a quantum computer.)

Our models follow the tradition of the original McCulloch and Pitts neural networks in that they are a mix of physical attributes (the information handling attributes of the nodes) and the logic associated with the global structure (Boolean, in the classical case). In our case, both the physics-like attributes and the structural logic are quantum-like. The physics-like attributes similarly invoke the internal structure of the ensemble of nodes while the quantum-like logic invokes the outer structural attributes. The physical analogue turns out to be a finite (but large) collection of fermion-like quanta, while the outer logic is a form of computational logic named for Gentzen. These two approaches to the same object(s) intermingle in interesting ways. For example, it turns out that our model is able to simulate the operation of *non-synaptic* signaling or connection modes, such as those implemented in actual brains via neurotransmitters and other substrate media, albeit in a rather general way. Such modes of substrate connection in actual brains are of course crucial to their function.

The table of contents displays the nature and course of the work, which is linear, with no footnotes, though there are digressions here and there. Each chapter after the first depends to a greater or lesser extent on results in the preceding chapter. The first chapter is rather more technical than a foundational chapter usually is. As explained there, it is necessary to rigorously rehearse the logical background, though the details may not be necessary for an understanding of the arguments to follow. For this reason most of the proofs and definitions required are relegated to an appendix explicitly to Chapter 1 (namely Appendix A). A reader not interested in these particular details may trip lightly over them: a guide for such a reader is supplied in the introduction to this first chapter.

The first three chapters develop the network model itself, while the second three develop applications of it.

Most of the chapters stay close to the model. An exception is the last chapter, Chapter 6. This entails an excursion somewhat beyond the strict confines of the model. It deals with a conjectured general syntax of retrieval that may underlie the syntax of human language as a special case. This is applied to the question of why we humans are the only primates to have innate recursive language, and to the uncovering of a language deficit that has been found to characterize the speech of schizophrenic patients. As with most journeys over thin ice — particularly well-trodden thin ice — I have thought it prudent to keep this one short.

There is a second appendix (Appendix B) which rehearses some topics from multilinear algebra which are fundamental to the models.

A disclaimer: what this work is not about. It is not about Artificial Intelligence, Deep Learning, nor any of the traditional applications of artificial neural nets. Although it may be hoped that there could be practical applications of these results in time, this aim was not the original impetus behind the work.

Acknowledgments

My thanks, above all, to Nadine Castro, whose depth of knowledge and insight first primed my interest in the logic of brains.

I will be forever unable to fully discharge a debt of gratitude to Robert A. Taylor MD, and the Santa Barbara Cottage Hospital staff, whose extraordinary skill and quick action saved my own neural networks from a fate not modeled in this work, and without which nothing would have been modeled in it.

A truly profound debt of gratitude is owed also to Joe Z. Tsien, whose penetrating insight, beautiful experiments and patient tutelage of a tyro was — and remains — a major inspiration.

An enormous debt of gratitude is owed to Felicitas Ehlen for a very long and very illuminating correspondence, which was directly responsible for Chapter 6, though any errors are mine alone.

I am tremendously grateful to many others whose help was instrumental. These include my collaborators on earlier work on some of the material treated here, namely: Gareth Owen, Gualtiero Piccinini and John Piers Rawling.

To legions of anonymous referees I extend my warmest thanks and apologies.

My no less grateful thanks for much additional help and insights, in the form of generous correspondence or word of mouth, to: C. Anthony Anderson, Wallace Arthur, Sonya Bahar, Jasper Brener, Peter Bruza, Nadine Castro, Felicitas Ehlen, Ortwin Fromm, James Hartle, Miriam Munson, Ronald Munson, Emmanuel Pothos, and Robert A. Taylor.

To Ivan Selesnick for manifold forms of help over a period of years — editorial, mathematical, biomedical and typographical — my grateful thanks. He is also responsible for Figure 5.4. My thanks to the multitalented polyglot Haifa Nouaime for her meticulous translations from the French of the Proust passage in Chapter 4, and elsewhere, and lessons in Arabic for Chapter 6.

The typesetting would have been impossible without the use of the splendid LaTeX editor Texpad. My thanks to the Texpad team: Jawad Deo and Duncan Steele.

With the exceptions of Figures 5.2, 5.3 and 5.4 the diagrams were made with the superb general drawing program OmniGraffle. My thanks to the Omni Group for this remarkable program.

To say that the literature on almost every neuroscientific, logical, linguistic and biolinguistic topic touched upon here is oceanic would be a vast understatement: galactic would be a more appropriate descriptor. Consequently, I must beg the forgiveness of those multitudes whose work should have been cited but has been overlooked.

My multimodal thanks to Shaun Tan Yi Jie of World Scientific for his apparently limitless patience and generous help.

S. A. S.
Santa Barbara, California, USA
March 30, 2022

About the Author

Dr. Stephen A. Selesnick is Professor Emeritus at the Department of Mathematics, University of Missouri–St. Louis, USA. He holds a PhD in Mathematics from the University of London, UK, and has over 50 years of experience in teaching and research. His area of interest is in cross-discipline interactions both within mathematics and outside of it. Thus his early work was on applying algebraic and topological methods in analysis, quantum logic, lattice theory, etc., and later to such things as surface definitions in CAD/CAM. Subsequently he became interested in physics, and worked in quantum field theory and applications to elementary particles. An attempt to build on the work of the late David Finkelstein led to a monograph in 2 editions, *Quanta, Logic and Spacetime*, purporting to derive the Lagrangians of fundamental physics from first principles. Some of this led to work on quantum computing and the logical provenance of the Lagrangians of the fundamental forces and gravity. More recently he has become interested in the mathematics of neural networks, particularly their mysterious quantum-like aspects.

Contents

Notes to the Reader

Semantic Overloads

In cross-disciplinary studies such as this one, there often arises the phenomenon of semantic overload: the same or similar words being used with different technical meanings in different contexts. Our subject is particularly prone to this source of confusion. To nip it in the bud, we shall draw attention to these possibilities sooner rather than later. To wit:

- **Projection**
  This term is used in mathematics to describe certain operators upon linear spaces, namely the ones that project every vector upon a subspace of the space. In neuroscience a projection generally refers to an axon or bundle of axons that extends from a neuron or ensemble of neurons, to a neuron or ensemble of neurons at some distance from the upstream neuron or neurons.
- **Convergence**
  The mathematical notion is assumed to be familiar to the reader. The neurological term generally applies to a number of axons emanating from different neurons meeting at, or converging upon, a common neuron.
- **Proof**
  In addition to its usual usage, this term is also applied in the area of proof theory as, epitomized by a Gentzen sequent calculus, another word for deduction.

- **Category**
  There is a strict mathematical meaning to this term and a very much less strict linguistic meaning to this term. We shall use both, but the differences will be clear from the contexts.

- **Clique**
  In graph theory a *clique* is formally defined as a graph in which every pair of distinct nodes are adjacent, i.e. are at the ends of a single edge. In Chapter 5 its usage is informal.

- **Motif**
  There is a formal graph theoretic definition which we shall not need in this work. In Chapter 5 its use is informal.

- **State**
  The word "state" has many connotations in the case of both classical and quantum-like systems. In classical systems states are usually unambiguously sets of points or elements in some ambient parameter space, such as the phase spaces of mechanics. In the case of actual quantum theory the word "state" has an ostensibly different connotation. A very general view of them is that they are certain functions defined on certain lattices, or algebras of operators: then deep theorems, such as that of Gleason, realize a correspondence between certain of these states — the so-called "pure" ones — and one-dimensional subspaces, or rays, in certain Hilbert spaces. In Chapter 1 we shall pursue an entirely logical approach to the issue of quantum-like behavior. The central class of objects that arises is the class of *ortholattices*. Thanks to the work of R. Goldblatt, and others, this apparently denatured logical theory has sufficient structure to provide us with built-in versions of "states" as elements of sets, *and* precise analogues of the "pure states", or one-dimensional subspaces ("rays"), of actual quantum theory. More generally the logical structure deals with "propositions" which are the precise analogues of the closed subspaces of Hilbert spaces, which have the status of propositions in *echt* quantum logic.

  Further confusion is threatened when we come to our main application, which is to "parameter windows" of complex systems: namely, subsets of Euclidean spaces, which consist of actual vectors. We promise to be vigilant in sorting out these issues as they arise.

At the outset, namely section 1.1, our use of the term "state" (of a system) shall be in a specific colloquial sense as a set of real parameters arranged in a list-like form, or vector. The linear algebraic structure will then follow from our logical considerations. The possible branching into quantum-like or classical behaviors will follow.

Later we shall understand the term "state" to refer to the one-dimensional subspace, or ray, determined by a vector. As noted, there is a one-to-one correspondence between such rays and "pure states", namely the unit vectors generating the respective rays.

In cases where confusion is risked we will be careful to make distinctions.

Some often used notation and terminology

- End of proof: □
- There exists: ∃
- For all: ∀
- Equal by definition: :=
- Transpose of the vector or matrix argument: In general $(\ )^T$, but in the case of our creation operators on a real exterior algebra, for reasons of tradition: $(\ )^\dagger$. Thus $a_i^\dagger := a_i^T$.
- Set complementation: $(\ )^c$
- Proportional to: $\propto$
- Natural log of ( ): $\ln(\ )$
- Set difference: $X \backslash Y := X \cap Y^c$
- Set element inclusions: we generally write for instance $x, y, z, \ldots \in S$ for individual element memberships $x \in S$, $y \in S$, $z \in S, \ldots$
- Binomial coefficient: $\binom{n}{k} := \dfrac{n!}{k!(n-k)!}$
- The field of real numbers: $\mathbb{R}$
- The field of complex numbers: $\mathbb{C}$
- If and only if: iff
- Injective (as in *injective* function or map): one-to-one

- Surjective (as in *surjective* function or map): onto
- Bijective: injective and surjective
- Kronecker delta: $\delta_{ij} = \begin{cases} 1 & \text{if } i = j \\ 0 & \text{if } i \neq j \end{cases}$
- For vector spaces V and W, $\text{Hom}(V, W)$ will denote the vector space of linear maps from V into W.
- For a vector space V over the field k, V^* will denote the dual space of V, namely $\text{Hom}(V, k)$.

Part I

Logic and networks

Chapter 1

Logical Foundations

There are few entries in the lexicon more abused than the word "quantum." It is one of the aims of this foundational chapter to precisely elucidate why and how we came to use this term in the context of a study of very complex systems such as those found in living organisms. This we attempt here by rigorously rehearsing the logic that arises when we try to cope with complexity by blocking or ignoring the differences between states of such systems which would produce negligibly discernible differences of effect. Roughly speaking, we regard such states as being *confusable* if their effects are indistinguishable. (We call this *hiding the variables*, which may be a misnomer since it is the differences in the variables that we are hiding.) We explain in this chapter how this leads to a quantum-like logic, quite independently of any considerations of quantum physics itself, except for the purposes of *post facto* comparisons and interpretations. This ultimately reduces to a consideration of subsets of Euclidean spaces from this point of view which will form the basis of the neuronal network model introduced in subsequent chapters.

For a purported introductory chapter this first chapter might appear to be technically daunting, particularly to readers not interested in the technicalities of the logic involved. For such readers a guide will be provided in the last paragraph of this section. The technical formalities are required, as mentioned, to justify the unfortunate necessity for the use of the often faddish, abused and possibly off-putting term *quantum* in *quantum-like*. Most of the formalities in this chapter may otherwise be ignored when reading the later chapters.

The layout of this chapter is as follows. In section 1.1 we informally introduce the idea of a system with confusable states, such as those that arise when considering the case of a biological cell, and its logic. This is treated a little more formally in section 1.2.1. In section 1.2.2 we develop the logic formally with most detailed proofs and definitions relegated to the chapter's appendix, Appendix A. The main result in section 1.2.2 is the Modal Embedding Theorem 1.2.4 which is used in section 1.3 to analyze the two archetype quantum paradigms, namely Schrödinger's unfortunate cat and the double slit experiment, entirely by means of the known B-modal system of logic. This analysis, together with a theorem characterizing the Booleanness of ortholattices, namely Theorem A.1.2, is used in section 1.4 to establish in some generality distinctions between quantum-like and non-quantum-like behaviors in our kind of system. The final section 1.4.2 exhibits and discusses various examples of the subsets of Euclidean spaces we have dubbed *parameter windows*: namely, the subsets constituting the supposed ranges of the systems' parameter values. Our further applications will be limited to the last such example, Example 5.

An attempt has been made to supply as much logical detail and background as possible where necessary in this chapter. However, only the last example is actually needed for what follows. Here is a possible tour through the chapter for readers not interested in the logical details: sections 1.1, 1.2.1, as much as possible of section 1.3, which is particularly important for our applications to follow, section 1.4, and a glance at the examples in section 1.4.2, particularly Example 5: we are interested in parameter windows with non-Boolean proposition lattices, typified by that example.

1.1   Irredeemable complexity: Biological systems

As noted, our model is based upon a possibly novel approach to dealing with extreme complexity. By way of example let us consider first the case of a single biological cell, prokaryotic or eukaryotic. It consists, roughly speaking, of a bi-lipid membrane, studded with receptors made of complex proteins, enveloping a complex fluid containing more proteins and other complex molecules such as RNA, DNA, organelles, cytostructures, etc. All of these are in a state of continuous and complex motion, involving diffusion in the fluid itself, along with the concomitant meeting and interaction of proteins and enzymes entailing the folding and unfolding of the proteins, their capture and release of phosphate groups, etc., which constitute the allosteric computations of unicellular life.

The specification of a *state* of such a cell at some point in time would entail some sort of numerical encoding into strings or vectors of real numbers of the variables parametrizing such things as chemical concentrations, temperature and pressure gradients, membrane voltages, the geometry and topology of the ever-changing protein complexes, the dynamics of the fluid medium, and so on. It would require a vector of real numbers with millions if not billions of entries. It is reasonable to suppose that there would be essentially little discernible or practical difference, when viewed from the outside, between one such state and another in which a relatively few of these microscopic variables are different. Thus, two such vectors could be far apart in Euclidean space, but negligibly different, or *confusable* insofar as the cell's behavior is concerned. For instance, vectors with billions of entries could have hundreds or thousands of differences and still be confusable. (Such tolerance of relatively small changes in its states would be a selective advantage to an organism). We shall regard this sort of confusability of states to be tantamount to the *hiding* of the variables involved, since their small variations are effectively suppressed.

The same idea applies *mutatis mutandis* to large conglomerations of cells such as those found in brains. The human brain is estimated to contain on average 10^{11} neurons. Later, in Chapter 2 we shall start the construction of our neuromorphic model at this level, thinking of the principal neurons as being the computational units and the confusable states as being those of a large collection of such neuron-like elements.

In this foundational chapter we shall pursue the logic of such systems in some detail. Systems of this type, some of whose internal states are not easily distinguished one from another by quiescent external observers as above, obey a variant of ordinary Boolean logic, called *orthologic* (**OL**). This logic has quantum-like attributes: actual quantum theory is surely not a theory of hidden variables, but a theory of hidden variables may be quantum-like and ours is one of those. Specifically, **OL** differs from ordinary logic in one respect only: namely the distributive law does not necessarily obtain. As a consequence the disjunct (OR, written $\sqcup$ in this logic) is not truth functional. That is to say, the proposition $p \sqcup q$ could be true without either p or q being true. This and some of the properties that follow from it are the essence of the quantum-like behavior such systems may exhibit and are the source of much of the disquiet experienced by classical thinkers when confronted by actual quantum theory. In the models of the logic of interest to us, this non-classical disjunct is realized as the *superposition* of states. Essentially, this process of hiding variables has traded their irredeemable complexity for the process of superposing states. Despite the fact that superpositions entail a continuum of possible states between those superposed, we contend that this trade is more than fair. For, although the quantum-like structure that emerges is not actual quantum mechanics, we have had enough experience of actual quantum mechanical paradigms to be able to extract meaningful information from the new setup.

1.2   The doctrine of hidden variables

The idea of *hidden variables* underlying actual quantum physics is anathema to most physicists. Subtle experiments thwart the idea that there are unknown but basically objective mechanisms underlying actual quantum physics. However, as we shall argue in this chapter and its sequels, for complex systems hiding the actual variables seems to be a good idea, and leads immediately to a logic and phenomenology very like actual quantum theory.

1.2.1 *Proximity and orthogonality spaces, ortholattices and the emergence of ortholgic*

Our interest is in systems whose sets of states may be characterized as *proximity spaces*. A proximity space $\langle W, \approx \rangle$ is a set W with a relation $\approx$ on it having the properties: reflexivity (for $w \in W$, $w \approx w$) and symmetry (for $w, v \in W$, $w \approx v$ iff $v \approx w$). Such a relation, called a *proximity*, is not generally transitive. Note that identity, $=$, is a proximity on any set. A proximity relation has the informal intuitive reading: $v \approx w$ iff the state v is *confusable* with the state w in the absence, say, of a direct observation or measurement. In other words, in the absence of any interference (such as a measurement), if such a system can be surmised to be in the state v, it may as well, for all intents and purposes, be surmised to be in the state w, and conversely. (For other similar interpretations and examples, see [11].)

This confusability of elements of W will now affect the practical operational matter of assembling subsets of W. If the proximity were identity, $=$, then a subset E is trivially assembled from its elements as

$$E = \bigcup_{v \in E} \{v\} \tag{1.2.1}$$

$$= \bigcup_{v \in E} \{w \in W : w = v\} \tag{1.2.2}$$

$$= \{w \in W : \exists\, v \in E \text{ such that } w = v\}. \tag{1.2.3}$$

In the case of a general proximity, $\approx$, this must be replaced by

$$\{w \in W : \exists\, v \in E \text{ such that } w \approx v\} =: \Diamond E. \tag{1.2.4}$$

This is the set constructible out of the elements of E to within the limits of confusability. Within the operational dictates of confusability, this is the closest to the set E one can get by assembling its elements. (Note that $E \subseteq \Diamond E$.) So the rule for reconstructing a subset E to within the limits of confusability is to place the modal "possibility" operator $\Diamond$ in front of it: $\Diamond E$. In the modal context these are the elements that are "possibly" in E. Cf. the discussion of the B-modal system in section 1.2.2.

Since, in the case of the identity proximity (and denoting set complementation by the superscript c) we have

$$E^c = \{w \in W : \exists v \in E \text{ such that } w = v\}^c, \qquad (1.2.5)$$

this becomes the subset $(\Diamond E)^c$. So, in the general case, and within the limits of confusability, we obtain, as the proper reconstruction of $(\ )^c$ according to the rule above, which entails placing the $\Diamond$ operator in front of the set to be constructed "up to confusability," the subset

$$(\Diamond E)' := \Diamond(\Diamond E)^c \qquad (1.2.6)$$

as the proper generalization of the Boolean complementation operation. For a proximity space $\langle W, \approx \rangle$ the sets of the form $\Diamond E$, for $E \subseteq W$, were shown by J. L. Bell [10, 11] to constitute a complete ortholattice, with join being the ordinary set union, the concomitant meet of two elements $\Diamond E$ and $\Diamond F$ being the largest subset of the form $\Diamond(\ )$ contained within $\Diamond E \cap \Diamond F$, and the complement given by $(\ )'$. An *ortholattice* is essentially a lattice with all the properties of a Boolean algebra except the distributive law: cf. Appendix A, section A.1.3 for definitions and first properties. Bell calls this lattice the lattice of "parts" of $\langle W, \approx \rangle$ and we shall follow him in denoting it by Part W since the proximity relation will never be ambiguous. (See Definition A.1.5 for more on the meet in this lattice.)

In the case in which the proximity is just $=$, then the lattice of parts is just the Boolean algebra of all subsets of W. The "associated" logic is what is known as the *Propositional Calculus* (**PC**) and the association is that Boolean algebras provide *models* for the logic, in the logical sense explained in section 1.2.2. There is a similar logic, mentioned above, that is modeled by the class of ortholattices: it is called *orthologic* (**OL**). It is a logic strictly weaker than standard **PC**, meaning that everything and more that can be proved in **OL** can be proved in **PC**. The difference is that **OL** does not require the presence of the distributive law.

The upshot is that the proper logic of propositions that deal "operationally" with the systems we are interested in, which generalizes ordinary **PC** in the Boolean case, is **OL**. Consequently we shall adopt **OL** as our overarching logic when dealing with the systems of interest to us here.

We shall therefore need first to rehearse the relevant material on this logic and its model theory. The logical machinations of this chapter are in service to the justification of the choice later of the extended spaces of states which form the basis of our network models. Some more logic, though of a different kind, will appear in Chapter 3 and its sequels. This "external" computational logic is simpler and our purely logical applications of it will be rather rudimentary. (The non-logical applications will be less rudimentary, though all mathematical background will be supplied.) A computable version of the logic **OL** is embeddable into this latter logic: a proof of this will given in the digressionary last section, section 3.4, of Chapter 3, which may be skipped.

The following diagram charts the logical development described in this chapter.

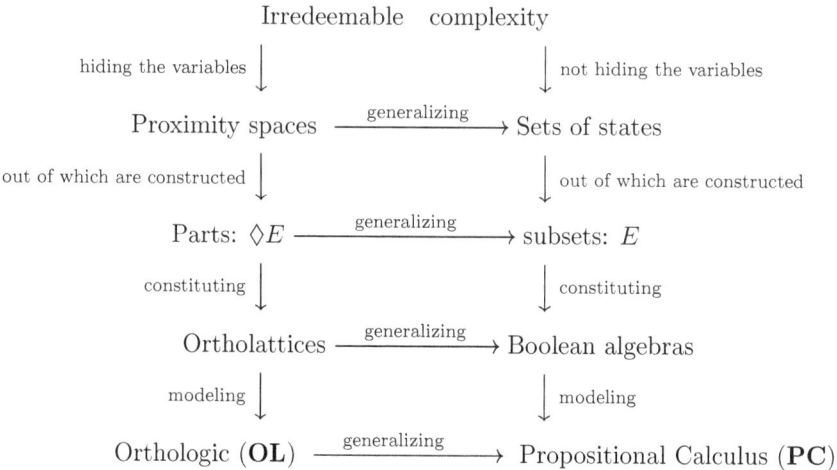

$$\text{Irredeemable} \quad \text{complexity}$$

hiding the variables $\downarrow$ $\downarrow$ not hiding the variables

Proximity spaces $\xrightarrow{\text{generalizing}}$ Sets of states

out of which are constructed $\downarrow$ $\downarrow$ out of which are constructed

Parts: $\Diamond E \xrightarrow{\text{generalizing}}$ subsets: E

constituting $\downarrow$ $\downarrow$ constituting

Ortholattices $\xrightarrow{\text{generalizing}}$ Boolean algebras

modeling $\downarrow$ $\downarrow$ modeling

Orthologic (**OL**) $\xrightarrow{\text{generalizing}}$ Propositional Calculus (**PC**)

The logic involved in the classical neural networks, such as those of MacCulloch and Pitts and their many descendants, and indeed several precursors, is the one described by the rightmost column in the diagram above. Our approach follows the leftmost column. It should be noted that we formulate no hypotheses concerning the origin or root cause of the "confusability" giving rise to our adoption of a proximity other than identity. Some or all of it could have arisen from actual quantum effects, such as those postulated to arise from the actual physics and chemistry of such things as microtubules or certain molecules, and other more general field-like effects: all of these would be covered by our orthological considerations. Similarly, the more recent neuromorphic models, involving complex new solid state materials and memristor-like circuits, would also be covered by these considerations.

1.2.2   *Orthologic and some of its models*

As mentioned above, **OL**, which is in fact the core logic underlying *echt* quantum logic, is a weakening of ordinary **PC**: one assumption is dropped, namely that conjunction distributes over disjunction, or equivalently, vice versa. It is perhaps surprising that the simple extension of Boolean logic to the non-distributive case should lead to a plethora of models, at least two of which lend themselves to quantum-like semantics. This writer knows of at least four such models, three of which we shall touch on here. In the Boolean case, they all collapse into one, namely the models provided by the category of Boolean algebras. The model we shall be aiming at involves modal logic, and we shall sketch this development here, relegating mathematical and logical details, where possible, to Appendix A.

In 1974 Goldblatt [41] introduced a deductive system for **OL** and proved completeness theorems for it in terms of certain models. These theorems are of the following type. A logical system may be specified by adopting a set of axioms and then deriving correct formulas, that is, "theorems," from these via sequences using specified rules of inference. If "models" exist for such a system then the validity of a formula may be defined in terms of conditions satisfied by the formula vis-à-vis the model. Then a *completeness* theorem would establish the equivalence of *theoremhood via sequences of inferences* (or *provability*) and *validity* in terms of the model. We shall here adopt the policy that the former type of assertion shall be symbolized by a simple turnstile ($\vdash$), usually bearing a subscript, while the model-based validity assertion shall be symbolized by a more elaborate turnstile ($\Vdash$, $\models$).

For example, ordinary **PC** may be characterized by morphisms of formulas into Boolean algebras: the completeness theorem in this case asserts that a formula is provable in **PC** if and only if its image under any morphism into any Boolean algebra is the top element. There is a similar completeness theorem for intuitionistic logic (**IL**) with Boolean algebras being replaced by Heyting lattices. Goldblatt proved a similar completeness theorem for **OL**, the target lattices in this case being ortholattices. (Since Boolean algebras are ortholattices it follows immediately from the respective completeness theorems that any orthotheorem is also a theorem of **PC**, but clearly not conversely: **OL** is strictly weaker than **PC**). As noted, in the **OL** case the model theory bifurcates in the sense that there is another kind of model that also characterizes **OL**, namely a Kripkean one. From the existence of such models, one finds a different sort of semantics arising solely from the peculiarities of disjunction, and it is this semantics that mimics quantum behavior. This is because in the case of the slightly stronger quantum logic itself, disjunction is exactly "quantum" superposition, the nexus of most if not all of the puzzlements classical thinkers experience when confronted with quantum theory. The existence of these Kripkean models of **OL** led Goldblatt to realize that **OL** itself may be embedded into a well-studied modal system, namely the *B-modal system* of Becker [9], B for "Brouwersche," which well predates the advent of quantum logic in 1936 [13]. The same result was obtained almost simultaneously but independently by Dishkant [28]: see also [24]. The associated Kripke models for this B-system also provide a semantics for probing the anomalies of disjunction and it is these we shall focus on, since this system is well known, and reveals the quantum-like behavior clearly and simply. In this section we shall give a simplified sketch of the path to the Modal Embedding Theorem for **OL**, leaving the details to Appendix A. (Cf. also [76, 81, 82].)

Goldblatt's realization of **OL** as a deductive system may be described as follows. The atoms, or primitive symbols, are

 (i) a denumerable collection Φ_0 of propositional variables $a_1, a_2, \ldots$;
 (ii) the connectives $\sim$ ("negation") and $\sqcap$ ("conjunction"); and
(iii) parentheses.

The set Φ of well-formed *orthoformulas* is constructed from these in the usual manner. Elements of Φ will be denoted by lowercase Greek characters α, β, ..., usually taken from the beginning of the alphabet. We shall generally try to reserve characters at the end of the Latin alphabet for elements of sets of various kinds.

Since there is no implication sign in Φ, a formal deductive calculus is based on *sequents* involving at most single formulas and written in the form

$$\alpha \vdash \beta \qquad\qquad (1.2.7)$$

for $\alpha, \beta \in \Phi$, the reading of which is that β may be inferred from α. Certain sequents are designated as *axioms*, and there are three *rules of inference,* namely, for any formula α, β:

AXIOMS

O1. $\alpha \vdash \alpha$
O2. $\alpha \sqcap \beta \vdash \alpha$
O3. $\alpha \sqcap \beta \vdash \beta$
O4. $\alpha \vdash \sim\sim\alpha$
O5. $\sim\sim\alpha \vdash \alpha$
O6. $\alpha \sqcap \sim\alpha \vdash \beta$

INFERENCE RULES

O7. $\dfrac{\alpha \vdash \beta \quad \beta \vdash \gamma}{\alpha \vdash \gamma}$

O8. $\dfrac{\alpha \vdash \beta \quad \alpha \vdash \gamma}{\alpha \vdash \beta \sqcap \gamma}$

O9. $\dfrac{\alpha \vdash \beta}{\sim\beta \vdash \sim\alpha}$

A disjunctive connective may be introduced through the following definition

$$\alpha \sqcup \beta := \sim ((\sim \alpha) \sqcap (\sim \beta)) \tag{1.2.8}$$

and dual forms of O2, O3, O6 and O8 follow.

A string $s_1; s_2; \ldots; s_n$ of sequents is called a *proof* of its last member s_n if each s_i is either an axiom or follows from some preceding sequent through the use of one of the rules of inference. If there exists a proof of a sequent $\alpha \vdash \beta$ then we write

$$\alpha \vdash_{\mathbf{O}} \beta \tag{1.2.9}$$

and say that β is *deducible from* α *in* **OL**. If $\alpha \vdash_{\mathbf{O}} \beta$ for *any* formula α we say that β is a *theorem of* **OL** or an *orthotheorem*, and we write

$$\vdash_{\mathbf{O}} \beta. \tag{1.2.10}$$

(Note that this is equivalent to $\alpha \sqcup \sim \alpha \vdash_{\mathbf{O}} \beta$.)

ALGEBRAIC ORTHOMODELS

The first characterization of **OL** is in terms of *ortholattices*. As mentioned, these are lattices $\langle L, \sqcup, \sqcap, 0_L, 1_L, (\ )' \rangle$ satisfying all the axioms for a Boolean algebra except the axiom purporting distribution of meet over join, or equivalently join over meet. Please see section A.1.3 for details. Given an ortholattice L a function $v_L : \Phi_0 \to L$ determines a *valuation* upon Φ via the recursive definitions

$$v_L(\alpha \sqcap \beta) = v_L(\alpha) \sqcap v_L(\beta) \tag{1.2.11}$$

$$v_L(\sim \alpha) = v_L(\alpha)'. \tag{1.2.12}$$

The algebraic characterization theorem for **OL** may then be stated as follows [41].

Theorem 1.2.1. $\gamma \vdash_{\mathbf{O}} \alpha$ *iff* $v_L(\gamma) \sqsubseteq v_L(\alpha)$ *for all ortholattices* L *and all valuations* v_L.

Then, because $\gamma \sqcup \sim \gamma \vdash_{\mathbf{O}} \alpha$ is equivalent to $\vdash_{\mathbf{O}} \alpha$, we have the following.

Corollary 1.2.1. $\vdash_O \alpha$ *iff* $v_L(\alpha) = 1_L$ *for all ortholattices L and all valuations v_L.*

Definition 1.2.1. A pair $\mathscr{A} := \langle L, v_L \rangle$ is called an *algebraic orthomodel* for **OL**, and a formula α deemed *valid in $\mathscr{A}$* if $v_L(\alpha) = 1_L$.

Then the last corollary may be restated as:

$$\vdash_O \alpha \text{ iff } \alpha \text{ is valid in every algebraic model.}$$

Since Boolean algebras are ortholattices, the following corollary follows immediately from the completeness theorem for **PC**.

Corollary 1.2.2. *If $\vdash_O \alpha$, then α is a theorem of* **PC**.

The converse is of course false.

KRIPKEAN ORTHOMODELS

These models were first posited by Goldblatt [41] and have been elaborated upon by Dalla Chiara [24] and others. The model in this case comprises a proximity space $\langle W, \approx \rangle$ and a function, called a *valuation* $\varrho : \Phi \to R(W)$ satisfying

$$\varrho(\alpha \sqcap \beta) = \varrho(\alpha) \cap \varrho(\beta) \tag{1.2.13}$$

$$\varrho(\sim \alpha) = \varrho(\alpha)^\perp. \tag{1.2.14}$$

Here, $R(W)$ denotes the complete ortholattice of propositions in $\langle W, \approx \rangle$ as described in section A.1.2, *et seq.*

Definition 1.2.2. A *Kripke orthomodel* $\mathscr{M} := \langle W, \approx, \varrho \rangle$ is a proximity space $\langle W, \approx \rangle$, with a valuation $\varrho : \Phi \to R(W)$.

We will say that a formula α is:

true at the "world" $w \in W$, and write $w \models_{\mathscr{M}} \alpha$ iff $w \in \varrho(\alpha)$;

true on the set E, and write $E \models_{\mathscr{M}} \alpha$ iff $w \models_{\mathscr{M}} \alpha$ for all $w \in E$: that is, iff $E \subseteq \varrho(\alpha)$;

true in the Kripke orthomodel $\mathscr{M}$ iff it is true at every world in W, that is iff $W \models_{\mathscr{M}} \alpha$;

Kripke valid, and write $\models \alpha$ iff it is true in *all* Kripke orthomodels.

Then we have the following simplified version of another completeness theorem for **OL**, also due to Goldblatt [41].

Theorem 1.2.2. $\vdash_{\mathbf{O}} \alpha$ *iff* $\models \alpha$.

A proof is given in section A.1.5.

THE B-MODAL SYSTEM

As noted, the main result we shall utilize in this chapter is the Modal Embedding Theorem [24, 28, 41]. The proof of this theorem of course entails first a statement of the completeness theorem for the B-system.

Here is a brief account of this modal system, namely the "Brouwersche" or B-system of Becker (cf. for example [19, 47, 48]). To describe it we introduce the ordinary Boolean connectives ¬ (negation), ∧ (conjunction), and the modal operator □ (necessity). Material implication → and the possibility operator ◊ are introduced through the usual definitions (for example $\Diamond := \neg \Box \neg$, or equivalently $\Box := \neg \Diamond \neg$). The axioms and inference rules include the usual ones for **PC** with *modus ponens*, and the modal additions (for formulas α and β):

$$\Box(\alpha \to \beta) \to (\Box\alpha \to \Box\beta) \tag{1.2.15}$$

$$\Box\alpha \to \alpha \tag{1.2.16}$$

$$\alpha \to \Box\Diamond\alpha \quad \text{(the "Brouwersche" axiom)} \tag{1.2.17}$$

If α is deducible then $\Box\alpha$ is deducible. (Necessitation) (1.2.18)

We denote the set of modal formulas by Φ_M. If a formula α can be deduced from a formula β by a chain of inferences from the axioms, we write $\beta \vdash_{\mathbf{B}} \alpha$. The theoremhood of a formula α is defined as usual: namely, α is a theorem of the B-system if $\beta \vdash_{\mathbf{B}} \alpha$ for all β, in which case we write

$$\vdash_{\mathbf{B}} \alpha. \tag{1.2.19}$$

(The origin of the odd nomenclature in equation (1.2.17) is that $\Box\Diamond := \Box \neg \Box \neg$ is like a strong form of double negation and the rule in equation (1.2.17) is then similar to the rule $p \to \neg\neg p$ in intuitionistic logic, the form of logic favored by L. E. J. Brouwer, whose converse is invalid in that system.)

Models for this system are described as follows, where $\mathbf{2}$ denotes the smallest Boolean algebra, which may also be written $\{\mathbf{0}, \mathbf{1}\}$.

Definition 1.2.3. A *Kripke B-model* is a triple $\mathscr{B} = \langle W, \approx, v \rangle$ where W is a set of worlds, $\approx$ is a proximity on W, and v is a function $v \colon \Phi_M \times W \to \mathbf{2}$ satisfying:

V1. For each $w \in W$, $v(\ , w) \colon \Phi_M \to \mathbf{2}$ is a Boolean valuation with respect to $\neg$ and $\wedge$. That is:
$v(\neg\alpha, w) = \neg\, v(\alpha, w)$ where $\neg$ on the right denotes complementation in $\mathbf{2}$ and
$v(\alpha \wedge \beta, w) = v(\alpha, w) \wedge v(\beta, w)$ where $\wedge$ on the right denotes the meet in $\mathbf{2}$;

V2. For any modal formula α, $v(\Box\alpha, w) = 1$ iff $v(\alpha, x) = 1$ $\forall x$ such that $x \approx w$.
It follows that:

V3. $v(\alpha \vee \beta, w) = v(\alpha, w) \vee v(\beta, w)$ where $\vee$ on the right denotes the join in $\mathbf{2}$, and

V4. For any modal formula α, $v(\Diamond\alpha, w) = 1$ iff $\exists\, x \approx w$ (i.e. $\exists\, x$ with $x \approx w$) such that $v(\alpha, x) = 1$.
A modal formula α is said to be:
true at the world w in the B-model $\mathscr{B}$, written $w \Vdash_{\mathscr{B}} \alpha$, iff $v(\alpha, w) = 1$;
true on the set $E \subseteq W$, written $E \Vdash_{\mathscr{B}} \alpha$, iff $w \Vdash_{\mathscr{B}} \alpha$ for all $w \in E$;
true in the B-model $\mathscr{B}$ iff $W \Vdash_{\mathscr{B}} \alpha$;
B-valid, written $\Vdash \alpha$, iff it is true in *all* B-models.

It is worth stating explicitly that, from V2 and V4 and the other definitions, we have:

$$w \Vdash_{\mathscr{B}} \Box\alpha \quad \text{iff} \quad \forall x \approx w,\ x \Vdash_{\mathscr{B}} \alpha \qquad (1.2.20)$$

$$w \Vdash_{\mathscr{B}} \Diamond\alpha \quad \text{iff} \quad \exists x \approx w,\ x \Vdash_{\mathscr{B}} \alpha. \qquad (1.2.21)$$

These models characterize the B-system, by the following completeness result:

Theorem 1.2.3. $\vdash_\mathbf{B} \alpha$ *iff* $\Vdash \alpha$.

This is proved in [19, 47, 48].

THE MODAL EMBEDDING THEOREM

As noted, a translation of **OL** into the B-system was found by Goldblatt and independently by Dishkant. The translation recursively assigns to each orthoformula $\alpha \in \Phi$ a modal formula $\alpha^\circ \in \Phi_M$ as follows:

T1. For atomic formulas a_i: $a_i^\circ = \Box \Diamond a_i$
T2. $(\alpha \sqcap \beta)^\circ = \alpha^\circ \wedge \beta^\circ$
T3. $(\sim \alpha)^\circ = \Box \neg \alpha^\circ$

Then we have the following Modal Embedding Theorem.

Theorem 1.2.4. *For* $\alpha \in \Phi \vdash_\mathbf{O} \alpha$ *iff* $\vdash_\mathbf{B} \alpha^\circ$.

An outline of the proof is given in section A.1.5.

The following sketch may be of help with the material above.

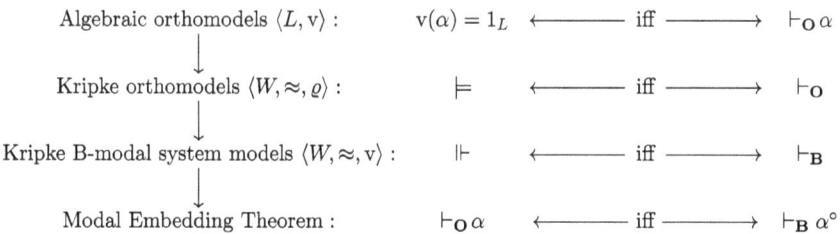

Algebraic orthomodels $\langle L, v \rangle$:	$v(\alpha) = 1_L$	$\longleftarrow$ iff $\longrightarrow$	$\vdash_\mathbf{O} \alpha$
$\downarrow$			
Kripke orthomodels $\langle W, \approx, \varrho \rangle$:	$\models$	$\longleftarrow$ iff $\longrightarrow$	$\vdash_\mathbf{O}$
$\downarrow$			
Kripke B-modal system models $\langle W, \approx, v \rangle$:	$\Vdash$	$\longleftarrow$ iff $\longrightarrow$	$\vdash_\mathbf{B}$
$\downarrow$			
Modal Embedding Theorem :	$\vdash_\mathbf{O} \alpha$	$\longleftarrow$ iff $\longrightarrow$	$\vdash_\mathbf{B} \alpha^\circ$

The translation of **OL** into the modal B-system is interesting and instructive, as we shall see in section 1.3, but the B-system is presumably a stronger logic.

Now, returning to section 1.2.1 in light of this model theory, our proximity spaces of states are seen to provide Kripkean orthomodels for **OL** in which the valuations ϱ take values in the ortholattice

$R(W)$ of propositions in an orthogonality space $\langle W, \perp \rangle$. But our argument in section 1.2.1 would seem to have identified the lattice Part W as the correct generalization of the lattice of Boolean propositions. Fortunately, these notions coincide thanks to the following theorem of J. P. Rawling. Here $\Diamond(\ )$ and $\Box(\ )$ are as defined in section A.1.1.

Theorem 1.2.5. *Given an orthogonality space* $\langle W, \perp \rangle$:

$$\Diamond \colon R(W) \to \text{Part}\, W \qquad\qquad (1.2.22)$$

is an isomophism of complete ortholattices. Its inverse is

$$\Box \colon \text{Part}\, W \to R(W). \qquad\qquad (1.2.23)$$

This is proved in section A.1.3 after Defintion A.1.5. We shall now generally relinquish the lattice Part W in favor of the completely isomorphic lattice $R(W)$ in those places where the Kripkean models are invoked, but see Example 3 in section 1.4.2 for an application of it.

1.3   Quantum-like behavior in the modal models

With the logical results in hand we can now investigate some semantics of orthotheorems in these models.

First, we recall some properties of actual quantum mechanics. In this theory a "quantum" (such as a particle like an electron or a photon) is not an object in the classical macroscopic sense since it does not have objective properties or attributes. These will generically depend upon the mode of probing or observing what attributes it does have. The observation, or measurement, of one property (for instance, the momentum of an electron) may render another property indefinite (the electron's position in this case) and vice versa. This of course flies in the face of classical tenets about objects. (Please see section 1.4.1 on the issue of objectivity.) This particular example epitomizes the Heisenberg Uncertainty Principle, whose real import is that quanta do not have objective states of being. Instead, certain repertoires of actions may be specified as possible to carry out upon a quantum, and it is the ensembles of these actions — which occupy a complex Hilbert space — that determine all possible propositions that can be asserted about the quantum. These Hilbert space elements may be thought of as descriptors of those possible elementary experiments that can be done on the quantum, and as such they are not inherent to the quantum, but rather inherent to the *episystem*, namely, everything involved in experimenting upon the system which is not the system itself, including the experimenter, apparatus, etc. (cf. [31]). "Observers" correspond to different sets of elements and that which is observed now depends on these sets, and are generically different among observers. (Technically, an "observer" corresponds to an algebra of commuting Hermitian operators on the Hilbert space in question, each such algebra determining an eigenbasis common to each of its elements.) This is how the non-objective nature of the quantum manifests itself in the formalism. The operational upshot of this is that a quantum can only be characterized per "observer" or experimenter or episystem, by certain combinations, called *superpositions*, of sets of measurable acts (namely, the eigenbasis mentioned above) which are in a sense simultaneously potential, or in *parallel*. (This is the property that is exploited in the optimistic science of quantum computing.) Only when a measurement or intervention occurs may the superposition collapse onto a suitable eigenvector representing that act corresponding to the measurement or intervention (whose eigenvalue is the value of the outcome). (This operational view is expounded for instance in [31]. See also [82], and for a treatment in the same spirit [80].)

The realization by Heisenberg of the non-objective nature of quanta constituted a radical departure from the ontology of macroscopic classical physics. For the systems of interest to us here, we do not posit such a radical ontology. For these systems, the states are actual, though they may not be directly accessible to "observers," by which we mean possible episystems, or other components of a family of systems. Thus, they reveal themselves, as in the case of actual quantum mechanics, in terms of the orthological version of superposition, namely *orthodisjunction*, via the Modal Embedding

Theorem (Theorem 1.2.4). As we have noted, this is how orthodisjunction is interpreted in actual quantum logic, whose models are certain lattices of closed subspaces of complex Hilbert spaces, which we do not discuss here. These non-classical properties are preserved in **OL** and its models, entirely in the absence of any physical trappings, as we shall now demonstrate.

Thus, returning to the B-modal system, suppose $\mathscr{B} = \langle W, \approx, \mathrm{v} \rangle$ is a B-model and α, β are two orthoformulas. We wish to simplify the modal translation of the formula $\alpha \sqcup \beta$. It will prove convenient to employ the formalism of *truth sets* introduced in Definition A.1.7. Namely, we write for each modal formula α:

$$\|\alpha\|_{\mathscr{B}} := \{w \in W : w \Vdash_{\mathscr{B}} \alpha\}. \tag{1.3.1}$$

Then from Proposition A.1.1 and the translation rules T1–T3 of the Modal Embedding Theorem:

$$\|(\alpha \sqcup \beta)^{\circ}\| = \|(\sim[(\sim\alpha) \sqcap (\sim\beta)])^{\circ}\|\| \tag{1.3.2}$$

$$= \|\Box [\Diamond\, \alpha^{\circ} \vee \Diamond\, \beta^{\circ}]\| \tag{1.3.3}$$

$$= \Box\,\Diamond\,(\|\alpha^{\circ}\| \cup \|\beta^{\circ}\|) \quad \text{from M1, etc.,}$$

$$\text{of Proposition A.1.1} \tag{1.3.4}$$

$$= \Box\,\Diamond\,(\|\alpha^{\circ} \vee \beta^{\circ}\|) \tag{1.3.5}$$

$$= \|\Box\,\Diamond\,(\alpha^{\circ} \vee \beta^{\circ})\|. \tag{1.3.6}$$

Thus, for $w \in W$,

$$w \Vdash_{\mathscr{B}} (\alpha \sqcup \beta)^{\circ} \quad \text{iff} \quad w \Vdash_{\mathscr{B}} \Box\Diamond(\alpha^{\circ} \vee \beta^{\circ}). \tag{1.3.7}$$

The "semantics" is now given by just unfolding the definitions in the last formula, namely:

In any state $w \in W$: $\forall u \approx w \; \exists x \approx u$ such that $x \Vdash_{\mathscr{B}} \alpha^{\circ}$ or $x \Vdash_{\mathscr{B}} \beta^{\circ}$.

The points to notice are:

- The validity or otherwise of either α° or β° may not be definite nor even defined in every state w. These may only be determined in at least one unknown daughter (i.e. w-proximal) state x, since $w \approx w$, and unknown granddaughter states. (Since the proximity relation is not transitive, these granddaughter states are not proximal, i.e. not confusable with, the current state w.) This indefiniteness of validity in the current state is the characteristic feature of (quantum-like) *superposition*. In the case of actual quantum mechanics, the system would be said to be *in a superposition* (of states).

- The conditions for the validity or otherwise of *both* branches of the disjunction are carried over to at least one unknown daughter state and many unknown granddaughter states and must, to a classical thinker, therefore be somehow available to them. So, to such a thinker, these potentialities must seem to coexist in some form over the span of pairs of states proximally connected to the current state w. This interpretation of superposition, in actual quantum theory, goes by the name *parallelism*.

After inspecting two examples we shall, in sections 1.3.1 and 1.4.1, delineate more precisely what we mean by quantum-like behavior in a model of this type.

EXAMPLE 1. SCHRÖDINGER'S CAT.

Our first application will be to Schrödinger's cat Ketzi. Here we take an atomic formula a to represent the proposition *Ketzi is alive*. The appropriate orthotheorem is then $\vdash_{\mathbf{O}} a \sqcup \sim a$. By Theorem 1.2.4 this translates into the B-theorem as $\vdash_{\mathbf{B}} (a \sqcup \sim a)°$. Our task is to simplify this translation and investigate its behavior vis-à-vis B-models.

From equation (1.3.6) with $\alpha = a$ and $\beta = \sim a$:

$$\|(a \sqcup \sim a)°\| = \|\Box\Diamond(a° \vee (\sim a)°)\| \tag{1.3.8}$$

$$= \|\Box\Diamond(\Box\Diamond a \vee \Box \neg \Box\Diamond a)\| \tag{1.3.9}$$

$$= \|\Box\Diamond(\Box\Diamond a \vee \neg \Diamond\Box\Diamond a)\| \tag{1.3.10}$$

$$= \|\Box\Diamond(\Box\Diamond a \vee \neg \Diamond a)\| \quad \text{from M9} \tag{1.3.11}$$

$$= \|\Box(\Diamond\Box\Diamond a \lor \Diamond\neg\Diamond a)\| \quad \text{from M1} \qquad (1.3.12)$$

$$= \|\Box(\Diamond a \lor \Diamond\Box\neg a)\| \quad \text{from M9} \qquad (1.3.13)$$

$$= \|\Box\Diamond(a \lor \Box\neg a)\| \quad \text{from M1.} \qquad (1.3.14)$$

We now consider the elements of W to represent "states" of a system in which the interpretation is carried out, and choose any $w \in W$. Then, from the last equation,

$$w \Vdash_{\mathscr{B}} (a \sqcup \sim a)^{\circ} \qquad \text{iff} \qquad w \Vdash_{\mathscr{B}} \Box\Diamond(a \lor \Box\neg a). \qquad (1.3.15)$$

As before, the semantics is given by just unfolding the definitions in the last formula, namely:

$$\text{In any state } w \in W: \ \forall u \approx w \ \exists x \approx u \text{ such that}$$

$$x \Vdash_{\mathscr{B}} a \text{ or } x \Vdash_{\mathscr{B}} \Box\neg a \qquad (1.3.16)$$

or

In any state $w \in W$: $\forall u \approx w \ \exists x \approx u$ such that $x \Vdash_{\mathscr{B}}$ *Ketzi is alive* or $x \Vdash_{\mathscr{B}} \Box$ *Ketzi is dead.*

If we interpret W as the set of Ketzi's states, in the absence of any external interference, such as releasing the poison or opening the box, then the above interpretation may be read:

In a given state w, $\forall u \approx w \ \exists x \approx u$ such that Ketzi is alive in state x or Ketzi is dead in state x and remains dead at x-proximal states (since $z \models_{\mathscr{B}} \neg a$ *for all* $z \approx x$). This occurs at every state w. (Note that since $w \approx w$, $\exists x \approx w$ such that the rest obtains, as before.)

Note here, as before, the appearance of the characteristic feature of "quantum" superposition: at every state w of the system the truth or otherwise of the proposition a may not be definite nor even defined, as far as an episystem, or "observer," is concerned. To a classical thinker the truth value at such states would seem to hover between the two alternative possibilities. This value is potential, not actual in such states, only determined at daughter states (i.e. proximal states, the u's in this case) and some granddaughter states (the x in this case), never at the state in question. Moreover, the apparent retention of both branches of a decision tree of potential outcomes over two generations of proximality seems to classical thinkers to imply some sort of inherent joint existence or storage of these two branches. As noted above, this is exactly ("quantum") *parallelism*. So to classical thinkers Ketzi seems to be dead and alive simultaneously in such states since to them both outcomes seem to be somehow stored and the truth value of the proposition a is always in abeyance, at least until the box is opened.

The reader may note that these conclusions have been reached just by working through the ineluctable algebra of modal logic, and that no metaphysical contortions concerning the actual state of Ketzi as an object or a non-object, her de Broglie wavelength, etc., were required.

The phenomena of "quantum" parallelism, and another quantum-like feature called *interference*, may be more plainly seen in our second example.

EXAMPLE 2. THE DOUBLE SLIT EXPERIMENT

In this case we concoct another orthotheorem in the form of a disjunct. For instance suppose the atomic formula a_i, $i = 1, 2$, denotes the proposition *the electron passes through slit i*. Then $\vdash_{\mathbf{O}} a_i \sqcup \sim a_i$ is the orthotheorem representing the proposition *the electron passes through slit i or it does not* and $\vdash_{\mathbf{O}} (a_1 \sqcup \sim a_1) \sqcup (a_2 \sqcup \sim a_2)$ is the orthotheorem (its orthotheoremhood following from the dual forms of either O2 or O3) representing the proposition *the electron passes through slit 1 or the electron passes though slit 2 or the electron passes through neither slit*. Writing this as $\vdash_{\mathbf{O}} \alpha_1 \sqcup \alpha_2$ it is an easy exercise using M1 and M9, and equations (1.3.6) and (1.3.14), to show that for an arbitrary B-model $\mathscr{B}$ and $w \in W$:

$$w \Vdash_{\mathscr{B}} (\alpha_1 \sqcup \alpha_2)^{\circ} \quad \text{iff} \quad w \Vdash_{\mathscr{B}} \Box \Diamond ((a_1 \vee \Box \neg a_1) \vee (a_2 \vee \Box \neg a_2)).$$
$$(1.3.17)$$

As before, this unfolds as

> For any state $w \in W$: $\forall u \approx w \, \exists x \approx u$ such that $x \Vdash_{\mathscr{B}} a_1 \vee \Box \neg a_1$ or $x \Vdash_{\mathscr{B}} a_2 \vee \Box \neg a_2$

which can be read:

> For any state $w \in W$: $\forall u \approx w \; \exists x \approx u$ such that in the state x: (the electron goes through slit 1 or stays away from it in all x-proximal states) or (the electron goes though slit 2 or stays away from it in all x-proximal states).

As before, there may be states w at which nothing concerning the truth or otherwise of the proposition can be determined or is defined. Moreover, since in general $(\alpha_1 \sqcup \alpha_2)^\circ \neq \alpha_1^\circ \vee \alpha_2^\circ$ we can have $w \Vdash_{\mathscr{B}} (\alpha_1 \sqcup \alpha_2)^\circ$ without either $w \Vdash_{\mathscr{B}} \alpha_1^\circ$ or $w \Vdash_{\mathscr{B}} \alpha_2^\circ$. That is to say, it might be the case that the proposition [*(the electron goes though slit 1 or does not) or (the electron goes though slit 2 or does not)*] is valid at w without it being the case that **either** the proposition *(the electron goes though slit 1 or does not)* is valid at w **or** the proposition *(the electron goes though slit 2 or does not)* is valid at w. Thus, if the electron could go through either slit independently (i.e. "classically"), the outcome would in general be different at such a state — namely the latter ordinarily disjunctive case above — from the former case. That is to say, the outcome for one slit is affected by the presence of the other slit. It would appear to a classical thinker that the advent of the second slit *interferes* with the state of affairs that obtains when the electron faces only one slit. This is "quantum" *interference*, which should not be confused with the entirely classical phenomenon of the interference of waves, but most often is. Although wave-like interference patterns do indeed emerge in experiments upon quanta like electrons in the two-slit experiment, they are not the result of the interference of actual waves.

This fundamental non-classical property — namely the non-truth-functionality of orthodisjunction, so disconcerting to classical thinkers — emerges more clearly in the Kripkean models appearing in the proof of Theorem 1.2.2 in section A.1.5. Namely, suppose we have an orthodisjunction of the form $\alpha \sqcup \beta$. Then in a Kripkean model $\mathscr{M}$:

$$\varrho(\alpha \sqcup \beta) = \varrho(\sim(\sim\alpha \sqcap \sim\beta)) \tag{1.3.18}$$

$$= \varrho(\sim\alpha \sqcap \sim\beta)^\perp \tag{1.3.19}$$

$$= (\varrho(\alpha)^\perp \cap \varrho(\beta)^\perp)^\perp \tag{1.3.20}$$

$$= (\varrho(\alpha) \cup \varrho(\beta))^{\perp\perp} \quad \text{from M10} \qquad (1.3.21)$$

$$\supseteq \varrho(\alpha) \cup \varrho(\beta) \qquad \text{from M5.} \qquad (1.3.22)$$

Thus, there may be states $w \in \varrho(\alpha \sqcup \beta)$ that are not in $\varrho(\alpha) \cup \varrho(\beta)$, or, in other words, there may be states w such that $w \models_{\mathcal{M}} \alpha \sqcup \beta$ but neither $w \models_{\mathcal{M}} \alpha$ nor $w \models_{\mathcal{M}} \beta$.

It is this property that confutes any attempt to build a Curry–Howard type correspondence for **OL** (or quantum logic).

1.3.1 *The hallmarks of quantum-like behavior, Born's Law and the manifestations of superposition*

We digress to recapitulate the characteristic signatures of quantum and quantum-like behavior.

- Superposition. Firstly, there is the primary ontological revolution instigated by Heisenberg: namely, the non-objective nature of quantum entities or *quanta*. Their states of being are not objective attributes. Cf. section 1.4.1. This manifests in actual quantum mechanics as the superpositional nature of the states of a quantum. The "actual" state of being is indefinite until a measurement operation or experiment is performed upon the quantum. In the logical context here, superposition is achieved by the orthodisjunct, whose apparently anomalous properties have been discussed.
- Parallelism. One interpretation or side-effect of superposition goes by the name of *parallelism*, discussed above: this paradigm has been found useful, particularly in application to quantum computing, though it is essentially just another name for superposition. Cf. the CAVEAT!! on page 27.
- Interference. Yet again, superposition is responsible for the phenomenon known as *interference* as in the double slit example. Here the situation obtaining for one slit is apparently altered by the advent of another slit. The orthodisjunct responsible in this interpretation is the one coming between the other orthodisjuncts in the rightmost assertion in (1.3.17). It has nothing to do with the interference of actual waves which unfortunately it resembles.

An important milestone in the path to quantum mechanics was the formulation of the so-called Born Law, which enables an interpretation of superposition. It says that if a quantum system is in a state ξ, an element of a Hilbert space, then the probability that it may be *found in* or *make a transition to* a state η, which we write as $\mathrm{prob}(\xi \to \eta)$, is given by

$$\mathrm{prob}(\xi \to \eta) = \left(\frac{|\langle \eta | \xi \rangle|}{\|\eta\| \|\xi\|} \right)^2. \tag{1.3.23}$$

For many decades this was accepted as an axiom of quantum theory. However, more recently it was seen to follow plausibly from other assumptions by Finkelstein [30] and independently and rigorously proved to so follow by Hartle [44]. For a sketch of one such argument see [82]. For a comprehensive discussion of the subsequent history of this important topic, see [45]. The proof entails assumptions about assembling quanta which we shall adopt since they do not fundamentally belong to physics *se ipse* but can be argued on mathematical grounds. This law enables us to interpret superpositions such as

$$\xi = \sum_i c_i \eta_i. \tag{1.3.24}$$

Thus, if the η_i are orthonormal and recalling that our c_is are *real* numbers, we have

$$\mathrm{prob}(\xi \to \eta_{i_0}) = \frac{c_{i_0}^2}{\sum_i c_i^2}. \tag{1.3.25}$$

That is to say, the probability that a superposition will "collapse" onto one of its component or Limbo-like states (see section 1.4.1 below), if these are orthonormal, is given in terms of the coefficients in the superposition by the right hand side of the above equation. We shall adopt this interpretation *mutatis mutandis* in our real cases. Thus the size of the coefficient c_i is a measure of the relative probability that if the system is in the state ξ, it will be found in the corresponding state η_i. In our paradigm, this is equivalent to the probability that the state ξ will be *confusable* with the state η_i, or, to all possible intents and purposes, *transition* to that state.

In our neuromorphic model to come, the basic units will turn out to be states of a system (a network) which are superpositions of basic states. So it is worth reiterating this interpretation of such a superposition. Such a superpositional state of a system is one in which, when the system is deemed to be in it, may be found in, or transition to, or be confusable with, one of the component states with the probability given by the Born Law, as above. The coefficients involved, being part of the collection of hidden variables, are generally unknown but may be surmised *ex post facto* by observation or more indirectly (cf. section 2.4.3).

The act of superposition will be found to have rather profound consequences in this context.

CAVEAT!!

The scope of this caveat extends to the rest of this section. The interpretations of superposition expounded here will be fundamental to our model(s) to come.

In our applications the coefficients in these superpositions will generally be variable in time. Such superpositional states will then manifest, via this interpretation, in recognizable ways which are very much less mysterious than the general interpretation of superposition. As a simple and not entirely artificial example, let us consider a time dependent superposition of the following form.

$$\xi(t) = (\cos t)\eta_1 + (\sin t)\eta_2 \qquad (1.3.26)$$

where the η_i are orthonormal. Then

$$\text{prob}(\xi(t) \to \eta_1) = \cos^2 t; \qquad (1.3.27)$$

$$\text{prob}(\xi(t) \to \eta_2) = \sin^2 t. \qquad (1.3.28)$$

Thus, $\xi(t)$ oscillates between a transition to the state η_1 and a transition to the state η_2. This is tantamount to *time division multiplexing* (or *duplexing* in this case) of the state $\xi(t)$. Time division multiplexing was invented in the 1870s by the pioneering

French telegraph engineer Émile Baudot (1845–1903) as a means of sending more than one message simultaneously along a single telegraph wire. It is a form of parallel processing and is still in use, though in sophisticated digital forms. It is apparent that our use of superposition will manifest as a very general form of time division multiplexing of neural behavior. Unlike the simple example given above, our coefficients — which may also be considered to be *weights* — will necessarily be unknown (though not unknowable) being among the ensemble of variables hidden from our theory. Since they will generally be time dependent, all possible time division multiplexing behaviors are consistent with it, from a simple change from one of the summand states to another, to oscillations among these states with any choice of frequency or frequencies. Our states will correspond to superpositions of subsets of neurons firing together. Since, in reality, these firings involve the movement of charges, they will generate electromagnetic fields. Given the possible oscillations described by our superposition paradigm, these fields would themselves interact and have their own oscillations. So we may expect oscillating electromagnetic fields to be associated inextricably with our notion of "firing pattern" which will be introduced in section 2.3.1. (This kind of alternation of states *à la* time division multiplexing in a superposition has been exploited in formal approaches to the ever hopeful discipline of quantum computing to effect exactly the sort of parallel processing that is its *raison d'etre* [103].)

To summarize: the phenomenon of superposition in our context and the applications to follow presents a very much less mysterious ontological problem than the superpositions of actual quantum physics. In our applications it is an encoding of the idea of parallel processing, and may manifest in such ways as time division multiplexing and other forms of cooperative behavior among cells and ensembles of cells. Actual oscillations among ensembles of firing cells may manifest as the familiar electromagnetic wave forms, since moving charges are involved.

1.4 Quantum-like vs. non-quantum-like behavior in the models

Although we have adopted **OL** as our overarching logic on the basis of the structure of propositional lattices in general proximity spaces, there are of course cases in which this lattice is in fact Boolean. In these cases there might be a non-trivial local modal semantics (as in Example 1 in section 1.4.2). However, in these cases the resulting semantics do not necessarily exhibit the orthodisjunctional anomalies to which we have attributed quantum-like behavior. The aim of this chapter, among others, is to establish, given the particular B-modal model specified by the parameter window of the complex system at hand, whether or not the lattice $R(W)$ is Boolean. Corollary A.1.4 will be most useful to us in this endeavor. However, it is important to pin down the precise distinction, which we attempt in the next section.

1.4.1 *The crucial difference: Greetings from Limbo*

Elucidating what it is that exactly distinguishes between classical behavior and quantum-like behavior in our context seems to be an elusive endeavor, as it is in actual quantum theory. Here we assume that classical behavior is associated with the Booleanness of our ortholattices and concomitantly quantum-like behavior is associated with their non-Booleanness. In the case of general ortholattices Theorem A.1.2 gives mathematical criteria for such distinctions, so we may try to interpret these conditions in more intuitive ways. To wit, let us take the Corollary A.1.1: an ortholattice L is not Boolean iff $\exists\, a \in L$ which is not uniquely complemented. That is to say, there must exist at least one $a^\sigma \in L$ such that $a \sqcap a^\sigma = 0_L$, $a \sqcup a^\sigma = 1_L$, and $a^\sigma \neq a'$. Of course there may be many such complements a^σ: in the simplest vector case, in dimensions greater than one, each pure state has an uncountable infinitude of such complements. In analogy with this case, we may think of these complements as being *non-orthogonal* complements, the *actual* complement a' generalizing the unique orthogonal one of the vector case. So we shall consider σ to act as an indexing of the set of these "non-orthogonal" complements to a, whatever this set is. We shall cavalierly interpret ⊓ as a type of AND, and ⊔ as a type of OR — a delicate point over which oceans of philosophical ink have been spilt and to which we shall not contribute anything other than handfuls of salt. Concomitantly we shall regard the elements of L as propositions, with a, for the sake of argument, taken as the proposition *Ketzi is alive*: then a' is the proposition *Ketzi is dead*. Then the a^σ are propositions akin to the proposition *Ketzi is dead* but none of them is that proposition. Neither could any of them be *Ketzi is alive*. So they are something like *Ketzi is in apartment σ of Limbo*. (Another descriptor for these complements that might also be appropriate is *zombie-like*: neither dead nor alive. However, zombies are not known to keep still, nor to occupy apartments, nor cat boxes, so we would not be able to pin hordes of them down.) Thus, from the two conditions on complements, we would have for each σ

Ketzi is alive   AND   *Ketzi is in apartment σ of Limbo*   is   FALSE;

Ketzi is alive   OR   *Ketzi is in apartment σ of Limbo*   is   TRUE.

(This is presumed to obtain in the absence of an external event such as an act of measurement or observation, for instance opening Ketzi's box.)

So for each σ all possible cases would seem to be covered, with possibly many of these Limbo-like propositions (the a^σ) none of which are the negation of the first one (a). This cannot happen in the Boolean case: the only proposition that, with the first one a, covers all possibilities in this way is the negated ("orthogonal") proposition itself: a'. There can be no Limbo-like propositions.

We can regard the pair a, a' as constituting a complete description of a putative underlying system (Ketzi in this case), relative to the proposition a, that is *aliveness*, in both the Boolean/classical case and the non-Boolean/quantum-like case: $a \sqcap a' = 0_L$, $a \sqcup a' = 1_L$. However, in the non-Boolean case, there are other, different, complete descriptions, perhaps very many of them, namely the pairs a, a^σ, the latter depending on a. The system described then has properties that depend on the property, or attribute, described by

a: namely the a^σ, with a and a^σ being mutually exclusive, but without a^σ denoting the contrary to a. This is *non-objectivity*. The complementary propositions a^σ could not pertain to an *object*, since no non-"orthogonal" properties of an object can *logically* depend upon the choice of some other mutually exclusive property. An object such as a tennis ball is either spherical or it is not. It has no Limbo-like unspherical properties that depend on the property of sphericity but are not its contrary. In this sense quantum-like systems have non-objective propositions whereas classical systems do not. (And Ketzi is therefore not an object, which is the least of her problems.)

The subjective nature of the ghostly Limbo-like propositions that must exist in the quantum-like case, possibly in great profusion, is another aspect of the presence of superposition, and is likely to contribute to the degree of disquiet experienced by classical thinkers when confronted by them.

The dichotomy is now clear. We shall attribute quantum-like behavior to those systems whose proximity spaces of states have non-Boolean proposition lattices.

We shall argue that the probability of such quantum-like behavior increases with complexity.

1.4.2 *Parameter windows*

The Kripkean orthomodels that arise in actual quantum physics are of the following type. Let $\mathfrak{H}$ denote a complex Hilbert space with inner product $\langle \, | \, \rangle$. We note that the family of closed subspaces of $\mathfrak{H}$ form an ortholattice with meet just the ordinary intersection of closed subspaces and join the closure of the algebraic direct sum of the summand spaces, the latter being the smallest subspace containing both the constituent subspaces, usually incorrectly denoted $\oplus$. (In the case of finite dimensional ambient spaces, this join is just the algebraic direct sum since all subspaces in this case are necessarily closed.) The orthogonality is just the usual orthogonal complement. Now put $\mathfrak{h} := \mathfrak{H}\backslash\{\mathbf{0}\}$, the set of non-zero elements of $\mathfrak{H}$, with $\xi \perp \eta$ iff $\langle \xi | \eta \rangle = 0$ for $\xi, \eta \in \mathfrak{h}$. Then $\langle \mathfrak{h}, \perp \rangle$ is an orthogonality space. If $E \subset \mathfrak{H}$ then the smallest closed subspace of $\mathfrak{H}$ containing E, denoted by $[E]$, is $E^{\perp\perp} \cup \{\mathbf{0}\}$ so if E is a proposition of $\langle \mathfrak{h}, \perp \rangle$, we have $E = [E]\backslash\{\mathbf{0}\}$. Conversely, if F is a closed subspace of $\mathfrak{H}$ it is easily seen that $F\backslash\{\mathbf{0}\}$ is a proposition of $\mathfrak{h}$. The bijective correspondence between the propositions of $\mathfrak{h}$ and the closed subspaces of $\mathfrak{H}$ given by $E \mapsto [E]$, the inverse map being given by $F \mapsto F\backslash\{\mathbf{0}\}$, is soon seen to be an isomorphism of the ortholattice $R(\mathfrak{h})$ and the lattice of closed subspaces of $\mathfrak{H}$. (Note that in this correspondence $\mathfrak{h} \supset \emptyset \mapsto [\emptyset] = \{\mathbf{0}\} \subset \mathfrak{H}$.)

In the cases of interest to us, namely the systems whose states may be described by vectors of real parameters, the relevant Hilbert-like lattices will arise from *real* Hilbert spaces of generally high but finite dimension. In other words, Euclidean spaces of the form $\mathbb{R}^n\backslash\{\mathbf{0}\}$. In these cases we shall write the vectors in bold face, $\mathbf{v}, \mathbf{w}$, etc., and write the inner product as for instance $\mathbf{v}.\mathbf{w}$: thus $\mathbf{v} \perp \mathbf{w}$ iff $\mathbf{v}.\mathbf{w} = 0$ and $\mathbf{v} \approx \mathbf{w}$ iff $\mathbf{v}.\mathbf{w} \neq 0$. Subsets of these spaces inherit their lattice operations and we shall be considering such subsets with this inherited structure. These subsets will be thought of as being defined by the parameter ranges of the systems of interest, hence the name *parameter windows*.

For formal definitions see section A.1.4. We note from Lemma A.1.4 that it is only necessary to investigate and classify to the extent possible the rays, or subspaces generated by single elements, in the orthogonality spaces in question, to gain access to the full lattice of propositions, and this is how most of our examples will be analyzed.

EXAMPLE 1

Take $n = 1$ and any subset $W \subset \mathbb{R}\backslash\{\mathbf{0}\}$. Then it is easy to see that $R(W) \cong \mathbf{2}$. For, any $\mathbf{w} \in W$ has $w^{\perp w} = \emptyset$ so $w^{\perp w \perp w} = W$. So $\emptyset$ and W are the only possible propositions. Note that for any $\mathbf{v}, \mathbf{w} \in W$, $\mathbf{v} \approx \mathbf{w}$. In this case we may imagine the underlying system to be a single cell or cell-like node, having only one parameter (such as a single membrane potential, in the case of nerve or neuron-like cell) whose range is W, which may be any subset of $\mathbb{R}$ avoiding the origin. Even in this simple case it is worth doing a modal analysis as follows.

Let us suppose that the atomic formula a stands, for example, for the proposition *the node fires* (or, as in the case of an actual neuron, *reaches the threshold potential*: cf. section 2.1. We could also say *the node turns ON*). Then the theorem $\vdash_{\mathbf{O}} a \sqcup \sim a$ translates via the Modal Embedding Theorem as $\vdash_{\mathbf{B}} (a \sqcup \sim a)^{\circ}$. Now with any choice of valuation v, $\mathscr{B} := \langle W, \approx, v \rangle$ is a B-model and we have for all $w \in W$

$$w \Vdash_{\mathscr{B}} (a \sqcup \sim a)^{\circ} \tag{1.4.1}$$

from which it follows as in (1.3.16) that for all $u \approx w$

$$\exists x \approx u \text{ such that } x \Vdash_{\mathscr{B}} a \quad \text{or} \quad x \Vdash_{\mathscr{B}} \Box \neg a. \tag{1.4.2}$$

Since $w \approx w$, for all w we have

$$\exists x \approx w \text{ such that } x \Vdash_{\mathscr{B}} a \quad \text{or} \quad x \Vdash_{\mathscr{B}} \Box \neg a. \tag{1.4.3}$$

Now, $x \Vdash_{\mathscr{B}} \Box \neg a$ iff $z \approx x$ implies $z \Vdash_{\mathscr{B}} \neg a$. So $W \Vdash_{\mathscr{B}} \neg a$ since $z \approx x$ for all $z \in W$. Thus we obtain finally:

In any state $w \in W$:

$$\exists x \in W \text{ such that } x \Vdash_{\mathscr{B}} a \quad \text{or} \quad W \Vdash_{\mathscr{B}} \neg a, \tag{1.4.4}$$

or

In any state $w \in W$:

$$\exists x \in W \text{ such that } \textit{the node fires at the non-zero value}$$
$$x \text{ or } \textit{the node does not fire at all.} \tag{1.4.5}$$

This result is not exactly surprising: if the node fires at all it must do so at at least one value, but it is perhaps worth stating it in the modal language. Namely, we note that in any state (w), or at any time, say, if the node is firable at all, there is some generically different state or *non-zero* value (x) at which this happens. Our formalism has forbidden inclusion of the 0 state, since it is not confusable with itself. (At another state w the node's condition may be different. Without further restrictions a node may be firable at one moment but not firable at the next.) This is of course classical behavior, though the value of x is not determined in the current state. This is indeterminacy, not parallelism. Of course, the value of x (the *threshold*) may be independent of w as it is in the case of actual neurons (cf. section 2.1).

We might summarize this non-quantum-like, or *classical*, 1-node case by saying that the node is generally in a state of potentially firing or not firing, i.e. potentially turning ON or not turning ON.

This behavior surprisingly mimics the similar all-or-nothing operation of an actual biological neuron: it can fire at some value of its membrane potential, though this value cannot be determined until the neuron is probed and/or stimulated. And when these cells are broken into, probed and stimulated, as they have been in laboratories over the course of centuries, it is found that the threshold potential value is indeed non-zero, in consonance with our model, though of course the actual potential does pass through the zero value during the cell's firing phase. In section 2.1, we give a vastly simplified account of the functioning of real neurons, and other excitable cells.

Although the behavior of a single node is classical as expected, the behavior of a cluster of more than one may not be, as we will see in Example 5 to follow.

We emphasize here that in this example and those to follow, an "observer" cannot just read the values in the nodes, since nothing is supposed to emanate from them until some event transpires, such as a firing or turning ON of the node.

EXAMPLE 2

Any $W \subset \mathbb{R}^n \backslash \{\mathbf{0}\}$ such that no pair in W are orthogonal must have $R(W) \cong \mathbf{2}$. For then, if $\mathbf{v} \in W$, $\mathbf{v}^\perp \cap W = \mathbf{v}^{\perp w} = \emptyset$, so $\mathbf{v}^{\perp w \perp w} = W$. So the only propositions are $\emptyset$ and W.

This covers the case when W is just a random sprinkling of a few points around the origin in $\mathbb{R}^n$. It seems unlikely that in this case pairs will be orthogonal and so while there are few points, the propositional lattice is likely to be Boolean. As the number of points increases this is less likely to occur and the lattices will become more probably non-Boolean. Please see the discussion following Example 4.

EXAMPLE 3

Now we consider what is actually the *locus classicus* of the traditional theory, namely *bit strings*: sequences of n 0s and 1s. We take W as the set of vectors represented by the non-null bit strings. Let us consider this for the case $n = 2$. Then $W = \{(1,0), (0,1), (1,1)\} := \{\mathbf{v}_1, \mathbf{v}_2, \mathbf{v}_3\}$. In this case, because $\mathbf{v}_1 \perp \mathbf{v}_2$, we have $\mathbf{v}_1^{\perp W} = \mathbf{v}_2$ so $\mathbf{v}_1^{\perp W \perp W} = \mathbf{v}_2^{\perp W} = \mathbf{v}_1$. Similarly, $\mathbf{v}_2^{\perp W \perp W} = \mathbf{v}_1^{\perp W} = \mathbf{v}_2$. And $\mathbf{v}_3^{\perp W} = \emptyset$, so $\mathbf{v}_3^{\perp W \perp W} = W$. Thus

$$R(W) = \{\emptyset, \mathbf{v}_1, \mathbf{v}_1^{\perp W}, W\} \cong \mathbf{2}^2. \tag{1.4.6}$$

It is interesting to see this by computing the lattice Part W in view of Rawling's Theorem 1.2.5, which is a laborious but mechanical process. First we list all the subsets of W:

$\emptyset,$
$\{\mathbf{v}_1\}, \{\mathbf{v}_2\}, \{\mathbf{v}_3\},$
$\{\mathbf{v}_1, \mathbf{v}_2\}, \{\mathbf{v}_1, \mathbf{v}_3\}, \{\mathbf{v}_2, \mathbf{v}_3\},$
$W.$

Then we apply $\Diamond_W$ to each set in turn:

$\Diamond_W \emptyset = \emptyset,$
$\Diamond_W \mathbf{v}_1 = \{\mathbf{v}_1, \mathbf{v}_3\},$
$\Diamond_W \mathbf{v}_2 = \{\mathbf{v}_2, \mathbf{v}_3\},$
$\Diamond_W \mathbf{v}_3 = W,$
$\Diamond_W \{\mathbf{v}_1, \mathbf{v}_2\} = \Diamond_W \{\mathbf{v}_1, \mathbf{v}_3\} = \Diamond_W \{\mathbf{v}_2, \mathbf{v}_3\} = W,$
$\Diamond_W W = W.$

Noting that

$$(\Diamond_W \mathbf{v}_1)' = \Diamond_W (\Diamond_W \mathbf{v}_1)^c \tag{1.4.7}$$

$$= \Diamond_W \mathbf{v}_2 \tag{1.4.8}$$

$$= \{\mathbf{v}_2, \mathbf{v}_3\} \tag{1.4.9}$$

we have

$$\text{Part } W = \{\emptyset, \Diamond_W \mathbf{v}_1, (\Diamond_W \mathbf{v}_1)', W\} \cong \mathbf{2}^2 \cong R(W). \tag{1.4.10}$$

This argument applies to the case of general n, reassuringly yielding the classical expectation $R(W) \cong \mathbf{2}^n$.

EXAMPLE 4

This will be our first non-Boolean example. Note that in this case, with four discrete parameter values permitted to each node of a two-node system, some work is required to cook up a non-Boolean example: as mentioned above, a random scattering of a few parameter values is more likely to yield a Boolean lattice: we have to work to confute this. We choose the vectors as follows. First choose an orthogonal pair of vectors $\mathbf{v}_1$ and $\mathbf{v}_2$. Then choose another vector $\mathbf{w}_1$ such that $\mathbf{v}_1 \approx \mathbf{w}_1$, and take $\mathbf{w}_2$ orthogonal to $\mathbf{w}_1$. Then $W = \{\mathbf{v}_1, \mathbf{w}_1, \mathbf{v}_2, \mathbf{w}_2\}$, $\mathbf{v}_1^{\perp w} = \mathbf{v}_2$ and $\mathbf{w}_1^{\perp w} = \mathbf{w}_2$. Thus

$$R(W) = \{\emptyset, \mathbf{v}_1, \mathbf{w}_1, \mathbf{v}_1^{\perp w}, \mathbf{w}_1^{\perp w}, W\}. \qquad (1.4.11)$$

This is a non-Boolean ortholattice by Corollary A.1.4 and construction: $\mathbf{v}_1 \approx \mathbf{w}_1$ and $\mathbf{v}_1^{\perp w \perp w} \cap \mathbf{w}_1^{\perp w \perp w} = \emptyset$. It is the double flanged Chinese lantern known as OM2 [56]. Please see Figure 1.1 for a depiction of an isomorph of this ortholattice.

These last two pointillist examples are somewhat artificial in the sense that with only a few parameter values available such systems

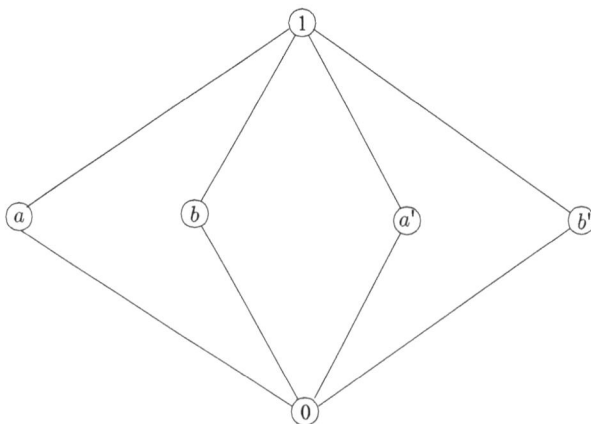

Figure 1.1 OM2.

could hardly claim complexity, but they illustrate the point that as complexity increases so does the probability of non-Booleanness. For, let us suppose that more points are added randomly to W. Initially, with few points, the chances are high that $R(W)$ remains Boolean, as in Example 2. As the density of points increases and they begin to fill up more space around the origin, it becomes more likely that vectors $\mathbf{v} \in W$ have orthogonal complements $\mathbf{v}^\perp$ that meet W so that $\mathbf{v}^{\perp w \perp w} \subset \mathbf{v}^{\perp\perp} \neq W$, and that there are more such $\mathbf{w} \in W$ with $\mathbf{v} \approx \mathbf{w}$, thus fulfilling the dictates of Corollary A.1.4 and forcing non-Booleanness upon $R(W)$. So probable classical behavior gives way to probable quantum-like behavior as complexity increases.

Our last class of examples addresses the applications we have in mind in modeling our neurons later (cf. Chapter 2).

EXAMPLE 5

Any $W \subset \mathbb{R}^n \backslash \{\mathbf{0}\}$ that "straddles" the origin has a non-Boolean $R(W)$. That is to say, any W that completely surrounds the origin will have a non-Boolean $R(W)$ isomorphic to the ortholattice of subspaces of $\mathbb{R}^n$. This follows from the application of Corollary A.1.4 as in the discussion following Example 4. Each ray will meet W, as will its orthogonal complement, and the Corollary then applies.

In what follows, our models will be built from nodes, modeling neurons or parts of neurons, that will be assumed to have parameter values (such as membrane potentials) that occupy intervals on the real line that straddle the origin, as seems to happen in nature. That is, they are of the form:

$$[s,t] := \{\lambda \in \mathbb{R} : s \leqslant \lambda \leqslant t\} \qquad (1.4.12)$$

where $s < 0 < t$. Thus, clusters of n of them will have parameter windows of the form

$$W := \prod_i^n [s_i, t_i] \backslash \{\mathbf{0}\} \subset \mathbb{R}^n \backslash \{\mathbf{0}\}. \qquad (1.4.13)$$

For $n > 1$ these will have non-Boolean proposition lattices and will therefore exhibit quantum-like behavior.

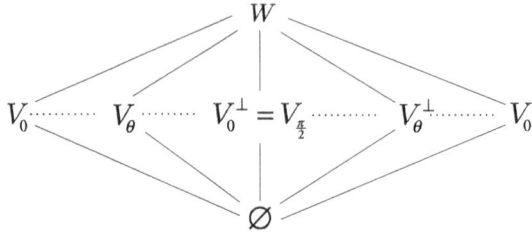

Figure 1.2   OM2$^{\aleph_0}$.

Again, the easily generalizable case of $n = 2$ will suffice to demonstrate the non-Boolean nature of such a parameter window. Note that each intersection of W with a line of slope $\tan\theta$ going through the origin — call it V_θ — is thus a proposition, and these are the only ones lying between $\emptyset$ and W. Thus, the lattice structure in this case is that of an infinite Chinese lantern as partially depicted in flattened-out form in Figure 1.2, with $\theta \in [0, \pi)$ and $V_{(\theta+\frac{\pi}{2})} = V_\theta^\perp$, with the meridian thought of as lying on a circle, identifying the leftmost subspace with the rightmost one. This is a non-Boolean ortholattice.

In this case we may expect to find quantum-like phenomena and indeed we do. With the parameter window of the form $W := [s_1, t_1] \times [s_2, t_2] \setminus \{(0,0)\}$ as in equation (1.4.13) with $n = 2$, let us now denote by the atomic formula a_i the proposition *the node i fires* (or *the node i is ON*) for $i = 1, 2$. Now the circumstances for each component node has changed since $R(W)$ is non-Boolean. We can define a valuation ϱ on the atoms a_i as follows

$$\varrho(a_1) = \{(\lambda, 0) \colon \lambda \in \mathbf{R}\} \cap W \qquad (1.4.14)$$

$$\varrho(a_2) = \{(0, \mu) \colon \mu \in \mathbf{R}\} \cap W \qquad (1.4.15)$$

with any propositions assigned to other atoms, and extend it inductively to all formulas.

Then it is obvious that there are an infinite number of $\mathbf{v} \in W$ such that in the Kripkean model $\mathcal{M} = \langle W, \approx, \varrho \rangle$

$$\mathbf{v} \models_\mathcal{M} a_i \sqcup \sim a_i \quad \text{but} \quad \mathbf{v} \not\models_\mathcal{M} a_i \text{ and } \mathbf{v} \not\models_\mathcal{M} \sim a_i$$

for $i = 1, 2$ (as in equation (1.3.22)). So each node now exhibits quantum-like behavior and can languish in Limbo-like, unobservable superpositions of firing (or being ON) and not firing (or being OFF), as in the case of Ketzi, as far as episystems are concerned. This is to be contrasted with the case of only one node considered above, which is entirely different: classical though indeterminate. The presence of the other node has made available more degrees of freedom, that is to say more superpositional states, and has therefore induced quantum-like behavior which had been entirely absent from the single node in isolation.

To see this more clearly, we apply Born's law, the "transition" relation in this case being confusability. Thus

$$\text{prob}(\mathbf{w} \approx \mathbf{v}) := \left(\frac{\mathbf{w}.\mathbf{v}}{\|\mathbf{w}\|\|\mathbf{v}\|} \right)^2 \qquad (1.4.16)$$

where the norms denote Euclidean lengths.

Let us test this in the case of a single node, with the model as above. Here clearly we have $\text{prob}(x \approx y) = 1$, which is consistent with the fact that it is certain that $x \approx y$ for every pair. This is a reflection of some observer's ignorance of the contents of the node until the node does something, like fire. (Of course in the case of an actual neuron, observers such as humans, equipped with appropriate instruments, may detect intracellular or membrane potentials when the cell is not firing, but the formalism must allow for all observers including, for example, other networks in the same brain. The same is true in the case of actual quantum mechanics.) Before such an event or observation, all possible non-zero values are mutually confusable.

Let us now choose a general state $\mathbf{v} = (v_1, v_2)$ in the last two-node example, and ask the question: what is the probability that $\mathbf{v}$ is confusable with a state in which the first node appears to behave as if it were alone, namely, that the first node is in its classical state of being potentially firable or not, that is, potentially ON or not? (Note that the W there is of the form $[a_1, b_1] \backslash \{0\}$ which in the present example lies along the axis in $\mathbb{R}^2$ determined by the first component of the **v**'s). This is tantamount to calculating $\text{prob}(\mathbf{v} \approx (v,0))$, to wit

$$\text{prob}(\mathbf{v} \approx (v,0)) = \left(\frac{(v_1, v_2).(v,0)}{\|\mathbf{v}\|\|(v,0)\|} \right)^2 \tag{1.4.17}$$

$$= \left(\frac{v_1 v}{(v_1^2 + v_2^2)^{\frac{1}{2}} |v|} \right)^2 \tag{1.4.18}$$

$$= \frac{v_1^2}{v_1^2 + v_2^2}, \tag{1.4.19}$$

a result which is independent of v. Thus, in a general state **v** of the two-node system, the first node's apparent behavior in isolation, though not quantum in itself, has only a certain probability (<1) of occurring. Likewise the second node. A similar effect obtains in a system of n nodes, of course, and we notice that the probability of any particular node appearing to behave in its isolated fashion when the system is in this general superpositional state **v** decreases as n increases, since the denominator in equation (1.4.19) will increase with n. This is certainly quantum-like behavior, and it increases with the number of nodes. Namely, the apparent classical behavior of each node in a general state of an n-node system becomes less probable with increasing n.

Put another way, as n increases, the number of superpositional possibilities, or families, increases (it turns out) at least exponentially. There is an increasing repertoire of states for the system to appear to be hovering among as the number of nodes goes up, while the probability of any one constituent node appearing to behave as it would by itself decreases: we note again that a human brain is estimated to have 10^{11} neurons. The upshot is that this quantum-like behavior again becomes more apparent as complexity increases: cf. the discussion following Example 4 above.

It may be noted also, from equation (1.4.19), that the probability that at least one node is in a classical state is unity, since this probability is the sum over the node index of those shown on the right hand side of that equation. (It is a simple matter to compute the probability of a subset of k nodes to appear to behave classically in a general state, and the result is generally <1.) So it is certain that such a system, in any state, will have at least one (unknown) classically behaving node while in that state, so that this node will either fire at a certain non-zero value or not fire at any value. This situation would seem evanescent, since the classically behaving node and its firing value could change with $\mathbf{v}$. But what we have been describing here is merely the kinematical template of all the possible outcomes and would require in reality the imposition of some sort of external principle or principles. For instance, it is known that in the case of biological neurons, the firing or threshold value (x in (1.4.5)) would be constant across many species (namely -55 mV). (In actual physics, it is usually a dynamical principle that confines a system's actual states to subsets of its state space. This is true of quantum physics also, but in this case it is the nature of the information determined by the state that is different from the classical case.)

It may seem counterintuitive that two nodes can be quantum-like while one node is not. However, even a small cluster of neurons, say, embedded within a mass of other neurons, is still a black box as far as our initial assumptions concerning these models are concerned. The analogy here is not to a cluster of neurons sitting on a laboratory bench hooked up to gauges and probes, but rather to a cluster of neurons embedded in a living brain.

This multiple node case is similar to, but apparently not identical with, the situation in actual physics in which there is a collection of non-interacting quanta: despite the lack of mutual interaction, the quantum assumptions enforce certain statistics upon the quanta. This raises the question of how to distinguish among such systems — for example a family of n one-node systems vs. an n-node system — which will be addressed in the subsequent chapters.

All these considerations (in the quantum-like cases) have been conducted in the absence of any interaction posited between the nodes which would be effected by assumptions concerning how they are networked and what characteristics the networks may have. These matters will be taken up in the following chapters.

Chapter 2

Neuronal Networks

In this chapter we introduce our network models and lay the ground-work for what might be described as their *physics-like* aspects. These notions will be applied and developed further in later chapters in concert with the other face of our model, namely the computational structure, which will be the subject of the next chapter.

The layout of the current chapter is as follows. Section 2.1 gives an abbreviated and simplified account of the structure and function of the "standard model" of a neuron. Section 2.2 introduces the basic standard neural network (upon which many variations have been historically performed) and we discuss very briefly the basic element of the Hopfield Ising-like model. In section 2.2.2 we introduce two combinators of network state spaces, namely the tensor product and the external direct sum. The former combinator is of central importance in the sequel: it will be shown to implement the kind of substrate-induced cooperative behavior exhibited by groups of biological neurons which transcends the purview of the classical synaptic-like models. The implementation of the external direct sum raises specifically quantum-like issues whose significance will emerge later. (This is not surprising since it is essentially interpreted as superposition, which is known to cause classical head scratching.) In section 2.3 we introduce the basic notion of a *bicameral* neuron (or *b-neuron*) and show that, in the quantum-like regime that naturally arises in view of our logical exertions in Chapter 1, such a b-neuron behaves like a spin-$\frac{1}{2}$ fermion. In section 2.4 we develop from first principles the "pseudo"-dynamics of such networks. That is to say, we derive the infinitesimal generator of the time translation operator for such systems. The states now represent possible subclusters of ON (or firing) b-neurons, and their superpositions, whose overall behavior mimics that of a finite system of fermions (though the Hilbert spaces in our cases are real). This infinitesimal generator thus mimics the Hamiltonian of such a system and we call it the *pseudo*-Hamiltonian (or *p-Hamiltonian*) of the network. It resembles the Hamiltonian of physical systems known as being of the *quasispin* type, and is similar, though in operator form, to the Hopfield Ising-like energy functional. In sections 2.4.2 to 2.4.4 eigenstates of such p-Hamiltonians are discussed and some general results proved. In section 2.5.1 we go into more detail concerning the significance of the new combinator, namely the ∧ or exterior product, that has arisen out of the fermionic structure. We also perform a "thought experiment" (section 2.5.2) to demonstrate the existence of the non-synaptic connection implemented by the exterior product. There is a final section 2.7, which may be skipped, delving into some aspects of the concomitant spin structure inherent in a b-neuron (or qubit).

2.1 Neurons: Structure and function

We give a very abbreviated account of the "standard" neuron and its behavior. Full accounts are too numerous to cite individually, and may be found in their multitudes online and elsewhere. For a couple of very good examples see [6, 29]. For a deep mathematical study of neuron dynamics, and to get some idea of the complexity of neuronal internal states, see [49].

A neuron is a specialized cell that functions in a node-like manner, is networked with other neurons, and communicates with them via the chemical mediation of electrical impulses in a manner to be described. A neuron generally has very many — generally numbered in the thousands — input channels *fanning in*, but only one signal is output, though generally to many recipients: i.e. *fanning out*. The inputs are branched projections, called *dendrites*, of the neuron body, or *soma*, that conduct electrochemical signals into the body of the neuron from other cells and local environmental factors such as the composition of the interstitial tissue and fluids. The outgoing branch is a single pseudopod-like projection, usually on the side opposite to the dendrites, called the *axon*. The active site where the axon connects to the soma is called the *hillock*, or sometimes the *trigger zone*. The axon conducts a single electrochemical output pulse or train of pulses, but may have many branching outputs at its end, called the axon *terminals*. These then provide the input signals to the dendrites of other neurons. (The axon can be immensely long relative to the size of the cell, as in nerve cells.)

The electrochemistry underlying this activity is extremely complex. In simple terms, electrical potentials are formed across the cell membranes via the bidirectional flows through the membrane of various types of ion: in this case mainly sodium and potassium ions, though there is a multitude of other ions in this environment. These ion flows are controlled by clusters of proteins embedded in the cell membranes (as they are in all cells) called *ion channels*, which act upon ion flows like gates, pumps and/or valves. There is a multiplicity of varieties of ion channel.

In its "resting" state, a neuron holds a potential of about $-70\,\mathrm{mV}$ (millivolts) reflecting a steady state of polarization between the internal and external ion flow states. If input signals arrive from the dendrites, or groups of them, the cell depolarizes and its potential goes up. Generally, the effect inside the soma is to add these inputs, which may increase (*incite*) or decrease (*inhibit*) the probability that the neuron will fire. Thus the cell may then rapidly repolarize without issue if the signal is not strong enough, in which case the signal does not penetrate very far into the soma. But if the stimulus is strong enough to enable a threshold potential to be reached (at about $-55\,\mathrm{mV}$), then positive feedback kicks in and there is a rapid depolarization and concomitant rise in the potential to a peak of about $60\,\mathrm{mV}$. At some point during this rise, the signal penetrates the soma and reaches the hillock, and the potential profile, or a train of copies of it, begins to be conducted along the axon via a spontaneously choreographed succession of opening and closing ion channels along the body of the axon. This is the point at which the neuron is said to *fire*. (This is a simplification: in fact some neurons may fire regardless of the value of the membrane potential, and some may not fire even if the membrane potential reaches the threshold.) Meanwhile, the cell body rapidly repolarizes and the *action potential* now plunges down to around $-90\,\text{mV}$, which has the salutary effect of preventing the signal from being conducted back up the axon toward the soma. After a refractory period in this territory, the potential rises again to the resting value of $-70\,\text{mV}$. The time scales are roughly as follows: the duration of the "spike" is about 0.5 ms (milliseconds); the entire duration of the action potential, including the refractory period, between the two rest states, has a duration of about 5 ms. It should be mentioned that it is only the frequency of the output signal train that is determined by the strength of the input signal, not the amplitude, which is independent of the input signal strength. A neuron cannot initiate firing when it is in the process of firing.

The mode of transmission of a signal from an axon terminal of one neuron, to a dendrite of another neuron, is generally not by direct contact but rather through a complex intermediary mechanism involving *synapses* (presynaptic bodies at the ends of axon terminals, with postsynaptic receptors at the ends of the receiving dendrites) and *neurotransmitters* which are complex molecules of a large variety of types, including serotonin, dopamine, tryptophan, histamine, etc. These neurotransmitters are released by the arrival of the spike at an axon terminal and enter a space between this presynaptic terminal and the postsynaptic dendrite of the receiving neuron, called the *synaptic cleft*. Upon reaching the dendrite of the postsynaptic neuron, the process is repeated in the postsynaptic neuron. This is again a simplification. As noted, neurons may fire in the absence of an incoming action, or may not fire even in the presence of an incoming action potential. (Our formalism will actually be able to take account of this characteristic *fragility* of life processes: cf. section 2.4.1). There also may be direct electrical contact between axon terminals and dendrites.

A set of equations describing the progress of the action potential, called the Hodgkin–Huxley model, was proposed in 1952 and was considered a great advance in biophysics, winning for its authors two-thirds of the Nobel Prize in Physiology and Medicine in 1963. The elucidation of the indirect neurotransmitter mechanism, by J. C. Eccles, garnered for its author the other third of the Prize.

The morphology described above may be dubbed the "standard model" of a neuron's structure, and in fact there is a vast collection of variations on it. However, the neurons in the cognitive areas of interest to us do conform to this model. Please see the second CAVEAT in section 2.3 and section 5.4.3 for more on this.

In addition to this complexity, mention must be made also of a plethora of other chemical and electrochemical elements in the environment of any collection of neurons, and the presence of other neuronal bodies within the interstices of the surrounding tissue. We will come to discuss some of these later.

It used to be thought that "computation" takes place in the somata of neurons, a reasonable and obvious conclusion which has colored the traditional approach to neuronal network models. This idea seems to be in conflict with another idea common to discussions of such neural networks, namely that "neurons have no memory." It is difficult to envisage computation without memory of some kind. In fact the issue of storage capability in a looser sense than memory *per se* will emerge as fundamental to the model we posit later. The view of somatic computation has slowly shifted towards the idea that the locus of computation, or what amounts to computation in such systems, actually lies elsewhere, specifically in the synapses. Our model favors this shift of focus.

2.2   Quantum-like networks and their combinations

2.2.1   *Standard neural networks*

The original such networks, for instance those of McCulloch and Pitts and their precursors, had single nodes playing the rôle of neurons. (For a superb account, with historical and philosophical perspectives, see [72].) These generally carried a single bit whose value, 0 or 1 denoting something like *firing* or *not firing*, was determined by addition (or *integration*) or other collective cumulative effects, of the input from the fanning in vertex structure. The single but fanned out output signal was determined by a neuronal threshold value or its analog mimicking the behavior of a biological neuron. There are many variations on this basic structure. The parameter windows of clusters of such nodes are then of the type considered in Example 3 of section 1.4.2. Their ortholattices of propositions are Boolean and so their "behavior" is classical.

An advance was made in 1982 [6, 46] by Hopfield who introduced what was essentially an energy parameter into such networks in analogy with the quantum mechanical system known as the *Ising* model. This allowed optimization and other schemes to be implemented relative to this energy function. As these models evolved it became clear that the techniques involved required by the AI type of applications became intractable if only binary data were involved. This led inevitably to more general continuous parameter ranges to be invoked for the neural units, essentially as in Example 5 of section 1.4.2 (cf. [4]).

For clusters of such neuronal units, it is clear from Corollary A.1.4 that even if only a couple of them are of the continuous, zero straddling interval type, as in Example 5 of section 1.4.2, then the parameter window of the whole cluster will be non-Boolean. So a quantum Ising model-inspired energy function, despite its apparently *ad hoc* nature, might indeed be viable and not obviously inconsistent. And indeed this seems to be the case for such models which have proved useful in their applications to artificial systems. However, when it comes to attempts to use them to explain the workings of actual biological systems, they fall short, for reasons both obvious and not so obvious. (Please see the penultimate paragraph of this section.) Since explaining biology was not the intent of such models, they are not to be taken to task for this.

In this section we shall attempt to derive a Hopfield Ising-like model from first principles for such networks as a limbering up exercise for what is to follow.

First we consider clusters of nodes and synaptic-like connections between such clusters. A node in this case represents the entire soma of a would-be neuron. These contain real values (*à la* membrane potentials). In Chapter 1 we argued that, under the hidden variable doctrine, it is only the state-based logic of propositions that is relevant to our considerations: the actual values inside the nodes at any moment are not relevant at the logical level, which is entirely structural. The state space for a collection of n nodes holding real values (read here as cell membrane — and later as parts of cell membrane — potential values) may be taken to be the product of (closed) intervals $W := \prod_{i=1}^{n}[s_i, t_i]$, where $[s_i, t_i] \subset \mathbb{R}$, the real numbers, with $s_i < t_i$ for all i, denotes the range of values allowed to the i-th node. The relevant associated logic of propositions is in this case modeled by the ortholattice of subsets of W that are intersections of W with subspaces of $\mathbb{R}^n$ with the origin removed. In case $n > 1$ the behavior of such a collection of nodes was shown in Chapter 1 to exhibit quantum-like behavior if the origin of $\mathbb{R}^n$ is an interior point of W. That is to say, within the space of states, there is a Euclidean open set containing the origin. (The origin itself is excluded as a state in these models.) In this case the relevant ortholattice is isomorphic with the lattice of subspaces of $\mathbb{R}^n$. This circumstance is seen to hold in that biological context in which neuron membranes, and portions of membranes, may sustain potentials which are both positive and negative.

Consequently, no logical harm is done in these cases if $\mathbb{R}^n$ is itself adopted as the space in which the pure states of the system are represented by rays, or equivalently, by normalized vectors which uniquely represent the rays they lie in. (Recall that a *pure* state is one which is not itself a convex superposition of other pure states). Here $\mathbb{R}^n$ is considered as a real Hilbert space with the Euclidean inner product. The quantum-like nature of such systems manifests as the existence of *superpositional* states, which realize disjunction in **OL**. Such a system, like Schrödinger's cat, may be in a superposition of pure states, such as *dead* and *alive*, or ON and OFF, without being "in" any component pure state.

Let us now specify that a cluster $\mathcal{N}_A$ of N nodes are labelled in a certain order $n_1^A, \ldots, n_N^A$. Then the real N-dimensional vector space of the corresponding (real) vector of nodal values $(v_1^A, \ldots, v_N^A)$ is spanned by the basis of vectors e_i^A, which is the vector with 1 in the i-th position and zeroes elsewhere. Thus a vector of the form

$v^A e_i^A$ is now interpreted as representing the presence of the value v^A in the node n_i^A. A possible vector of values may be expressed as

$$\mathbf{v}^A = \sum_{i=1}^{N} v_i^A e_i^A \qquad (2.2.1)$$

where v_i^A is the value *in* the node n_i^A. In general these values will depend on time. We shall for the time being denote this "state" space by $\mathfrak{H}_A \cong \mathbb{R}^n$ and regard it as a real Hilbert space with the usual Euclidean inner product as above.

CAVEAT!
There is a hidden assumption in this formulation which will occasionally be retracted in future. Namely we have assumed that each node may be labeled by a unique state, such as e_i^A, which is not taken to be a superposition of other states of the node.

Now, with $\mathcal{N}_B$ denoting another such cluster, with M, say, nodes, we may connect $\mathcal{N}_A$ "synaptically" to $\mathcal{N}_B$ by connecting each node n_j^A to certain nodes of $\mathcal{N}_B$ (or none of them) by single links. (We denote this informally by $\mathcal{N}_A \rightarrow \mathcal{N}_B$.) We shall suppose that the synaptic connections are such that as a result of n_j^A firing at some time t a value is added to the contents of the connected $\mathcal{N}_B$ nodes a short time τ later. This presynaptic value, $v_j^A(t)$, in n_j^A when it is firing is assumed to be non-zero. We shall assume that the postsynaptic graded potential induced upon the connected $\mathcal{N}_B$ node n_i^B, via a single synaptic link, at the later time $t+\tau$ may be expressed in the form $\tau \alpha_{ij}^{AB}(t) v_j^A(t)$, linear in τ for small enough τ. (It is worth noting here the "wrong way around" indexing convention, which we will be adhering to: namely $(\ )_{ij}$ signifies the direction $j^{\text{th}} \rightarrow i^{\text{th}}$. Its choice comports with the standard conventions of the matrix formulations to follow.) Here $\alpha_{ij}^{AB}(t)$ is some real scaling factor, or weight. It must be multiplied by the number of single links from n_j^A to n_i^B at time t which we shall denote by $\Lambda_{ij}^{AB}(t)$: thus the value to be added to the contents of n_i^B at time $t+\tau$ is $\tau \Lambda_{ij}^{AB}(t) \alpha_{ij}^{AB}(t) v_j^A(t)$. We note that $\Lambda_{ij}^{AB}(t)$ can be zero. Our assumptions require that the value $\alpha_{ij}^{AB}(t)v_j^A(t)$, associated with a single synaptic link $n_j^A \rightarrow n_i^B$, shall be the value of the graded potential in the postsynaptic node n_i^B at a later time. If this value is denote $p_{ij}^{AB}(t)$ we require $p_{ij}^{AB}(t) = 0$ when $v_j^A(t) = 0$: if the presynaptic value is zero, there should be no subsequent contribution to the postsynaptic potential. Then we may take

$$\alpha_{ij}^{AB}(t) = \begin{cases} \dfrac{p_{ij}^{AB}(t)}{v_j^A(t)} & \text{if } v_j^A(t) \neq 0 \\ 0 & \text{if } v_j^A(t) = 0 \end{cases} \tag{2.2.2}$$

which will reproduce the desired case.

We assume that all the postsynaptic values arrive within the interval τ. Thus, taking into account all the connections from $\mathscr{N}_A$ to $\mathscr{N}_B$, we have the following increment or decrement

$$v_i^B(t+\tau) - v_i^B(t) = \tau \sum_{j=1}^{N} \Lambda_{ij}^{AB}(t)\alpha_{ij}^{AB}(t)v_j^A(t) \tag{2.2.3}$$

or, in an obvious notation

$$\mathbf{v}^B(t+\tau) - \mathbf{v}^B(t) = \tau J^{AB}(t)\mathbf{v}^A(t). \tag{2.2.4}$$

Thus, the connection is characterized by the matrix

$$J^{AB}(t) = (\Lambda_{ij}^{AB}(t)\alpha_{ij}^{AB}(t)). \tag{2.2.5}$$

CAVEAT!
From now on, we shall assume our Λ_{ij}, α_{ij} and hence the matrices J are constant in time.

Thus equation (2.2.3) now reads

$$v_i^B(t+\tau) = v_i^B(t) + \tau \sum_{j=1}^{N} J_{ij}^{AB}v_j^A(t). \tag{2.2.6}$$

Here, the J_{ij} $(i, j \geqslant 1)$ act as *scaling factors* or *weights*, which multiply or scale the nodal values that are operated upon.

The matrix J^{AB} defines a linear map (for which we shall use the same symbol) $J^{AB} \colon \mathfrak{H}_A \to \mathfrak{H}_B$. (The linear structure emerges from the fanning in and out of the basic neural connection but it also preserves the ortholattice structure (and therefore also the logic) of the state spaces.)

This $\mathbf{v}^B(t + \tau)$ is a superpositional state of $\mathcal{N}_B$ which omits the additive, or superpositional, interference of other possible inputs into $\mathcal{N}_B$ from sources other than $\mathcal{N}_A$ during the period between t and $t+\tau$. This extraneous input is often simulated by adding a stochastic term, as in [46]. This is certainly consistent with a quantum-like underlying model and we shall return to it later.

In summary, we have interpreted the case of a cluster $\mathcal{N}_A$ being connected to a cluster $\mathcal{N}_B$ via irreducible (single) axonal/synaptic connections in terms of a linear map $\mathfrak{H}_A \to \mathfrak{H}_B$ at time t. That is, both the wiring "diagram" and the values transferable at the time t are encoded in equation (2.2.6) (while ignoring other sources of input to $\mathcal{N}_B$).

Definition 2.2.1. A *unicameral neural net* is a cluster $\mathcal{N}_A$, equipped with a connection $\mathcal{N}_A \to \mathcal{N}_A$ as above.

The matrix $\Lambda^{AA} = (\Lambda_{ij}^{AA})$ is in this case called the *adjacency matrix* of the network. In what follows, if the network is unambiguous we shall generally drop the superscripts. If we deal with a labeled network we shall use the unrepeated label as a superscript where necessary, as in Λ^A, J^A, etc. The matrix J is often called the *connection* matrix of the network. Its entries are not to be considered among the variables to be hidden but are rather part of the structure of the network.

Now let $\mathcal{N}$ denote such a network. In standard treatments the *efficacy* of a synaptic link from node n_j to node n_i is a real number corresponding to the "amount of penetrating current per synaptic spike" going from the node n_j to the node n_i along the synapse. It is usually denoted by J_{ij} [6]. Our choice of the same notation above is not coincidental, as we shall see later (section 2.4.1).

This matrix J of efficacies would seem to encapsulate a complete description of the network, and the operator it defines upon the state space would be irresistibly reminiscent to a physicist of the matrix of *single particle exchange integrals* of the Ising model. This model describes the quantum behavior of regularly arranged ensembles of magnetic dipole moments having only two spin states. There is not very much further one can go with any actual quantum mechanical manipulation of this network model for reasons to be explained. About the only obvious thing to do is to consider the *expectation value* of this operator in any state $\mathbf{x}$. Namely

$$\langle \mathbf{x} \,|\, J \,|\, \mathbf{x} \rangle = \mathbf{x}^T J \mathbf{x} \tag{2.2.7}$$

$$= (x_1, x_2, \dots) \begin{pmatrix} J_{11} & J_{12} & \cdots \\ J_{21} & J_{22} & \cdots \\ \vdots & \vdots & \ddots \end{pmatrix} \begin{pmatrix} x_1 \\ x_2 \\ \vdots \end{pmatrix} \tag{2.2.8}$$

$$= \sum_{i,j} x_i J_{ij} x_j \tag{2.2.9}$$

$$= \sum_{i,j} J_{ij} x_i x_j. \tag{2.2.10}$$

This expression, preceded by a minus sign, is the basis of the Hopfield model. It is reminiscent, though in non-operator form, of the kinetic part of the Hamiltonian of the Ising model which is known to be rather intractable computationally even in its simpler physical incarnation. In our network case there are generally no geometrical symmetries such as those that ameliorate matters in the physics case, nor are there known analogues in the network case of the known physical force laws that structure the J_{ij} in the physics case, not to mention other simplifying attributes. However, biologically non-viable conditions may be imposed to ameliorate the computations, if the aim is to enable artificial computational paradigms, which historically it largely was. Among these conditions are the requirements of full connectivity of the network, symmetry of the J_{ij}, and no self-connections: $J_{ii} = 0$. All of these are clearly not viable in a model that seeks to mimic the behavior of actual neuronal networks, though workarounds can be concocted [6].

As we saw in Example 1 of section 1.4.2, a state space of one real dimension must describe classical behavior but ensembles of them can and do have state spaces that describe quantum-like behavior, as in Example 5 of that section. So it does not strain the credibility of the Hopfield-like models to apply quasi-quantum methods to them. However, there is a price to be paid. And it is paid in the lack of an inherent statistics of many such systems when taken together as in a network or cluster. In fact there is no non-trivial way we can arrange the combination of one-dimensional real spaces in a quantum-like manner. (For, anticipating further developments somewhat, the general recipe to combine quantum-like systems is to take the tensor product of their state spaces and then impose upon it an algebraic structure depending on the statistical nature ascribed to the quanta. In this case $\bigotimes^n \mathbb{R} \cong \mathbb{R}$ and this cannot reproduce the assumed state space $\mathbb{R}^n$ whatever statistics we ascribe to the underlying systems.) For this reason we will not further pursue possible dynamics for such networks.

As we shall see, by expanding this model and taking care to be consistent with its inherent statistics, we will be rewarded by finding collective effects that are apparently present in actual brains.

2.2.2 *Combinations of networks and their state spaces*

We shall shortly introduce an expanded version of the basic neuronal unit which more closely resembles the biological case. Before doing so we consider two ways of combining network states spaces.

Both are of course quantum-like, and are beyond the remit of classical discussion. This is both a handicap to classically chauvinist thinkers but also a liberation in certain respects, as we shall see.

2.2.3 *The tensor product $\otimes$*

In nature there are various mechanisms whereby ensembles of neural cells, or parts of cell membranes (which we will also model by nodes shortly), become associated co-operatively with each other, either by sharing the same neuromodulatory chemical environment, volumetric medium, or because they are jointly orchestrated or tuned, for instance by a shared substrate of interneurons or other cells, exosomes, and other extracellular bodies implementing a level of electrochemical communication independently of the synaptic structure [87]. In this kind of combination (or pairing in the case of two networks) a node n_i^A of a network $\mathcal{N}_A$ may be associated with a node n_j^B of a network $\mathcal{N}_B$ in such a way that the pair (n_i^A, n_j^B) may act as a unit, which contains a single value. The state of such a unitized pair is well modeled by the tensor product $e_i^A \otimes e_j^B$ of the corresponding states (cf. section B.1.1 for a brief account). A value assigned to this paired state cannot be attributed to any one constituent state since $\lambda e_i^A \otimes e_j^B = e_i^A \otimes \lambda e_j^B = \lambda(e_i^A \otimes e_j^B)$, $\lambda e_i^A \otimes \mu e_j^B = \mu e_i^A \otimes \lambda e_j^B = \lambda\mu(e_i^A \otimes e_j^B)$ and similarly for any number of tensorial tuples. That is to say, in keeping with the doctrine of hidden variables, such a combined tuple of nodes *hides* the location of a value λ: for instance the state $\lambda(e_i^A \otimes e_j^B) = \lambda e_i^A \otimes e_j^B = e_i^A \otimes \lambda e_j^B$ cannot distinguish between the circumstance that the value λ is in the i-node and the circumstance that it is in the j-node. Thus the space generated by all possible such pairs is the tensor product of the individual spaces,

$$\mathfrak{H}_A \otimes \mathfrak{H}_B. \tag{2.2.11}$$

Although the basic indecomposable tensors represent discernible possible classical unitized combinations of the nodes of the component networks, the general elements of this space, being linear combinations of them, of course represent superpositions and as such defy classical interpretation. Such superpositions are not available to classical constructs such as standard networks. The ability of this model to take such *substrate* or interstitial liaisons into account is another way in which the model differs from the standard neural net model.

In more detail, an element of $\mathfrak{H}_A \otimes \mathfrak{H}_B$ is a superposition of states of pairs of nodes tightly bound by a substrate connection, of the form $e_i^A \otimes e_j^B$. Thus an element in this space is of the form

$$\theta = \sum_{i,j} \lambda_{ij} e_i^A \otimes e_j^B \tag{2.2.12}$$

where some or all of the real factors λ_{ij} may be zero. These are possible quantum-like states of the combined system, which may never find itself in any of them. When in such a state, the combined system will be in a state confusable with the state $e_{i_0}^A \otimes e_{j_0}^B$, in which the node $n_{i_0}^A$ is paired with node $n_{j_0}^B$, with probability

$$\text{prob}(\theta \to e_{i_0}^A \otimes e_{j_0}^B) = \frac{\lambda_{i_0 j_0}^2}{\sum_{i,j} \lambda_{ij}^2}. \qquad (2.2.13)$$

In this way the details of the microscopic computations proceeding inside each node, and among an ensemble of co-acting nodes, are hidden, in keeping with our doctrine of hidden variables, while intracellular effects are taken into account. This closely mimics the way in which interneurons among the other mechanisms mentioned above orchestrate the activity of principal neurons. Moreover, should one of these values momentarily become zero, then the probability of the combined state being available (or firing) falls to zero. This is both synchronizing and inhibiting. It simulates in a vastly simplified but effective manner the kind of inhibitory synchronization effected by biological interneurons, a hugely diversified group including basket cells and chandelier cells, which connect local principal neurons and induce very rapid inhibitory signalling between networks of them, having synapses of both chemical and electrical types. Of course, the joint value could go up, and in that case there are problems for the neural applications we have in mind. Cf. sections 2.3 and 2.5 for further discussion.

In order for the tensor product of state spaces of nodes to reflect the actual states of an underlying system it is necessary for such substrate connections to be present in the underlying system. If they are not, then there may be no state in the tensor product that refers to an actual state of the system. This could happen in a biological system, in which case we may claim that the tensor product may not be taken. This differs from the case of actual quantum physics in most of its applications.

Moreover, since the phenomenon of superposition occurs in our quantum-like systems, tensor products will inevitably produce quantum-like *entanglement* among tensor product states.

(For an account of entanglement in the actual quantum case see [69].) And just as superpositional states may be regarded as substitutes for the multiplicity of hidden complex microscopic chemical states, so entangled states may be regarded as stand-ins for the complex hidden microscopic chemical and/or volumetric extrasynaptic communication channels, both internal and external.

We reiterate that the tensor product of state spaces is the state space of the combined system if the general substrate connection is available to "glue" the pairs of nodes together. And these pairs are evanescent, since the glue may ebb and flow and change its targets in time. (This would be reflected in the change in the coefficients in expansions such as equation (2.2.12)). It is to be noted that because of this evanescence we are not positing a new *network* structure on these unitized pairs other than the ones already in place. Such a structure could be described if we were willing to posit connections between the unitized pairs if and when they happen to exist, but there will be no need for us to do so.

There will be other reasons to delimit the use of the $\otimes$ combinator, which we shall address as the need arises.

2.2.4 *The external direct sum $\oplus$ and multimodality*

The second combinator presents certain difficulties in application, some of which can be ameliorated to good effect within the quantum-like paradigm. One problem that remains with it is that, unlike the previous case, no known mechanism implements it, though there must be one in actual brains, as we shall see.

Let us first recall the definition of the *external direct sum*, $\mathfrak{H}_A \oplus \mathfrak{H}_B$ of the two vector (or Hilbert) spaces $\mathfrak{H}_A$ and $\mathfrak{H}_B$. It is simply the Cartesian product $\mathfrak{H}_A \times \mathfrak{H}_B$ endowed with the coordinate-wise operations:

$$\lambda(v, w) := (\lambda v, \lambda w); \tag{2.2.14}$$

$$(v_1, w_1) + (v_2, w_2) := (v_1 + v_2, w_1 + w_2); \tag{2.2.15}$$

and with inner-product

$$\langle (v_1, w_1)|(v_2, w_2)\rangle := \langle v_1|v_2\rangle + \langle w_1|w_2\rangle. \tag{2.2.16}$$

If $\mathfrak{H}_A$ has a basis $\{e_1^A, \ldots, e_m^A\}$ and $\mathfrak{H}_B$ has a basis $\{e_1^B, \ldots, e_n^B\}$ then it is immediate that $\mathfrak{H}_A \oplus \mathfrak{H}_B$ has a basis $\{(e_1^A, 0), \ldots, (e_m^A, 0),$ $(0, e_1^B), \ldots, (0, e_n^B)\}$ so that $\dim(\mathfrak{H}_A \oplus \mathfrak{H}_B) = \dim \mathfrak{H}_A + \dim \mathfrak{H}_B$. Maps between spaces determine maps between their direct sums in an obvious way.

(We shall reserve the use of the notation $\oplus$ for the *external* direct sum of vector spaces. When subspaces of a vector space are the summands in such a direct sum, this direct sum will coincide with the algebraic direct sum — namely the smallest subspace containing all the summands, which we denoted $\sqcup$ in Chapter 1, the join in the ortholattice of subspaces — only if these subspaces satisfy a disjointness condition. Cf. [62], Proposition 4.6.)

Now suppose that we have two (non-trivial) networks $\mathcal{N}_A$ and $\mathcal{N}_B$ with corresponding state spaces $\mathfrak{H}_A$ and $\mathfrak{H}_B$. Of course the space $\mathfrak{H}_A \oplus \mathfrak{H}_B$ can always be formed, but is it the state space of any entity materially related to the original networks? That is to say, is it the state space of any entity other than a cluster of nodes in one-to-one correspondence with a basis, which exists trivially for all finite dimensional vector spaces? Clearly if the networks are disjoint, having no nodes in common, we may, *ceteris paribus*, consider the two networks taken together, with no further connections being posited between the nodes respectively of $\mathcal{N}_A$ and the nodes of $\mathcal{N}_B$, as a single network with $\mathfrak{H}_A \oplus \mathfrak{H}_B$ as its state space and $J^A \oplus J^B$ as its connection matrix, with the nodes in one-to-one correspondence with the elements in the basis shown above. (In later applications even this possibility will be called into question.)

But what if a node is shared? A simple example may suffice.

Two networks are shown in Figure 2.1. The state space of $\mathcal{N}_A$, with our conventions in place, is $\mathfrak{H}_A \cong \mathbb{R}e_1^A \oplus \mathbb{R}e_2^A \oplus \mathbb{R}e_3^A$, where we have explicitly written the generator of each one-dimensional component space. And similarly for the state space $\mathfrak{H}_B$ of $\mathcal{N}_B$.

If the networks are disjoint, as shown in Figure 2.1, then the two taken together, *ceteris paribus*, with no other connections posited, constitute a single network with state space $\mathfrak{H}_A \oplus \mathfrak{H}_B$.

But what if n_3^A and n_2^B coincide and are actually the same node? Would $\mathfrak{H}_A \oplus \mathfrak{H}_B$ now represent the state space of a non-trival network?

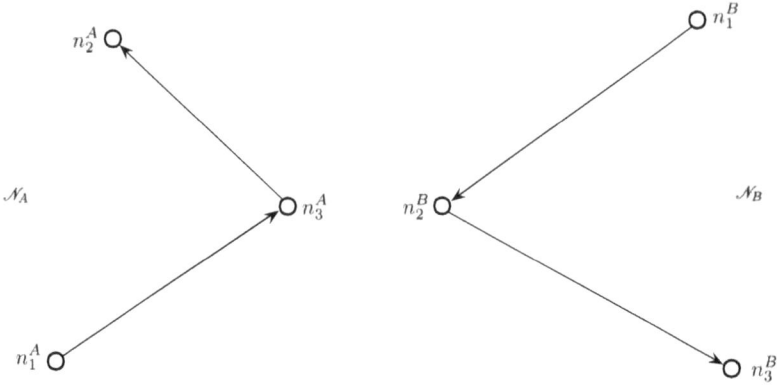

Figure 2.1 Two disjoint networks.

It has a basis $\{(e_1^A,0),(e_2^A,0),(e_3^A,0),(0,e_1^B),(0,e_2^B),(0,e_3^B)\}$. Nodes $n_1^A, n_2^A, n_1^B, n_3^B$ each have unique corresponding (pure) states, but the common node corresponds to *two* different (pure) states $(e_3^A,0)$ and $(0,e_2^B)$. To say they are different means saying the vectors are linearly independent. (They remain independent even if $e_3^A = e_2^B$, which is another problem with this construct.) Consequently, if we allow certain nodes to possess *at least* two states so that they can be in superpositions of these two states, then $\mathfrak{H}_A \oplus \mathfrak{H}_B$ does in fact represent the state space of the network depicted in Figure 2.2, if the common node is one of these *at least bimodal* nodes. In future we shall describe such nodes as just being *bimodal*, the *at least* conditional being understood.

The shared node itself has a state space spanned by the states $(e_3^A,0)$ and $(0,e_2^B)$. The connection matrix is still $J^A \oplus J^B$. As noted the shared node is bimodal: it could exist in a superposition of (at least) two states, one by virtue of its membership of $\mathcal{N}_A$ and one by virtue of its membership of $\mathcal{N}_B$. If it were a neuron or part of a neuron, it would presumably respond to two different *modalities* and, since the real scalar coefficients in a superposition are changing in time, so would the probabilities of the node being in one or other of these two states. It may appear to fluctuate between the states and, as explained in section 1.3.1, achieve an analogue of the effect known to communications engineers as time division multiplexing of

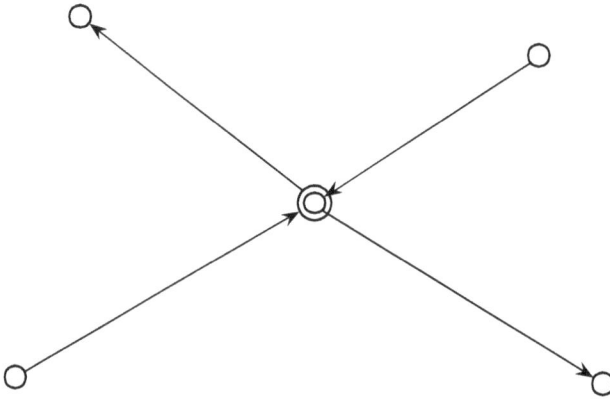

Figure 2.2 The central node has two states.

spiking behaviors. Such cells, should they exist, could achieve high efficiencies since they could implement sparse neural coding. The existence of such cells has been conjectured for at least a couple of decades and there is mounting evidence that such cells do in fact exist in mammalian cortices: cf. [18]. (The reference [67] is also likely to be relevant.) However, this property of multimodality, also called polymodality, seems to be confined to the sensorium, namely those areas concerned with physical sensations, such as visual perception, touch, taste, smell, etc., not to the cognitive areas, although there must be some relations between these areas. (The case for more general multimodality has been mooted: cf. [18]).

The upshot of these considerations is that if two networks are not disjoint and their shared nodes are all bimodal then, *ceteris paribus*, the external direct sum of their state spaces may be taken as the state space of a non-trivial network, though one in which the shared nodes have two-dimensional state spaces. Similar conclusions hold for direct sums of many networks, in which case *multimodal* would be the term of art for the shared nodes. If at least one node is not multimodal, the direct sum is not *implemented*, meaning that it cannot be realized as the state space of a relevant network. In this case the status of the set of ostensibly different networks must be reappraised: perhaps it was a single network.

It may appear to be too good to be true, or an unlikely happenstance, that a family of networks may have an intersection all of whose nodes have exactly the right multiplicity of modalities, but in reality it is likely that the multimodal nodes are common to the various networks *because* they are multimodal. It is a question of *propter hoc*. (Please see section 5.4.1 for the discussion of a special case.)

This option of multimodality is of course not open to classical networks and probes the essential difference between classical and quantum-like ontologies, namely the possibility of superposition of several possible states in the latter case.

This discussion seems to exclude the simple possibility that two networks just overlap in shared nodes of any modality. The $\oplus$ construction does in fact exclude this case, which we shall regard as a case of a single network, requiring reassessment as above, since a mere renumbering of the whole ensemble, which entails a numbering of the union of sets of constituent nodes, would produce this single network of which the two original networks are both subnetworks. The state space of the ambient network in this case is the Hilbert space whose nodal basis is the union of the original two nodal bases. As noted, this is in general neither the internal nor the external $\oplus$.

This $\oplus$ is the construct we shall use in implementing the logic introduced later (Chapter 3), where it will be interpreted as a sort of disjunct. These often seem to be problematical from a logical point of view, as in intuitionistic logic, quantum logic, *et al.*, and our case is no exception. (For readers interested in general issues of disjunction the book [51] is strongly recommended.) Among the subtle issues raised here is in the actual implementation of $\oplus$ even in the case of disjoint networks. How may it be determined that the two networks do or do not belong to the "same" system? If it is determined that they cannot both be regarded as subsystems of some discrete system, then the direct sum may not be used. As we shall see (first in section 5.6.2) this possibility has deep ramifications.

Definition 2.2.2. In case the $\oplus$ of two state spaces A and B, $A \oplus B$, of two networks $\mathcal{N}_A$ and $\mathcal{N}_B$ may be properly implemented as the state space of a canonical network as above, we shall denote that network by $\mathcal{N}_{A \oplus B}$.

2.3 The bicameral neuron

CAVEAT!
Unless explicitly stated otherwise, all nodes henceforth will be considered to be unimodal. Multimodality will be considered separately and well advertised.

Our main model of a neuron will comprise a system of two, more primitive, value-bearing nodes whose individual parameter window, a rectangle in the real plane, may be such that it exhibits quantum-like behavior, unlike the case of a single node model of a neuron, which must remain classical.

Actual neurons are generally extended in space, some of them having for instance very long axons, others having large somas, such as pyramidal cells, and this has consequences which are generally ignored in neural net models. In this work we shall simulate such an extension by resolving our single node, or *unicameral*, neuron model into one having two "chambers." The first is the input chamber corresponding to that part of the soma to which the dendrites are attached (depicted by the sub-oval on the left in Figure 2.3), while the second is the output chamber, corresponding to the hillock (or trigger zone)–*cum*–axon system, or axon initial segment, along which the action potential is conducted upon its firing (depicted by the sub-oval on the right in Figure 2.3). In this way we can accommodate the values of the membrane potential at the two extreme ends of the neuron, which will generally be different. Once the graded input potential achieves the threshold value, assuming it does, the action potential is triggered and occupies the output chamber, depicted by the rightmost oval in the diagram. We shall assume that when, and very shortly after, the action potential is triggered, the entire soma, and hence both chambers, achieve the same potential. After this very short period, as the action potential begins to move down the axon, the two nodes are gain independent of each other. This arrangement includes the case of neurons that fire independently of the dendritic inputs, such as pacemaker cells.

CAVEAT!

There is a very wide diversity of neuron cell types. Not all of them conform to this standard model, as was noted by Cajal himself. An important class of such cells is epitomized by dopaminergic cells: these have axons emanating directly from the dendrites, and so cannot be resolved into a bicameral structure [39]. The logic appropriate to a single such neuron, which is effectively unicameral, would therefore seem to remain classically Boolean. Clusters of them may be prone to a collective logic that is not Boolean, but in this case, possibly because of their rôle as reward determiners, each bestowing or not bestowing rewards, the parameter windows of such clusters are likely to remain in the positive sector of the space of states with the consequence that the collective logic is Boolean. There is experimental evidence that this is the case for dopaminergic neurons: cf. section 5.4.3.

The diagram in Figure 2.3, and the model itself, is of course schematic and simplified. Real synapses and dendrites may often impinge anywhere on a neuron's soma.

The vertical and approximately vertical dotted lines on the left and right are meant to indicate that *many* more arrows might be involved: possibly in the many thousands. Since the basic neuron now comprises two nodes, its state space is two-dimensional, isomorphic to $\mathbb{R}^2 \cong \mathbb{R} \oplus \mathbb{R}$, where the first summand accommodates the state space of the input node on the left of Figure 2.3 and the second summand accommodates the state space of the output node on the right, and will be interpreted as a two-node system. Since we are assuming that the parameter window is of the type described in Example 5, section 1.4.2, straddling the origin, we shall assume quantum-like behavior in the general case. As such we have something very like a *qubit*, which is the quantum two-state analogue of the classical *bit*: the complex version is generally taken as the basic computational unit in the hopeful discipline known as quantum computing.

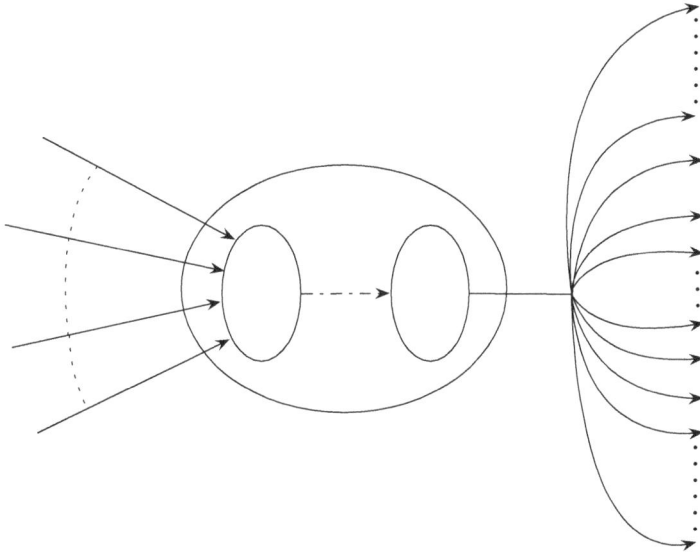

Figure 2.3 Bicameral neuron schematic.

We shall call a node of this type a *bicameral neuron* or *b-neuron*.
Denoting the generic vector in the state space $\mathbb{R}^2$ of such a two-node system by $\binom{\lambda}{\mu}$ we define a pair of operators

$$\mathfrak{a}\binom{\lambda}{\mu} = \begin{pmatrix} 0 & 1 \\ 0 & 0 \end{pmatrix}\binom{\lambda}{\mu} = \binom{\mu}{0} \tag{2.3.1}$$

$$\mathfrak{a}^T\binom{\lambda}{\mu} = \begin{pmatrix} 0 & 0 \\ 1 & 0 \end{pmatrix}\binom{\lambda}{\mu} = \binom{0}{\lambda} \tag{2.3.2}$$

where $(\ )^T$ denotes operator transpose or adjoint relative to the Euclidean inner product. These operators satisfy the following *anticommutation relations*:

$$\mathfrak{a}\mathfrak{a}^T + \mathfrak{a}^T\mathfrak{a} = 1, \tag{2.3.3}$$

$$\mathfrak{a}^2 = 0, \tag{2.3.4}$$

$$(\mathfrak{a}^T)^2 = 0. \tag{2.3.5}$$

It is quickly checked that these operators are indeed mutually adjoint. These relations identify the system as a fermion-like one, since it can be shown that they represent respectively the annihilation and creation of a fermion-like quantum of one type.

The operator $\mathfrak{a}^T$ represents the internal transfer of the current value in the input node from that node to the output node and thus represents a *preparation* to fire the neuron: if λ has reached the threshold value then, at that moment, it becomes a value of the action potential and moves into the trigger zone or hillock, replacing any value μ that may be there (equation (2.3.2)). We note that the operator $\mathfrak{a}^T$ is not an observable of the system, but rather an internal unobservable operation that in a sense stands for the electrochemical ionic flows, etc., that physically transfer the potential across the cell body, while hiding the hugely complex details. The dotted horizontal arrow in the diagram is meant to represent it.

The anticommutation relations listed above are formally the same as the relations satisfied by the *creation* operator, corresponding to our $\mathfrak{a}^T$, and the *annihilation* operator, corresponding to our $\mathfrak{a}$, for a fermionic (spin-$\frac{1}{2}$) quantum of one type, although acting in a real space (which may be complexified). The analogy may be misleading, since there are no actual physical fermions involved, but it reveals the underlying quantum-like behavior inherent in the logic. We shall adopt the deeply ingrained convention of choosing the un-transposed operator as the corresponding annihilation operator for these *virtual* fermions. Although there are no actual physical particles implied by our picture we note that the fermionic *number operator* $\mathfrak{a}^T \mathfrak{a}$ has the two eigenvalues 0 (with eigenspace the set of vectors $(\lambda, 0)$ having the output node at zero: the OFF state when normalized), and 1 (with eigenspace the set of vectors of the form $(0, \mu)$ having the input node at zero: the ON, or firable, or excited state when normalized). In general a b-neuron will be in a superposition of these two states, not necessarily in one or the other. An array of many such b-neurons may be expected to exhibit large scale behavior similar to that of an Ising-like model, or more generally, that of a *quasispin* model [64]. For a single b-neuron, let us suppose, as in ordinary quantum mechanics, that the (kinetic) energy observable, or (kinetic) Hamiltonian, is proportional to the number operator. Then the two energy eigenspaces are exactly the "classical" state spaces of the individual nodes, and this is consistent with the quantum mechanical view that these eigenstates are exactly the ones with definite energy values, which the system collapses into when the energy is measured or observed (possibly by another neuron). Note that $\mathfrak{a}^T \mathfrak{a}$ is a projection. Moreover, the minimal energy states of a b-neuron are the ones in which the output node is at *zero*. Consequently, under the usual physical rubric of minimizing energy, b-neurons would tend towards a state of having released their action potentials, or having *fired*. However, the overall behavior of a network of many b-neurons will be very different, as in the case of a physical many-quanta system. In section 2.4.1 we will derive the analogue of the Hamiltonian for such a network.

As a space of *states* of the underlying b-neuron this space has additional algebraic structure which captures an important aspect of actual neurons and reveals an improved mathematical model of the associated connective in the presence of a co-activating substrate that overcomes the problem we found with the $\otimes$-product in section 2.2.3. To see this, first note that insofar as the operator $\mathfrak{a}^T$ represents a preparation to fire the b-neuron, that fact that $(\mathfrak{a}^T)^2 = 0$ means that this preparation cannot be executed twice at the same time: and indeed a biological neuron cannot fire while it is already firing. Now consider the subspace of the algebra of operators on $\mathbb{R}^2$, namely $\operatorname{End}\mathbb{R}^2$, generated by the identity operator I and $\mathfrak{a}^T$, which may be written

$$\mathcal{A} := \mathbb{R}I \oplus \mathbb{R}\mathfrak{a}^T \subset \operatorname{End}\mathbb{R}^2. \tag{2.3.6}$$

Here we have explicitly written in the two generators I and $\mathfrak{a}^T$. It is immediately seen that $\mathcal{A}$ is in fact a (commutative) *subalgebra* of $\operatorname{End}\mathbb{R}^2$ in view of equation (2.3.5). Now note that the inclusion map $\iota : \mathbb{R}\mathfrak{a}^T \hookrightarrow \mathcal{A}$ is easily seen to satisfy the UMP of Appendix B.1.3 so that $\mathcal{A} \cong E(\mathbb{R}\mathfrak{a}^T)$ as an algebra, where $E(V)$ denotes the exterior algebra of the vector space V, also known as the Fermi-Dirac space of V. If we make explicit the state representing the b-neuron's output node, e say, then we may express the state space of the b-neuron in the form

$$E(\mathbb{R}e) = \mathbb{R} \oplus \mathbb{R}e \tag{2.3.7}$$

where $e^2 = e \wedge e = 0$, which will be seen to express the fact that the neuron cannot fire twice simultaneously and at the same time realizes the exterior product $\wedge$ as essentially the unique combinator modeling this property mathematically.

In the last equation we have explicitly written in the generator of the state space of the output node. When it comes to deciding on a symbol to use for the generator of the input node's state space we encounter a notational problem of the overloaded sort quite familiar in cross-disciplinary studies. Namely, the state space $E(\mathbb{R}e)$ is an *algebra* in which the generator of the input node is the unit, or identity, which may of course be identified with the number $1 \in \mathbb{R}$, the vacuum or 0-th grade component of $E(\mathbb{R}e)$. On the other hand, in the *occupation number* formalism we shall soon be adopting, which is the paradigm of interest to us here as it is in quantum computing, this state, namely the algebra unit, is the vacuum or state of zero occupancy and is usually denoted by $|\mathbf{0}\rangle$ (the OFF state). It is unfortunate that the state of zero occupancy must be represented by the scalar we denote by 1, but the mathematics of the situation forces our hand. To confuse matters further, e, which represents the state of single occupancy of the b-neuron, is in this formalism denoted by $|\mathbf{1}\rangle$ (the ON or firing state, or state of single occupancy). In this work we shall employ both notations, with warnings, depending upon which context is more compelling. Thus, to reiterate, $|\mathbf{0}\rangle = 1 \in \mathbb{R}$ is the OFF, or vacuum state, or state of zero occupancy of the b-neuron, and also the unit of the exterior algebra $E(\mathbb{R}e)$, while $e = |\mathbf{1}\rangle$ is the ON state, or state of single occupancy, or firing state of the b-neuron. And now we have $E(\mathbb{R}e) = \mathbb{R}1 \oplus \mathbb{R}e = E(\mathbb{R}|\mathbf{1}\rangle) = \mathbb{R}|\mathbf{0}\rangle \oplus \mathbb{R}|\mathbf{1}\rangle$, with $e \wedge e = |\mathbf{1}\rangle \wedge |\mathbf{1}\rangle = 0$ and $1 \wedge e = e \wedge 1 = 1.e = e = |\mathbf{0}\rangle \wedge |\mathbf{1}\rangle = |\mathbf{1}\rangle$.

We must admit the possibility that the output node of a b-neuron may be multimodal. We shall assume this possibility only for the output node, since we assume this is where the multimodality would be manifested. Then we would still have a bicameral neuron but the output node may have many states. That is to say, there will be different modes of being ON. Moreover, these modes can mix. For example, let us consider a b-neuron with a bimodal output node having two states e and e'. Then the state space of the b-neuron would be $E(\mathbb{R}e \oplus \mathbb{R}e') \cong \mathbb{R} \oplus \mathbb{R}e \oplus \mathbb{R}e' \oplus \bigwedge^2(\mathbb{R}e \oplus \mathbb{R}e')$ and there is a possible new state $e \wedge e'$ in which both states of the output node are occupied. This is possible but, as with all states, it may not ever be manifested.

2.3.1 *Bicameral networks*

The nodes of a network of such b-neurons must now be treated as (real) fermion-like systems and, as such, the corresponding space of states must be considered first.

Specifically, a b-neuron as specified in section 2.3 is equivalent to a (real) fermionic system with two pure states we have characterized as ON, corresponding to it having some non-zero value in its output node so that it is in a state of firing, and 0 in its input node, and OFF, in which there is a 0 in its output node: the state of not firing. A collection of such fermionic elements may then exist in states in which some are ON and some are OFF. We may think of such ON states as states of *occupancy*. For a collection (or network) of b-neurons, n_i, $i = 1, \ldots, N$, the states in which *single* b-neurons are ON, or occupied, is $\mathfrak{H} := \bigoplus_i \mathbb{R}e_i \cong \mathbb{R}^N$, where e_i is the state corresponding to the n_i^{th} b-neuron's *output* node. The space of states of two b-neurons being ON, or occupied, is $\bigwedge^2 \mathfrak{H}$, because they are fermionic, and so on. The space of no b-neuron being on, or zero occupancy, is the space generated by that single state also called the *vacuum* state. Thus the space of all occupancies of the whole ensemble or network of b-neurons is

$$\mathbb{R} \oplus \overset{2}{\bigwedge} \mathfrak{H} \oplus \cdots \oplus \overset{N}{\bigwedge} \mathfrak{H}.$$

That is to say it is the exterior algebra of $\mathfrak{H}$. The existence of the fermionic structure of the individual b-neurons has determined the choice of the exterior algebra on the space $E(\mathfrak{H})$. As noted in section B.1.3, this makes the vector space isomorphism $E(\mathfrak{H}) = E(\bigoplus_i \mathbb{R}e_i) \cong \bigotimes_{i=1}^N E(\mathbb{R}e_i)$ (equation (B.1.15)) an isomorphism of algebras if the *graded product* is used on the right. The Fermi-Dirac space/exterior algebra represents the space of states of *co-acting* or *co-spiking* subcollections of b-neurons. We shall interpret this space also as the space of *firing patterns*. We should emphasize here that these spaces of states are *possibility* spaces: they describe what is possible. An actual system (i.e. network) may or may not fulfill the requirement for the actual existence or otherwise of certain such states.

We denote by a_i, $a_i^\dagger$, $i = 1, \ldots$, respectively the usual annihilation and creation operators on this space. The dagger here denotes transpose since we are working over the real numbers. Thus

$$a_i^\dagger(e_{j_1} \wedge \cdots \wedge e_{j_m}) = e_i \wedge e_{j_1} \wedge \cdots \wedge e_{j_m} \qquad (2.3.8)$$

from which it follows that

$$a_i(e_{j_1} \wedge \cdots \wedge e_{j_m}) = \delta_{ij_1}(e_{j_2} \wedge \cdots \wedge e_{j_m}) - \delta_{ij_2}(e_{j_1} \wedge e_{j_3} \cdots \wedge e_{j_m}) + \cdots$$
$$\cdots + (-1)^{m+1}\delta_{ij_m}(e_{j_1} \wedge \cdots \wedge e_{j_{m-1}}). \qquad (2.3.9)$$

The 0 index will be reserved for the vacuum, or unit in the exterior algebra. Thus $e_0 = 1$. Summations will generally be over the non-zero indices. (Note that these entail $a_i^\dagger(1) = e_i$ and $a_i(e_j) = \delta_{ij}1$.)

The last equation (2.3.9) may be described as follows: a_i has the effect on an element $e_{j_1} \wedge \cdots \wedge e_{j_m}$ of giving 0 if e_i does not appear. If it does appear, shuffle it to the front keeping track of the sign changes, and then remove it. These operators obey the *canonical anticommutation* relations:

$$\{a_i, a_j^\dagger\} = a_i a_j^\dagger + a_j^\dagger a_i = \delta_{ij}I, \qquad (2.3.10)$$

$$\{a_i, a_j\} = 0, \qquad (2.3.11)$$

$$\{a_i^\dagger, a_j^\dagger\} = 0, \quad \text{from which it follows that}$$

$$a_i^2 = (a_i^\dagger)^2 = 0. \qquad (2.3.12)$$

These circumstances lead to a useful paradigm known as the *occupation number* formalism [64]. The idea is to take account of the right hand side of the identity (B.1.15) to develop a compact vector-like notation in which each i-entry may take values characterizing the ON and OFF states in $E(\mathbb{R}e_i)$. For, each *number operator* $a_i^\dagger a_i$ has the eigenvalues 0 and 1. The eigenvalue 0 has as eigenvector the state $|\mathbf{0}\rangle := 1 \in \mathbb{R}$ this $\mathbb{R}$ being the first summand in $E(\mathfrak{H})$, which is regarded as the state of zero occupancy, containing no "particles," also called, as noted, the *vacuum* state. In our context, $|\mathbf{0}\rangle$ is the state in which no b-neurons are firing (though there may be "hidden" values, such as a resting potential, in the input nodes). The other eigenstate is the ith ON state, in which there is a (necessarily) single i^{th} occupancy, corresponding to the state of firing of the ith b-neuron.

Let us write a general state of occupancy of the network system in the form

$$|\nu_1, \nu_2, \ldots, \nu_N\rangle \tag{2.3.13}$$

where

$$\nu_i = \begin{cases} \mathbf{1}_i & \text{if b-neuron } n_i \text{ is ON, or occupied or firing;} \\ \mathbf{0}_i & \text{if b-neuron } n_i \text{ is OFF, or unoccupied or not firing.} \end{cases} \tag{2.3.14}$$

If all the b-neurons are OFF in (2.3.13), we have

$$|\nu_1, \nu_2, \ldots, \nu_N\rangle = |\mathbf{0}\rangle. \tag{2.3.15}$$

We shall, further, adopt the convention that any $\mathbf{0}_i$ will simply be dropped from the list $\{\nu_1, \ldots, \nu_N\}$ in (2.3.13), since their inclusion just amounts to multiplying an element of $E(\mathfrak{H})$ by instances of the algebra unit $1 \in \mathbb{R}$. Then we note that for instance,

$$|\mathbf{1}_{j_1}, \mathbf{1}_{j_2}, \ldots, \mathbf{1}_{j_m}\rangle = a_{j_1}^\dagger a_{j_2}^\dagger \ldots a_{j_m}^\dagger |\mathbf{0}\rangle \tag{2.3.16}$$

$$= e_{j_1} \wedge e_{j_2} \wedge \cdots \wedge e_{j_m}. \tag{2.3.17}$$

All of these notations will prove helpful in what follows. (We shall generally use lowercase Greek characters to denote generic elements of $E(\mathfrak{H})$ in order to distinguish them from elements in the original extended state space $\mathfrak{H}$. *States* of the now multipartite system, comprising N b-neurons, will, as usual, be associated with rays in $E(\mathfrak{H})$, classified by their normalized elements.) Note that the alternating nature of the $\wedge$-product precludes more than one simultaneous firing of any b-neuron.

We have dubbed the basic states depicted in the last set of equations *firing patterns* since they denote the "classical" states of network occupancies, namely the set or pattern of b-neurons that are ON and therefore co-firing at an implicit time. As noted earlier we shall extend this terminology to all the states in $E(\mathfrak{H})$. Thus, superpositions of "classical" firing patterns will also be considered to be firing patterns.

We may now define a network of b-neurons, to be called a *b-network*, as follows.

Definition 2.3.1. Let $\mathcal{N}_A$ denote a unicameral network, with matrix J^A defined as above. Then the *b-network, denoted by $\mathcal{N}_A^b$, associated with $\mathcal{N}_A$* is defined as the network obtained by replacing each node of $\mathcal{N}_A$ by the b-neuron whose output node is the node being replaced, and whose links are implemented by the map $E(J^A)\colon E(\mathfrak{H}_A) \to E(\mathfrak{H}_A)$.

Here the last stated map is the unique map of exterior algebras that extends J^A. It is determined by the assignments $E(f)(e_{i_1}^A \wedge \cdots \wedge e_{i_k}^A) = f(e_{i_1}^A) \wedge \cdots \wedge f(e_{i_k}^A)$ (cf. section B.1.3). Thus, the map $E(J^A)$, which coincides with J^A on the first grade elements of $E(\mathfrak{H}_A)$, connects the basic *singleton firing patterns* e_i^A of the b-network to its *singleton firing patterns*. That is to say, it maps the single ON b-neurons (generating the first grade subspace $\mathfrak{H}_A$) of the b-network's state space $E(\mathfrak{H}_A)$ to superpositions of single ON b-neurons of the network. This is exactly the effect to be expected when putative synaptic connections have been posited and are operating. So this simulates the effect of an actual synaptic wiring being in place but *hidden*, revealing itself only in the mapping of firing patterns to firing patterns which implements the net effect such a connection would have.

Given two b-networks $\mathcal{N}_A^b$ and $\mathcal{N}_B^b$, say, connections $f\colon \mathfrak{H}_A \to \mathfrak{H}_B$ from the first underlying unicameral network $\mathcal{N}^A$ to the second may be extended uniquely as above to the map $E(f)\colon E(\mathfrak{H}_A) \to E(\mathfrak{H}_B)$. As in the definition of the self-connections which constitute the structure of a b-network, the map f connects the basic *singleton firing patterns* e_i^A of the A-network to *singleton firing patterns* e_i^B of the B-network. That is to say, it maps the single ON b-neurons (generating the first grade subspace $\mathfrak{H}_A$) of the A-network to superpositions of single ON b-neurons of the B-network. Again, this is exactly the effect to be expected when putative synaptic connections between the networks have been posited and operate. So again this simulates the effect of an actual synaptic wiring being in place, revealing itself only in the mapping of firing patterns to firing patterns which implements the net effect such a connection would have. Therefore we shall refer to this type of connection as *synaptic-like*.

The further options that are introduced by positing a quantum-like structure allows also the possibility of modelling *non*-synaptic-like connections, as noted in section 2.2.2. We can accommodate this possibility by allowing connections between b-networks, via maps between their state spaces of firing patterns, that are more general than these synaptic-like ones.

Again, we must consider the possibility of b-networks with multimodal members. Here we note that the assumptions would require the underlying unicameral network to contain the would-be multimodal output nodes. Then the definition above of a b-network applies *mutatis mutandis* where only the state spaces need to be modified by the inclusion, via tensoring, of exterior algebras generated by the extra states of the multimodal output nodes. However, as noted in the next section, there are problems involved in this attempt to incorporate multimodal nodes into our model.

2.3.2 *Multimodal nodes*

It is noteworthy that the introduction of the disjunctive combinator $\oplus$ into the context of state spaces of underlying networks — these spaces being, needless to say, at the center of our interest — leads to the necessary existence of multimodal nodes or neurons. Admittedly this is a line of reasoning reminiscent of those employed in generally spurious ontological arguments: however, it does seem that evolution has in fact produced such neurons. They are a problem for our model, in that should a would-be b-neuron have a multimodal output node, then the b-neuron is no longer a two-state quantum-like system and our argument that it is fermionic falls to the ground. On the other hand since our nodes seem to be fermionic in themselves, we could just declare such a b-neuron to be bicameral with the output chamber now having a multiplicity of states, all of them being fermionic.

For instance, there could be gates, ion channels, synapses, etc., which are sometimes quiescent and sometimes not: this could be simulated via a superposition of possible states. Let us assume this possibility only for the output node, since we assume this is where the multimodality would be manifested. Then we would still have a bicameral neuron but the output node may have many states. That is to say, there will be different modes of being ON, which is the point, of course, of being multimodal. Moreover, these modes will in general be in a superposition: this possibility lies behind the implementation of a form of time division multiplexing. For example, let us consider a b-neuron with a bimodal output node having two states e and e' (for instance a single cell being "broadly tuned" to be responsive to high frequencies of sound — one mode — and at the same time responsive to low frequencies — the other mode. Then the state space of the b-neuron would be $E(\mathbb{Re} \oplus \mathbb{Re}') \cong \mathbb{R} \oplus \mathbb{Re} \oplus \mathbb{Re}' \oplus \bigwedge^2(\mathbb{Re} \oplus \mathbb{Re}')$ and there is a possible new state $e \wedge e'$ in which both states of the output node are occupied. In the sequel we shall avoid such multimodal states as far as is possible: they have, after all, only been found in nature to occur in the sensorium, though their possible existence will have to be taken into account in the considerations in Chapter 6, since they obviously have an effect upon cognition: *vide* Proust's narrator's taste of the madeleine. A thorough treatment of such neurons *per se* will be postponed until another time and place.

2.3.3   CAVEAT! *Tensor products of b-network state spaces*

We will be dealing with many tensor products of spaces of the form $E(\mathfrak{H}_A) \otimes E(\mathfrak{H}_B)$ where these are supposed to be the state spaces of the indicated b-networks. While it is always necessarily true that $E(\mathfrak{H}_A) \otimes E(\mathfrak{H}_B) \cong E(\mathfrak{H}_A \oplus \mathfrak{H}_B)$ as Hilbert spaces, we have noted that it may not be the case that $\mathfrak{H}_A \oplus \mathfrak{H}_B$ represents the state space of an underlying network. In our applications to networks, some caution is therefore required in these cases, since although $E(\mathfrak{H}_A) \otimes E(\mathfrak{H}_B)$ may represent the state space of a network — namely the two component networks with the possible all-to-all substrate connections among their nodes taken into account and considered as a compound system — it may not, under these circumstances, represent the state space of a b-network according to our definition. That is to say, it is not necessarily the exterior algebra of the state space of a unicameral network.

2.3.4 Relabeling the nodes of an external direct sum

It will be convenient at this point to introduce, by way of illustration, a labeling scheme for the case of the multimodal networks that might arise upon successful applications of $\oplus$ to a number of different networks. We illustrate this for the case discussed in section 2.2.4, namely $\mathfrak{H}_A \oplus \mathfrak{H}_B$ for the networks $\mathcal{N}_A$ and $\mathcal{N}_B$. We shall simply retain the old numbering for the unshared nodes and introduce double superscripted indices for the shared nodes, assuming there are any, with an arbitrary enumeration via the subscripted indices. Thus the new network whose state space is $\mathfrak{H}_A \oplus \mathfrak{H}_B$, which we shall denote by $\mathcal{N}_{A \oplus B}$, has nodes labeled n_1^A, n_2^A, n_{12}^{AB}, n_1^B, n_3^B. Here n_{12}^{AB} labels the bimodal shared node in the center of Figure 2.2. The corresponding basis for $\mathfrak{H}_A \oplus \mathfrak{H}_B$ is

$$\{e_1^A, e_2^A, e_1^{AB}, e_2^{AB}, e_1^B, e_3^B\}. \tag{2.3.18}$$

Here the node n_{12}^{AB} has two states labeled e_1^{AB} and e_2^{AB}. Note that if this is the network of output nodes of the associated b-network, we can now form basic firing patterns of the form $e_1^A \wedge e_1^{AB} \wedge e_2^A$, $e_3^B \wedge e_2^{AB}$, etc., because of the fundamental algebra isomorphism $E(\mathfrak{H}_A \oplus \mathfrak{H}_B) \cong E(\mathfrak{H}_A) \otimes E(\mathfrak{H}_B)$. This scheme can be extended to any number of component networks in an obvious way.

2.4 Bicameral network dynamics

In elementary non-relativistic quantum mechanics the action of time upon states is taken to be an operator, $U(t)$, say, necessarily unitary since normalized states are assumed to develop into normalized states, the Hilbert spaces in this case all being complex. A state ξ develops in time $t \in \mathbb{R}$ to

$$\xi(t) = U(t)\xi. \tag{2.4.1}$$

Under mild constraints, a theorem of Stone and von Neumann applies
to yield the representation

$$U(t) = \exp(-itH), \tag{2.4.2}$$

where H is an Hermitian operator, and this leads immediately to the
Schrödinger equation

$$\frac{\partial}{\partial t}\xi(t) = -iH\xi(t). \tag{2.4.3}$$

Here H is identified with the (time independent) Hamiltonian of the
system in units in which $\hbar = 1$.

In section 2.4.1 we consider this Schrödinger-like dynamics and
derive an operator implementing the time evolution of the states of
such a network, which is the analogue of the Hamiltonian in physics
proper. Our dynamics differs fundamentally from that of ordinary
quantum mechanics in at least the following respects:

- the state spaces are real finite dimensional spaces. And here we
  should point out the often overlooked fact that the theory of real
  vector spaces is much more difficult than the theory of complex
  vector spaces. One reason is that every complex vector space is
  at the same time also a real vector space so that there are more
  real vector spaces than complex ones; another is that complex
  eigenvalues and their corresponding eigenstates must be discarded
  in the real case;
- inner products between vectors in these spaces are not conserved
  since the vectors represent actual values in the network nodes,
  and these can change in time. The *state* represented by a vector is
  the ray it determines, which, as in ordinary quantum mechanics,
  may be associated uniquely with its normalized element. Since
  probabilities of transitions $\mathbf{w} \to \mathbf{v}$ are given by inner products of
  the associated states or normalized vectors, as in

$$\mathrm{prob}(\mathbf{w} \to \mathbf{v}) = \left(\frac{\mathbf{v}.\mathbf{w}}{\|\mathbf{v}\|\|\mathbf{w}\|}\right)^2, \tag{2.4.4}$$

the tautologous statement that the transition $\mathbf{v} \to \mathbf{v}$ is certain still
obtains at all times. It is clear from this that it is not necessarily

the case that $\mathrm{prob}(\mathbf{w}(t) \to \mathbf{v}(t)) = \mathrm{prob}(\mathbf{w} \to \mathbf{v})$. In fact the time evolution operator we shall derive (section 2.4.1) is not orthogonal;

- the simplifications inherent in most of the physical analogues of our system due to spatial symmetries, known force laws and their symmetries, etc., are not available here.

Finally, we note that the connection between the quantities of interest in a quantum theory of many degrees of freedom, namely certain amplitudes, and the quantities of interest in the associated statistical theory, namely partition functions, etc., is given by the replacement of the time variable (t, say) in the quantum theory by the corresponding imaginary variable (it). This leads to expressions of the type we obtain here. We shall not pursue this statistical connection in this work.

2.4.1 *Derivation of the pseudo-Hamiltonian for a b-network*

Before considering the dynamics of b-networks we shall digress briefly to discuss an aspect of their state spaces that is highly relevant to their computational capacity. Thus consider a system, such as the one just described above in section 2.3.1, as a classical system: that is, its classical states are not subject to superposition. Then the number of (classical) subsystems, or (classical) firing patterns, is 2^N. In order to store such patterns, perhaps for future use in computations, we would require a store whose size increases exponentially with N. In view of the fundamental identity equation (B.1.15), however, we have

$$E(\mathfrak{H}) \cong E \left(\bigoplus_{i=1}^{N} \mathbb{R}e_i \right) \cong \bigotimes_{i=1}^{N} E(\mathbb{R}e_i) \qquad (2.4.5)$$

so that the number of b-neurons required to house these firing patterns in the presence of quantum-like superposition (and, in this case, also concomitant entanglement) only increases linearly in N. This benefit presumably underlies some of the efficacy of Ising-like model simulations *à la* [6, 46].

Basic to the construction of a pseudo-Hamiltonian, or infinitesimal generator of time translation, are the analogues of those expressions which, in the case of the physics of many-fermion systems, are called the *(one particle) exchange integrals,* which express the interaction between the fermions, such as electrons, which are exchanged between the nodes or sites of a lattice. In our case, they are not of course actual particles, and, moreover, the sites do not constitute a lattice.

To motivate their choice in the case of a b-network, we note that equation (2.2.6), which applies here for the case of the output network to which the b-network is associated, tells us what the infinitesimal generator of time translation must reduce to when applied to the singleton or basic firing patterns corresponding to the single ON nodes of the b-network. Thus it becomes a problem of finding an operator that does this job.

This is not difficult to do. For, reverting to equation (2.2.6), we have

$$v_i(t+\tau) = v_i(t) + \tau \sum_{j=1}^{N} J_{ij} v_j(t), \qquad (2.4.6)$$

$i = 1, \ldots, N$. Thus for a firing pattern of the form

$$\xi(t) = \sum_{i=1}^{N} v_i(t) e_i \qquad (2.4.7)$$

we have

$$\sum_i v_i(t+\tau) e_i = \sum_i \left(v_i(t) + \tau \sum_j J_{ij} v_j(t) \right) e_i \qquad (2.4.8)$$

$$= \sum_i v_i(t) e_i + \tau \sum_i \sum_j J_{ij} v_j(t) e_i \qquad (2.4.9)$$

so, interchanging the indices in the double summation

$$= \sum_i v_i(t) e_i + \tau \sum_j \sum_i J_{ji} v_i(t) e_j \qquad (2.4.10)$$

$$= \sum_i v_i(t) e_i + \tau \sum_i v_i(t) \left(\sum_j J_{ji} e_j \right). \qquad (2.4.11)$$

Now we note that

$$\sum_j J_{ji} e_j = \sum_{j,k} J_{jk} \delta_{ki} a_j^\dagger |\mathbf{0}\rangle \qquad (2.4.12)$$

$$= \sum_{j,k} J_{jk} a_j^\dagger \delta_{ki} |\mathbf{0}\rangle \qquad (2.4.13)$$

$$= \sum_{j,k} J_{jk} a_j^\dagger a_k e_i \qquad (2.4.14)$$

so equation (2.4.11) becomes

$$\sum_i v_i(t+\tau) e_i = \sum_i v_i(t) e_i + \tau \left(\sum_{j,k} J_{jk} a_j^\dagger a_k \right) \sum_i v_i(t) e_i \quad (2.4.15)$$

that is

$$\xi(t+\tau) = \xi(t) + \tau \left(\sum_{j,k} J_{jk} a_j^\dagger a_k \right) \xi(t). \qquad (2.4.16)$$

Thus it is clear that the kinetic-like *pseudo-Hamiltonian*, or *p-Hamiltonian*, namely the infinitesimal generator of time translations, for the b-network in isolation is

$$H := -\sum_{i,j} J_{ij} a_i^\dagger a_j. \qquad (2.4.17)$$

The time *development* or *evolution* operator is then

$$T(t) := \exp(-tH) \qquad (2.4.18)$$

$$= \exp\left(t \sum_{i,j} J_{ij} a_i^\dagger a_j \right), \qquad (2.4.19)$$

where it will generally be understood that such sums are over $i > 0$ and $j > 0$.

The sign of H is moot, since we have not specified the signs of the J_{ij}. However, some justification for this sign choice will be posited shortly (section 2.4.3).

In the complex case of actual physics the analogous expression describes the kinetic term for a class of many-fermion systems going under the rubric *quasispin* [64]. It can be seen as an operator version of the Hopfield energy functional. We note that H does not give rise to an orthogonal time evolution operator unless $J_{ij}(t) = J_{ji}(t)$, a completely unreasonable restriction upon real neuronal networks if one has a viable biological model in mind, and one we shall not impose. This allows for the presence of *autapse*-like terms in which $i = j$, corresponding to the cases in which neurons may be synaptically/axonally connected to themselves, a frequent occurrence in real neural networks. But, more significantly, the non-orthogonality of the time evolution operator brings to the fore a property of *fragility* that is encountered in living systems, though not usually in non-living systems [22]. That is to say, something that happens at some point in time may fail at another even when the circumstances necessary for its occurrence are the same. Thus, the firing of a neuron may or may not always entail the safe arrival of neurotransmitter at a dendrite of a downstream neuron. Our model differs significantly from ordinary quantum physics in precisely this aspect. Namely, in ordinary quantum physics it is a more or less sacrosanct principle that probabilities of transitions with respect to the dynamics of the system remain invariant in time: the dynamical observables, in particular the time evolution operator itself, is *unitary*: i.e. it preserves probabilities of transitions in time. In our pseudo dynamics acting upon real spaces, the analog of unitarity is orthogonality and as noted our evolution operator is generally not orthogonal. So the probability of a transition between states at one time may be different at another time. This precisely simulates fragility in the sense described above.

We now digress briefly to point out an important mathematical feature of our expression for the p-Hamiltonian that does not seem to be made explicit in physics texts, though parts of it are used in discussions of variational principles. To motivate the discussion we note first that for small t the operator $T(t) - I := \delta_t$ produces a small change, or deformation, in any state, as in $T(t)\xi - \xi = -tH\xi$. Therefore it is not surprising that the operator H may in fact be realized as a deformation in its own right. To see this, we apply it to a general basic occupied state, written for simplicity as $e_1 \wedge \cdots \wedge e_m$, to obtain

$$\sum_{i,j} J_{ij} a_i^\dagger a_j (e_1 \wedge \cdots \wedge e_m)$$

$$= \sum_i \left(\sum_j J_{ij} a_i^\dagger a_j (e_1 \wedge \cdots \wedge e_m) \right) \tag{2.4.20}$$

$$= \sum_i \left(J_{i1} a_i^\dagger a_1 (e_1 \wedge \cdots \wedge e_m) + J_{i2} a_i^\dagger a_2 (e_1 \wedge \cdots \wedge e_m) + \cdots \right) \tag{2.4.21}$$

since any a_j with j outside the range we have
labeled $1, \ldots, m$ yields 0

$$= \sum_i \left(J_{i1} (e_i \wedge e_2 \wedge \cdots \wedge e_m) + J_{i2} (e_1 \wedge e_i \wedge e_3 \wedge \cdots \wedge e_m) + \cdots \right) \tag{2.4.22}$$

$$= \left(\sum_i J_{i1} e_i \right) \wedge e_2 \wedge \cdots \wedge e_m$$

$$+ e_1 \wedge \left(\sum_i J_{i2} e_i \right) \wedge e_3 \wedge \cdots \wedge e_m + \cdots \tag{2.4.23}$$

$$= \left(\sum_{i,j} J_{ij} a_i^\dagger a_j e_1 \right) \wedge e_2 \wedge \cdots \wedge e_m$$

$$+ e_1 \wedge \left(\sum_{i,j} J_{ij} a_i^\dagger a_j e_2 \right) \wedge e_3 \wedge \cdots \wedge e_m + \cdots . \tag{2.4.24}$$

Thus, the operator $\sum_{i,j} J_{ij} a_i^\dagger a_j$ acts as a deformation of the state $e_1 \wedge \cdots \wedge e_m$ since if we put

$$\delta_t = \sum_{i,j} J_{ij} a_i^\dagger a_j \tag{2.4.25}$$

we find that equation (2.4.24) yields

$$\delta_t (e_1 \wedge \cdots \wedge e_m) = \delta_t e_1 \wedge e_2 \wedge \cdots \wedge e_m$$

$$+ e_1 \wedge \delta_t e_2 \wedge e_3 \wedge \cdots \wedge e_m + \cdots . \tag{2.4.26}$$

This is to say that the operator δ_t acts on the exterior algebra to satisfy the Leibnitz relation for differentials. This last equation is easily seen to imply that $\delta_t(\xi \wedge \eta) = \delta_t\xi \wedge \eta + \xi \wedge \delta_t\eta$ for any firing patterns ξ and η, δ_t being linear. Such an operator upon an algebra is called a *derivation*: cf. section B.1.3.

The derivational nature of operators of this form have the consequence that the corresponding time evolution operator is an *automorphism* of the exterior algebra $E(\mathfrak{H})$ ([50], chapter 1, section 2). This property will be of significance later.

With H as in equation (2.4.17), the time evolution operator is

$$\exp(-tH) = I - tH + \frac{t^2}{2!}H^2 - \cdots . \tag{2.4.27}$$

Its effect on a state of single occupancy, which we now write $|1_{i_0}\rangle$ say — to produce the state evolving from the firing of the single b-neuron n_{i_0} — is given for small t by the superposition

$$\exp(-tH)|1_{i_0}\rangle - |1_{i_0}\rangle + t\sum_{i,j} J_{ij}a_i^\dagger a_j|1_{i_0}\rangle$$

$$+ \frac{t^2}{2!}\sum_{i,j,k,l} J_{ij}J_{kl}a_i^\dagger a_j a_k^\dagger a_l|1_{i_0}\rangle + \cdots . \tag{2.4.28}$$

Note that since t is small, contributions to the superpositions associated with higher powers of t are progressively less likely. To interpret this, let us unfold the term in $t^2/2!$, namely:

$$\sum_{i,j,k,l} J_{ij}J_{kl}a_i^\dagger a_j a_k^\dagger a_l|1_{i_0}\rangle = \sum_{i,j,k,l} J_{ij}J_{kl}a_i^\dagger a_j a_k^\dagger \delta_{li_0}|0\rangle \tag{2.4.29}$$

$$= \sum_{i,j,k} J_{ij}J_{ki_0}a_i^\dagger a_j|1_k\rangle \tag{2.4.30}$$

$$= \sum_{i,j,k} J_{ij}J_{ki_0}a_i^\dagger \delta_{jk}|0\rangle \tag{2.4.31}$$

$$= \sum_{i,j} J_{ij}J_{ji_0}|1_i\rangle \tag{2.4.32}$$

$$= \sum_{i} (J^2)_{ii_0}|1_i\rangle \tag{2.4.33}$$

where $(J^2)_{ii_0}$ represents the appropriate entry in the matrix J^2. This term represents the superposition of the states of single occupancy which have been reached by firings of n_{i_0} along 2-step axonal paths. The next term is similar, involving 3-step paths, and so on. Thus the formalism picks out sequential firings along paths of an increasing number of steps leading from n_{i_0} consequent upon its firing: these contribute to the superposition with probability decreasing as path length increases. Thus short paths are favored and this implies a certain nearest neighbor behavior. We note here that this behavior, where nearest neighbor effects dominate, is built into the Ising-like models by fiat.

Let $J : \mathfrak{H} \to \mathfrak{H}$ denote the linear map represented by the matrix $(J)_{ij}$. Then it is immediate that, restricted to $\mathfrak{H}$, $H = -J$, and that on $E(\mathfrak{H})$, H is the unique derivational extension of $-J$ since H is itself a derivation.

The p-Hamiltonian H in equation (2.4.17) has been derived on the assumption that external influences acting on the network at hand have been excluded. We shall simplistically divide such influences into two types. First, a field-like term which is supposed to collect the external influences such as neuromodulation, substrate connections, etc., coming from external non-synaptic sources. Such influences can only have the effect of turning groups of neurons on or off or leaving them untouched, so in their simplest from, up to first order in the $a, a^\dagger$, we may take them to be of the form

$$\sum_{i=1}(l_i(t)a_i^\dagger + h_i(t)a_i) + \mathscr{E}(t)I \qquad (2.4.34)$$

where $l_i(t)$ (respectively, $h_i(t)$) subsumes the effects of external influences to switch b-neurons ON (respectively OFF), and $\mathscr{E}$ denotes a "self-interaction" term. Some, or all, of these terms may be zero. The other term would depend on the synaptic connections with other networks and would be a complex expression involving the "hopping" operators $J_{ij}a_i^\dagger a_j$ of the kind appearing in the internal p-Hamiltonians. In our single application of an external p-Hamiltonian, which is the subject of Chapter 5, we will deal mainly with external stimuli of the above non-synaptic type.

We note also that each hopping operator $J_{ij}a_i^\dagger a_j$ summand in H maps each subspace $\bigwedge^n \mathfrak{H}$ into itself, so that H assumes block diagonal form, and so basic firing patterns represented by superpositions of the basic states of the form $|1_{i_1}, \ldots, 1_{i_m}\rangle = e_{i_1} \wedge \cdots \wedge e_{i_m}$ are mapped to states of the same grade (or occupation number): occupation number is preserved in time. This is because H by itself represents the network's infinitesimal time evolution in isolation. In order for occupation number to vary in time, terms of the form equation (2.4.34), representing external fields or stimuli, must be added to H.

Classical theories involving spaces of states can deal only with states of single occupancy, in our terminology. Namely, the elements in the first grade summand $\mathfrak{H} \subset E(\mathfrak{H})$. The states of higher occupancy, namely those in $\bigwedge^n \mathfrak{H}$ for $n > 1$, will be unavailable to such treatments. Classical treatments, moreover, cannot make operational or semantic distinctions between pure states and super-positional ones, even in principle. The decoherence or otherwise of superpositional states could have no significance in such theories. (In [84] we argued that, even in the simplest non-trivial case of 3-neuron motif classes, both higher occupancy states and their superpositional nature have a discernible bearing upon their observed behavior. In that work we adopted the positive sign for the p-Hamiltonian and the adjoint of the exchange matrix J, neither of which materially alters the conclusions.)

If there are multimodal b-neurons involved these considerations change only in the addition of extra states and alternative (i.e. superposed) firing patterns, some including one set of output modes and others including other sets of output modes.

2.4.2 *Firing patterns and eigenstates*

A general element ξ of $E(\mathfrak{H})$ at time t may be written

$$\xi(t) = \sum_{p \geqslant 0} \sum_{i_1 < \cdots < i_p} \alpha_{i_1 \ldots i_p}(t) e_{i_1} \wedge \cdots \wedge e_{i_p} \qquad (2.4.35)$$

$$= \alpha_{i_0}(t)e_{i_0} + \sum_{i_1} \alpha_{i_1}(t)e_{i_1} + \sum_{i_1 < i_2} \alpha_{i_1 i_2}(t)e_{i_1} \wedge e_{i_2} + \cdots \qquad (2.4.36)$$

where the coefficients $\alpha_{i_1...i_p}(t)$ are real numbers and the $p = 0$, or 0-th grade component, is a multiple of the vacuum state $e_{i_0} = |0\rangle = 1 \in \mathbb{R}$. By the linearity property of the exterior product each basic *firing pattern* in this superposition is similar to the tensor $e_{i_1} \otimes \cdots \otimes e_{i_p}$ in that the ON or firing output nodes corresponding to the e_{i_k} are unitized and share the value $\alpha_{i_1...i_p}$. Denoting by $\hat{\xi}(t)$ the unit vector corresponding to $\xi(t)$ we remark that the probability that the network will transition from the general state $\xi(t)$ in equation (2.4.35) to the basic firing pattern state $e_{i_1} \wedge \cdots \wedge e_{i_k}$ is given by

$$\text{prob}(\xi(t) \to e_{i_1} \wedge \cdots \wedge e_{i_k}) = |\langle e_{i_1} \wedge \cdots \wedge e_{i_k} \,|\, \hat{\xi}(t) \rangle|^2 \qquad (2.4.37)$$

$$= \frac{\alpha_{i_1...i_k}(t)^2}{\sum_p \sum_{i_1 < \cdots < i_p} \alpha_{i_1...i_p}(t)^2}. \qquad (2.4.38)$$

A basic firing pattern of grade greater than one, such as $e_{i_1} \wedge \cdots \wedge e_{i_p}$, thus represents the state of the corresponding *co-spiking, unitized* set of corresponding b-neurons being ON. Such a state may or may not be realized by the b-network.

Thus, a general firing pattern (equation (2.4.36)) represents a state of the b-network which is a superposition of basic firing patterns of co-acting subsets of b-neurons. The α coefficients determine the probability that, when the network is in such a state, the corresponding basic firing pattern is the one that is active (equation (2.4.38)). These coefficients, hence these probabilities, change in time according to the dynamics derived above. When in such a state, the network might appear to be in one of these basic firing patterns (i.e. a subset of co-spiking neurons) according to the probability inherent in the coefficients. These probabilities change in time, giving rise to changing patterns of sets of co-spiking neurons, as in the phenomenon described in section 1.3.1.

The states of single b-neurons, which are not part of a co-acting unitized set, are fully described by the first grade elements of the exterior elements, namely those of the original space generated by the e_i. Thus for one of these grade one states, namely

$$\xi(t) = \sum_i \alpha_i(t)e_i, \qquad (2.4.39)$$

the probability that for instance two b-neurons, namely n_{i_1} and n_{i_2} are simultaneously ON, or firing at time t, but not necessarily unitized, is

$$\text{prob}((\xi(t) \to e_{i_1}) \text{ and } (\xi(t) \to e_{i_2})) = \text{prob}(\xi(t) \to e_{i_1})\text{prob}(\xi(t) \to e_{i_2})$$
(2.4.40)

$$= \frac{\alpha_{i_1}(t)^2 \alpha_{i_2}(t)^2}{(\sum_i \alpha_i(t)^2)^2}$$
(2.4.41)

and so on for any number of simultaneously but not necessarily unitized sets of b-neurons. The probability of these being ON while all the others are OFF is left as an easy exercise.

In actual quantum mechanics the (non-zero) eigenvectors of a time independent Hamiltonian play a fundamental rôle. The normalized ones represent stationary states and are the ones that are observable with definite energies (namely the eigenvalues). Only their unitary phases change in time. In our case the eigenstates do not have complex phases, but eigenvectors of our pseudo-Hamiltonians still represent stationary states of *occupancy*, which we have dubbed *firing patterns*. The states correspond to the rays such vectors determine which remain fixed in time, though their "phases" are real scaling factors that vary exponentially in time. Namely, if

$$H\xi(0) = \lambda\xi(0)$$
(2.4.42)

then for small t and $\lambda \neq 0$,

$$\xi(t) = \exp(-tH)\xi(0)$$
(2.4.43)
$$= \exp(-\lambda t)\xi(0).$$
(2.4.44)

(It is quite possible and often happens that there are no non-zero eigenstates.)

Thus, over short time intervals, the *states* (i.e. the rays) determined by such eigenvectors remain stable, while the scaling value of each basic firing pattern in $\xi(0)$ changes by the *same* scalar factor $\exp(-\lambda t)$. Let us suppose an eigenvector $\xi(0)$ is written in the form shown in equation (2.4.35) at time zero. Then at time t, for small t, we have

$$\xi(t) = \sum_{p \geqslant 0} \sum_{i_1 < \cdots < i_p} \exp(-\lambda t)\, \alpha_{i_1 \dots i_p}(0) e_{i_1} \wedge \cdots \wedge e_{i_p}. \qquad (2.4.45)$$

(In future we will generally write $\alpha_{i_1 \dots i_p}$ for $\alpha_{i_1 \dots i_p}(0)$ the time dependence being carried by the dynamical operator, in this case the exponential scaling factor.)

The normalized version of $\xi(t)$ represents the same state or firing pattern regardless of the exponential factor, as long as it is not zero, so in this sense they are stable, at least for short times. These states are often considered to be the ones of cognitive significance in this neural network context since they are locally stable, and, in the quantum-like context, are exactly the observable ones.

Although, as just noted, over short time intervals, the *states* (i.e. the rays) determined by such eigenvectors remain stable, while in them *each* b-neuron's output value scales by the *same* factor $\exp(-\lambda t)$. This can be seen by considering the single occupancy, or first grade, component of the eigenstate ξ, whose coefficients represent the values in the output nodes of those b-neurons which are firing if the system were to enter that state at time 0. (As noted the space of these states is mapped into itself by the p-Hamiltonian.) Equation (2.4.45) asserts that over the next short time interval the nodal values scale by the exponential factor. Thus, for this possibly very short time interval after entering the eigenstate, this scaling behavior is the same for all the possibly different co-spiking subsets of b-neurons involved in this state, and this globalized behavior dominates any local synaptic activity. These eigenstates thus represent a sort of non-local coherence in the intensity of the possibly different basic firing patterns of the network, though this coherence may be fleeting.

For all networks having p-Hamiltonians of the form H the two states of extreme occupancy, namely the vacuum state $|0\rangle$ in which no b-neurons are firing, and the state of full occupancy $|1_1, \dots, 1_N\rangle = e_1 \wedge \cdots \wedge e_N$ in which all b-neurons are firing, have the following properties:

$$H|\mathbf{0}\rangle = 0, \qquad\qquad (2.4.46)$$

$$H|\mathbf{1}_1,\dots,\mathbf{1}_N\rangle = -\mathrm{tr}J\,|\mathbf{1}_1,\dots,\mathbf{1}_N\rangle, \qquad\qquad (2.4.47)$$

where $\mathrm{tr}J := \sum J_{ii}$ denotes the trace of the matrix $J(t)$. The first of these is obvious while the second follows at once from equations (2.4.25) and (2.4.26). The first of these equations asserts that a network in which no b-neurons are firing will remain in that state if no external influences are applied, as expected. The second asserts that if all b-neurons are firing then the short term future behavior in the absence of external influences is entirely determined by the presence and behavior of autapses. If there are none, or if they are such that $\mathrm{tr}J = 0$, then the network remains in its state of full firing, at least for the short term. If $\mathrm{tr}J > 0$ the values in all the output nodes will start to increase exponentially with the same scaling factor at time t, namely $\exp(-t\,(-\mathrm{tr}J)) = \exp(t\,\mathrm{tr}J)$, with firing activity increasing synchronously and independently of synaptic activity, leading perhaps to seizure or some other catastrophic failure of the network. If $\mathrm{tr}J < 0$ the values will decrease exponentially, leading ultimately to the opposite catastrophe of coma or death. This behavior is consistent with the J_{ij} terms being thought of as increments to the value of the depolarizing postsynaptic potential value and is some evidence in support of the correctness of the sign adopted in equation (2.4.17). Note that it would take only a single excitatory autapse to precipitate such a short term explosion. In reality, the complexity of the behavior of J would yield mediating behaviors. Thus, for instance, we might expect to find increased autapse involvement in epileptic tissue *in vitro*, with increased non-synaptic synchronized firing activity as seizure is approached. There seems to be some support for both of these conclusions: cf. [52, 63]. (In most cases the eigenvalues are not so simply related to the J_{ij} values, and of course in many cases there may be no non-trivial eigenstates at all. Note also that equation (2.4.47) applies, of course, to the case of a single autapse, i.e. a single b-neuron synaptically connected to itself, and shows that such a b-neuron either dies or blows up.)

2.4.3 *A pair of toy b-networks*

To get some further idea of the function of the *exchange* terms or *efficacies* J_{ij} let us perform a couple of little thought experiments.

First we consider the simplest case of a b-neuron n_1, say, connected by a single axonal link to another b-neuron, n_2, say. The p-Hamiltonian for this b-network in isolation is then $H = -J_{21} a_2^\dagger a_1$. It is quickly seen that it has no non-zero eigenstates: it is unstable. Let us see what happens in time if n_1 fires at time $t = 0$. That is to say, it is in the state $|1_1\rangle$ at $t = 0$. Then

$$T(t)|1_1\rangle = \exp(-tH)|1_1\rangle \tag{2.4.48}$$

$$= \exp(tJ_{21} a_2^\dagger a_1)|1_1\rangle \tag{2.4.49}$$

$$= (I + tJ_{21} a_2^\dagger a_1 + \frac{t^2}{2!} J_{21}^2 a_2^\dagger a_1 a_2^\dagger a_1 + \cdots)|1_1\rangle \tag{2.4.50}$$

$$= |1_1\rangle + tJ_{21}|1_2\rangle. \tag{2.4.51}$$

Thus, after time $t > 0$, the probability $p_2(t)$ that the system is in a state in which n_2 is ON or *has entered* the firing state, is:

$$p_2(t) = \frac{t^2 J_{21}^2}{1 + t^2 J_{21}^2} \tag{2.4.52}$$

which is less than 1, so is not certain, but approaches 1, i.e. certainty, as $t \to \infty$ as is to be expected. This illustrates the fragility mentioned earlier: the firing of the downstream b-neuron is not certain, even when the upstream b-neuron has fired. The probability that n_1 is still ON at time t is

$$p_1(t) = \frac{1}{1 + t^2 J_{21}^2}. \tag{2.4.53}$$

This goes to zero as $t \to \infty$ as is to be expected. Note also that, for a fixed time t, the probability of n_2 being ON, or having entered the firing state, goes up with increasing $|J_{21}|$ and down with decreasing $|J_{21}|$. Thus large values of $|J_{21}|$ are associated with *excitation* and small values of it are associated with *inhibition*.

If either one of these probabilities are known at some time $t_0 \neq 0$, the constant J_{21} can be computed up to sign. Namely,

$$J_{21} = \pm \frac{1}{t_0} \sqrt{\frac{p_2(t_0)}{1 - p_2(t_0)}} \qquad (2.4.54)$$

$$= \pm \frac{1}{t_0} \sqrt{\frac{1 - p_1(t_0)}{p_1(t_0)}}. \qquad (2.4.55)$$

Note that if we are considering the J_{21} to be proportional to the increment induced upon the postsynaptic n_2 when excitatory, then the positive signs in the above equations should be chosen. Moreover, if $J_{21} > 0$ then the value updating the n_2 output node scales linearly and positively in time according to equation (2.4.51), while its probability of firing increases with time. This increase again speaks to the correctness of the sign chosen in equation (2.4.17).

Our second toy will be a pair of b-neurons, n_1 and n_2, each connected by a single axonal link to a third b-neuron n_3. Then $H = -(J_{32}a_3^\dagger a_2 + J_{31}a_3^\dagger a_1)$. The temporal development of a general state $\xi = \lambda|1_1\rangle + \mu|1_2\rangle$ of the two upstream b-neuron output nodes is then given by

$$T(t)\xi = \exp(-tH)\xi \qquad (2.4.56)$$

$$= (I + t(J_{32}a_3^\dagger a_2 + J_{31}a_3^\dagger a_1) + \frac{t^2}{2!}(J_{32}a_3^\dagger a_2 + J_{31}a_3^\dagger a_1)^2 + \cdots)\xi \qquad (2.4.57)$$

$$= \lambda|1_1\rangle + \mu|1_2\rangle + t(\lambda J_{31} + \mu J_{32})|1_3\rangle. \qquad (2.4.58)$$

$$\qquad (2.4.59)$$

In this state the probability that n_3 has entered the ON state is:

$$p_3(t) := \frac{t^2(\lambda J_{31} + \mu J_{32})^2}{\lambda^2 + \mu^2 + t^2(\lambda J_{31} + \mu J_{32})^2}. \qquad (2.4.60)$$

We note that:

For fixed λ, μ, Js,

$$p_3(t) \rightarrow 1 \text{ as } t \rightarrow \infty.$$

For fixed λ, μ, t,

$$p_3(t) \to 1 \text{ as } |\lambda J_{31} + \mu J_{32}| \to \infty;$$
$$p_3(t) \to 0 \text{ as } |\lambda J_{31} + \mu J_{32}| \to 0.$$

This shows again the rôle of the efficacies and also that our b-neurons are integrators. Again the firing of n_3, after a firing state of the upstream two-neuron system is entered, is not certain but approaches certainty in time. And again the influence upon n_3 is excitatory if the absolute value of the integrated input is large, and inhibitory if this quantity is small.

Note that the internal mechanisms of the neuron, namely those by which the resultant incoming potentials either rise above the threshold and trigger the action potential, or do not, are not available in this picture. It should not be a surprise that, having gone to a lot of trouble to hide these complexities, the brush produced is too broad to paint them. Instead our model represents a broader picture depicting the results of these internal machinations, rather than the machinations themselves. In fact it seems to have captured the pertinent fact of the mesoscale life of standard neurons: namely that their probability of firing depends upon the magnitude of their integrated inputs.

The last toy example had a pair of b-neurons converging on a single one. Readers might like to try the simplest diverging example of a single upstream b-neuron connected to each of a pair of downstream b-neurons. With equal efficacies on both axon terminals they will find that as time goes on after the firing of the single upstream b-neuron, the probability of one of either of the two downstream b-neurons firing approaches $1/2$.

Of course none of these examples takes account of the possible external or substrate influences which abound *in vivo*.

2.4.4 *Symmetry promotes stability*

As noted, Hamiltonians analogous to those of the form shown in equation (2.4.17) appear in many-body (or condensed matter) physics, where they are known to be extremely recalcitrant computationally

[64]. In [84], section 9.4, we found the eigenvectors for many of
the 13 three-neuron motifs and noted that some of the eigenspaces
were more than one-dimensional. In physics this sort of degeneracy
is associated with greater stability against decay and this would
have a bearing upon the survival of such states as representers
of memoranda, or other cognitive attributes. (For, if H has an
eigenspace of dimension greater than one, then there are eigenvectors
which are superpositions of other eigenvectors. Such superpositions
are then stable in the sense of equation (2.4.44) and therefore resist
decay via decoherence.)

Now, eigenvalue degeneracies of H are associated with *symmetries* of H. Thus suppose an operator S on $E(\mathfrak{H})$ were such that

$$[H, S] := HS - SH = 0. \tag{2.4.61}$$

Then, if ξ is an eigenvector we have $H\xi = \lambda\xi$, for some real λ, so
that

$$HS\xi = SH\xi \tag{2.4.62}$$
$$= S\lambda\xi \tag{2.4.63}$$
$$= \lambda S\xi. \tag{2.4.64}$$

Thus $S\xi$ is also an eigenvector. If $S\xi$ and ξ are linearly independent
then the eigenvalue λ is degenerate. Such an S is called a *symmetry*
of H. Although we cannot specify the eigenspectrum of a general H,
we can claim that the presence of a symmetry increases the likelihood
of the existence of degenerate eigenstates which are more stable than
otherwise.

We seek symmetries of the p-Hamiltonian as follows. As noted
(section B.1.3) any linear map $\phi : \mathfrak{H} \to \mathfrak{H}$ extends uniquely to an
algebra map $E(\phi) : E(\mathfrak{H}) \to E(\mathfrak{H})$. In what follows, we recall that
the T superscript denotes the transpose of its argument.

Theorem 2.4.1. *Suppose* $\phi : \mathfrak{H} \to \mathfrak{H}$ *is a linear map. Then*
$E(\phi)H = HE(\phi)$ *if and only if* $\phi J = J\phi$ *where* J *denotes the
exchange matrix* (J_{ij}).

Proof. The proof is a brute force calculation and uses the fact that H is a derivation on the exterior algebra [84]. With notation as above, put

$$\phi(e_k) = \sum_s \lambda_{sk} e_s, \qquad (2.4.65)$$

for $\alpha_{sk} \in \mathbb{R}$.

Now suppose $E(\phi)H = HE(\phi)$. Then, for all k,

$$E(\phi)\left(-\sum_{i,j} J_{ij} a_i^\dagger a_j\right)(e_k) = \left(-\sum_{i,j} J_{ij} a_i^\dagger a_j\right)\phi(e_k). \qquad (2.4.66)$$

The left hand side soon yields

$$E(\phi)\left(-\sum_{i,j} J_{ij} a_i^\dagger a_j\right)(e_k) = -\sum_s \left(\sum_i J_{ik}\lambda_{si}\right)(e_s) \qquad (2.4.67)$$

while the right hand side yields

$$\left(-\sum_{i,j} J_{ij} a_i^\dagger a_j\right)\phi(e_k) = -\sum_i \left(\sum_s J_{is}\lambda_{sk}\right)(e_i). \qquad (2.4.68)$$

Equating coefficients of the basic elements gives $\phi J = J\phi$. Conversely, if this last equation holds, the above steps reversed show that $E(\phi)H(e_k) = HE(\phi)(e_k)$ for all k. Then, H being a derivation and $E(\phi)$ being an algebra map,

$$E(\phi)H(e_{i_1} \wedge \cdots \wedge e_{i_k})$$
$$= E(\phi)(He_{i_1} \wedge \cdots \wedge e_{i_k} + e_{i_1} \wedge He_{i_2} \wedge \cdots \wedge e_{i_k} + \cdots) \quad (2.4.69)$$
$$= E(\phi)H(e_{i_1}) \wedge \phi(e_{i_2}) \wedge \cdots \wedge \phi(e_{i_k}) + \phi(e_{i_1}) \wedge E(\phi)H(e_{i_2}) \wedge \cdots$$
$$(2.4.70)$$
$$= HE(\phi)(e_{i_1}) \wedge \phi(e_{i_2}) \wedge \cdots \wedge \phi(e_{i_k}) + \phi(e_{i_1}) \wedge HE(\phi)(e_{i_2}) \wedge \cdots$$
$$(2.4.71)$$
$$= H\phi(e_{i_1}) \wedge \phi(e_{i_2}) \wedge \cdots \wedge \phi(e_{i_k}) + \phi(e_{i_1}) \wedge H\phi(e_{i_2}) \wedge \cdots$$
$$(2.4.72)$$
$$= H(\phi(e_{i_1}) \wedge \cdots \wedge \phi(e_{i_k})) \qquad (2.4.73)$$
$$= HE(\phi)(e_{i_1} \wedge \cdots \wedge e_{i_k}) \qquad (2.4.74)$$

which proves the assertion. □

So $E(J)$ itself is a symmetry. If J is a *normal* matrix, namely if it commutes with its transpose, $J^T J = J J^T$, then $E(J^T)$ is also a symmetry. We may interpret this in a significant special case. Namely, let us suppose that our network has at most a single link, or edge, connecting each pair of nodes. Then for this network, $\Lambda_{ij} = 1$ or 0. Let us suppose further that the α_{ij} scaling factors (equation (2.2.5)) are all the same non-zero constant α, say. (This is not an unreasonable assumption for a network of neurons selected by evolution to perform a common task.) Then

$$J_{ij} = \Lambda_{ij}\alpha \qquad (2.4.75)$$

and the normalcy condition reads

$$(J^T J)_{ij} = (J J^T)_{ij} \qquad \text{or} \qquad (2.4.76)$$

$$\sum_k \Lambda_{ik}^T \Lambda_{kj} \alpha^2 = \sum_k \Lambda_{ik} \Lambda_{kj}^T \alpha^2 \qquad \text{or} \qquad (2.4.77)$$

$$\sum_k \Lambda_{ki} \Lambda_{kj} = \sum_k \Lambda_{ik} \Lambda_{jk}. \qquad (2.4.78)$$

So, with $i = j$ we obtain

$$\sum_k \Lambda_{ki}^2 = \sum_k \Lambda_{ik}^2 \qquad \text{or} \qquad (2.4.79)$$

$$\sum_k \Lambda_{ki} = \sum_k \Lambda_{ik} \qquad (2.4.80)$$

since $\Lambda_{ij} = 1$ or 0. This says that the number of links *from* any node n_i — the left hand side of the last equation — is equal to the number of links *to* that node — the right hand side. Such a network is said to be *balanced*. So networks of this type with normal exchange matrices are balanced. (The converse is false. Balanced networks do not have to have normal adjacency matrices. From above a necessary and sufficient condition for a network of the above type to be balanced is that the diagonal of the matrix $\Lambda^T \Lambda - \Lambda \Lambda^T$ be zero.) In this case, since there is likely to be greater degeneracy, we may expect the associated network's stability against noise, decoherence, etc., to be greater than the average. Examples include *cliques* which are networks in which

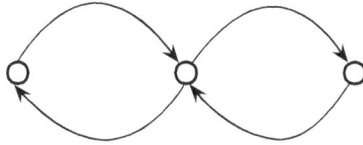

Figure 2.4 The dual dyad motif: M9 in the Sporns listing of three-node motifs [90].

every node is connected (singly) to every other node (also termed *all to all* connectivity). They have adjacency matrices which are about as symmetric as matrices can get. In these we would expect to find unusual stability (and synchronicity) which would bestow a selective advantage by providing efficient receptacles for memoranda, and possibly other functional attributes. An abundance of such cliques has in fact been found in the neocortex of the rat, and they are also present in even such a primitive nervous system as that of the nematode *C. elegans* [77]. The simplest non-trivial example is the mutually connected two-neuron network, the *dyad*, which has also been called a *resonance pair*, and has similar functional attributes, namely stability and synchrony promotion [42].

A significant symmetric (hence normal) example that is not a clique is afforded by the three-node motif known as the *dual dyad*, depicted in Figure 2.4.

Its eigenspectrum was investigated in [84] and it was found to have the most stability of all the 13 three-node motifs. This comports with the experimental finding that it is the most frequently found such motif across studies (of cat and macaque brains). This is presumably because, among other things, its stability bestows a selective advantage.

2.5 Remarks on the nature of the substrate connections

In this section we review the relevance of the substrate connection implemented by the ∧-product, and probe its nature for the case of the simplest network consisting of two b-neurons.

2.5.1   *The significance of the exterior product*

We have interpreted a tensor of the form $\lambda e_i \otimes e_j = e_i \otimes \lambda e_j$ as a state of a pair of corresponding nodes bound together by a substrate of some kind so that the bound pair registers the single value $\lambda \in \mathbb{R}$. If this value decreases to zero, the combined state contributes nothing to any superposition: the probability of firing, in the applications to follow, falls to zero. Thus the substrate connection admits *inhibitory* input into nodes, and for this reason we identified it with the substrate provided by generally inhibitory effects in actual brains such as an interconnection via interneurons or other substrate mechanisms. This kind of connection would also admit excitatory inputs since the algebra admits any real value for λ so it may increase as well as decrease. In this case, a problem would arise for self-connection of nodes via this substrate, whose basic state would be of the form $\lambda e_i \otimes e_i$ for, if a node were connected to itself in this manner and admitted excitatory input, then it would run the risk, in the application to b-neurons, of precipitating the forbidden state of double firing of the b-neuron.

The exterior product, $\wedge$, which arose spontaneously out of our adoption of the operator $\mathfrak{a}$ in equation (2.3.1) or equation (2.3.2), expressly excludes this possibility, since $\lambda e_i \wedge e_i = 0$, while at the same time also admitting inhibitory substrate connections, since it has the same multilinear property of sharing scalar values as the $\otimes$-product. Since $\xi \wedge \xi = 0$ for any firing pattern, no firing pattern can co-spike, or unitize, with itself via this substrate bond. Whatever sort of bond this $\wedge$-combinator represents, it does not allow self-connection. Given b-neurons, n_i and n_j say, the $\wedge$ bond presumably acts only in one direction: either $n_i \rightarrow n_j$ or $n_j \rightarrow n_i$ but not both. Then such a $n_i \rightarrow n_i$ bond is impossible, which is the quantum-like import of the statement $e_i \wedge e_i = 0$. This is in contradistinction to the $\otimes$-combinator, which allows for instance self-connection via $e_i \otimes e_i (\neq 0)$. We note that in the case of b-neurons, this tensor ($\otimes$) combinator ignores the fermionic structure, and is promiscuous in that it is defined between any two vectors belonging to different vector spaces.

The exterior product may reflect either inhibitory binding substrates (such as those provided by a plethora of GABAergic interneurons) or excitatory substrates (such as those provided by a glutamatergic background as in mammalian cortices) or both. (In a sense, it provides a stronger bond than $\otimes$ since it resists self-interaction and concomitant blowup.) This has a bearing upon memory formation. Although the chemical environment is exceedingly complex, it seems clear that a flux of the excitatory neurotransmitter glutamate is involved in memory formation as is the ambiguously modulating neurotransmitter dopamine. We can crudely simulate this in a formal fashion by regarding the influx of some excitatory influence, such as extracellular glutamate, as provoking a non-synaptic transition of the form $\xi_1 \otimes \cdots \otimes \xi_k \mapsto \xi_1 \wedge \cdots \wedge \xi_k$. Here the $\otimes$-product is associated with an inhibitory substrate while the $\wedge$-product is associated with both inhibitory and excitatory substrates such as glutamate and dopamine. In a sense this transition represents a *consolidation*, since the $\wedge$ combinations are somewhat more resistant to dissolution than are the $\otimes$ combinations tolerating, as they do, excitatory modulation. This transition will play a rôle in our deduction of early long term potentiation in Chapter 4. (Note that this map is just the multiplication map $\bigotimes^k E(\mathfrak{H}) \to E(\mathfrak{H})$ of the exterior algebra.)

Let us briefly review the differences between the connectors $\otimes$ and $\wedge$ in the context of firing patterns of b-networks.

- The $\otimes$ product $\xi \otimes \eta$, where ξ and η may represent firing patterns of any networks (for the time being) or the same network, represents a possible state of a unitized pairing of each b-neuron of one with each b-neuron of the other, via a substrate connection or connections, which enables the simultaneous firing of both networks. Such states are not necessarily stable if the connection is excitatory;

- The $\wedge$ product $\xi \wedge \eta$, where ξ and η are now understood to be firing patterns of the *same* b-network, represents *another* firing pattern of the b-network. Since it does not allow the possibility of simultaneous double firing, it tolerates both excitatory and inhibitory substrate connections and will be associated with long term memory consolidation consistently with the intuition that it is a more stable connective.

A somewhat subtle and counterintuitive point is that although the $\otimes$ product does not obviously implement a directional "flow" of substrate connectivity, as does the $\wedge$ product, it is not the case that $\xi \otimes \eta$ represents the same state as $\eta \otimes \xi$. In contradistinction, although the "flow" implementing the $\wedge$ product *does* have a directional attribute, $\xi \wedge \eta$ and $\eta \wedge \xi$ *do* represent the same state, since $\xi \wedge \eta = -\eta \wedge \xi$.

A transition of the form $\xi \otimes \eta \mapsto \xi \wedge \eta$ thus represents the change of a possible state of paired co-acting firing patterns to a state of a single firing pattern if there is no risk of double firing in the pair. Otherwise the transition is impossible. (In a sense this is a change from a somewhat unstable configuration to a more stable one.) Thus, for instance, a self-connection of a b-neuron via a substrate, with firing state $e_i \otimes e_i$, cannot make this transition to a firing pattern since $e_i \wedge e_i = 0$. The same remark applies to any tuples of basic firing patterns which have repeats on both sides of a $\otimes$ sign. Such transitions, from insecure $\otimes$-products of firing patterns to a single more secure firing pattern, will play a basic rôle in our approach to memory formation. (Long term memories cannot form properly in the presence of such overlaps.)

2.5.2   *Probing the exterior product connection*

In this section we will conduct another thought experiment that exposes some aspects of the substrate connection we have postulated to be represented by the $\wedge$-product.

We consider the simple case of a degenerate b-network consisting of two b-neurons n_{b_1} and n_{b_2} with no synaptic connection between them. Let us consider the case in which n_{b_2} is subjected to an external flux of the kind we have considered above. The *external* or *interaction* p-Hamiltonian in this case is then $H_I = la_2^\dagger + ha_2$ (cf. equation (2.4.34)). Here we have dropped the time dependence for notational simplicity. We seek the eigenstates of H_I. The matrix representation of this operator, relative to the basis $\{|\mathbf{0}\rangle, |\mathbf{1}_1\rangle, |\mathbf{1}_2\rangle, |\mathbf{1}_1, \mathbf{1}_2\rangle\}$ of $E(\mathbb{R}|\mathbf{1}_1\rangle \oplus \mathbb{R}|\mathbf{1}_2\rangle)$, is soon seen to be

$$\begin{pmatrix} 0 & 0 & h & 0 \\ 0 & 0 & 0 & -h \\ l & 0 & 0 & 0 \\ 0 & -l & 0 & 0 \end{pmatrix}. \tag{2.5.1}$$

If $lh > 0$ (so that also $h/l > 0$) there are two real non-zero eigenvalues, namely $\lambda = \pm\sqrt{lh}$, and four eigenstates, with each eigenvalue having two eigenvectors, namely for: $\lambda = +\sqrt{lh}$,

$$|\mathbf{0}\rangle + \sqrt{\left(\frac{h}{l}\right)}|\mathbf{1}_2\rangle, \qquad \text{and} \tag{2.5.2}$$

$$|\mathbf{1}_1\rangle - \sqrt{\left(\frac{h}{l}\right)}|\mathbf{1}_1, \mathbf{1}_2\rangle, \tag{2.5.3}$$

and for $\lambda = -\sqrt{lh}$,

$$|\mathbf{0}\rangle - \sqrt{\left(\frac{h}{l}\right)}|\mathbf{1}_2\rangle, \qquad \text{and} \tag{2.5.4}$$

$$|\mathbf{1}_1\rangle + \sqrt{\left(\frac{h}{l}\right)}|\mathbf{1}_1, \mathbf{1}_2\rangle. \tag{2.5.5}$$

(The notation may be confusing: for instance the column vector corresponding to $|\mathbf{1}_1, \mathbf{1}_2\rangle$ is $\begin{pmatrix} 0 \\ 0 \\ 0 \\ 1 \end{pmatrix}$, etc.)

So in this case the external flux impinging upon b-neuron n_{b_2} produces four stable states, *two of which involve the activation of b-neuron n_{b_1}* though only n_{b_2} is directly affected by the external flux, and there is no synaptic connection between them. This shows that the presumed connection via the substrate represented by the combinator $\wedge$ actually exists.

(Moreover, writing the eigenstates in the form

$$|\mathbf{0},\mathbf{0}\rangle \pm \sqrt{\left(\frac{h}{l}\right)}\,|\mathbf{0},\mathbf{1}_2\rangle, \qquad \text{and} \qquad (2.5.6)$$

$$|\mathbf{0},\mathbf{1}_1\rangle \mp \sqrt{\left(\frac{h}{l}\right)}\,|\mathbf{1}_1,\mathbf{1}_2\rangle \qquad\qquad (2.5.7)$$

it is apparent that all of these eigenstates involve some form of entanglement.)

The existence of tensor or exterior product states allows for the simulation of *neuromodulation* in our model. In the example above the firing behavior of a system comprising two b-neurons is modulated by the presence of the external "field" effecting the interaction we have denoted by H_I, regardless of whether there are synaptic connections or not. Similar effects occur in larger systems. Such effects, though implemented in an entirely different manner, have recently been shown to be effective in AI applications [99]. This example will be greatly generalized in section 5.2.

2.6 Multimodal b-networks

How do the dynamical considerations change if a b-network is multimodal? The main difference is that there are now more nodes, concomitantly more a, $a^\dagger$ operators, more J efficacy or exchange terms, and a more complex indexing scheme. Each output node formerly written as e_i^A, for instance, is now replaced by a super-position of "modal" states, with a more complex indexing scheme as illustrated for the bimodal case in section 2.3.4. The state spaces of such networks are still Fermi-Dirac/exterior algebras, with similar p-Hamiltonians (internal and external). A b-network state space is in this case equivalent to a system of (more) b-neurons, the only essential difference being in the new indexing scheme. A single n-modal b-neuron behaves in essentially the same way as a set of n unimodal b-neurons: this is the essence of their wished-for usefulness.

2.7 Digression: Spin, the Hopfield model and a free-floating tangent

As noted, our model resembles the attractor network model of Hopfield [6]. This was adapted from the quantum theory of many bodies, the paradigm being the so-called Ising model, which describes the quantum behavior of regularly arranged ensembles of magnetic dipole moments having only two spin states. The Hopfield model, though based on a quantum paradigm, does not posit operators such as our creation and annihilation operators, but, among other differences, uses ordinary scalar quantities instead.

Our model does not explicitly involve the quantum notion of spin, which is one of notorious subtlety, but in fact every two-state quantum system may be regarded as a spin system with two possible spin states. There are at least two routes to understanding this.

2.7.1 *The Jordan-Wigner paradigm*

The first was noted by Jordan and Wigner [55] in the early years of Heisenberg's "matrix" mechanical revolution. Their observation requires us to extend the base field from the real field $\mathbb{R}$ to the complex field $\mathbb{C}$. This can be accomplished by tensoring with $\mathbb{C}$ over $\mathbb{R}$, which is possible since $\mathbb{C}$ is a vector space over $\mathbb{R}$. Thus a real vector space V is replaced by $V_{\mathbb{C}} := \mathbb{C} \otimes_{\mathbb{R}} V$, the *complexification* of V, which is a complex vector space with the obvious action of $\mathbb{C}$. Operators on V now inherit this complex action. Now let V represent the two-dimensional real vector space of one of our b-neurons, with $V_{\mathbb{C}}$ its complexification. Recall our basic operators $\mathfrak{a}$ and $\mathfrak{a}^T$:

$$\mathfrak{a} = \begin{pmatrix} 0 & 1 \\ 0 & 0 \end{pmatrix} \tag{2.7.1}$$

$$\mathfrak{a}^T = \begin{pmatrix} 0 & 0 \\ 1 & 0 \end{pmatrix}. \tag{2.7.2}$$

In the Jordan-Wigner paradigm, these are regarded as the spin-lowering and spin-raising operators respectively. Then, by analogy with quantum mechanical spin systems, we can form the (Hermitian) spin *operators*, namely:

$$S_x := \frac{1}{2}(\mathfrak{a} + \mathfrak{a}^T) \qquad (2.7.3)$$

$$S_y := \frac{1}{2i}(\mathfrak{a} - \mathfrak{a}^T) \qquad (2.7.4)$$

$$S_z := \frac{1}{2}I - \mathfrak{a}^T \mathfrak{a}. \qquad (2.7.5)$$

Using equations (2.7.1) and (2.7.2) we find immediately that

$$S_x := \frac{1}{2}\begin{pmatrix} 0 & 1 \\ 1 & 0 \end{pmatrix} = \frac{1}{2}\sigma_1 \qquad (2.7.6)$$

$$S_y := \frac{1}{2}\begin{pmatrix} 0 & -i \\ i & 0 \end{pmatrix} = \frac{1}{2}\sigma_2 \qquad (2.7.7)$$

$$S_z := \frac{1}{2}\begin{pmatrix} 1 & 0 \\ 0 & -1 \end{pmatrix} = \frac{1}{2}\sigma_3 \qquad (2.7.8)$$

where the σ_i denote the Pauli matrices. These are (up to a constant) the usual representations of the spin (or quantum mechanical angular momentum) operators of ordinary quantum mechanics. (Extending this correspondence to many-fermion systems is not straightforward: cf. [55].) It is these operators that are involved in such many-body physical situations as the Ising model of which the Hopfield network model is a reduced classical version.

We shall not pursue the spin aspects any further in this work.

2.7.2 *A mathematical subdigression*

The above argument is clearly rather *ad hoc* and requires a fore-knowledge of actual physics. There is another purely mathematical argument, known to algebraic geometers, that reveals the orientable nature of such a qubit-like system and does not require any physics knowledge. We give an abbreviated sketch of this result.

Let k denote an arbitrary field. Then the quotient ring $k[\epsilon]/(\epsilon^2)$ is called the ring of *dual numbers*. Its elements are of the form $a + b\varepsilon$ where $a, b \in k$ and $\varepsilon := \epsilon \bmod (\epsilon^2)$ with $\varepsilon^2 = 0$. Thus, $k[\epsilon]/(\epsilon^2)$ is clearly isomorphic with $E(k)$. A *local* ring is one having a unique (proper) maximal ideal. The ring $k[\epsilon]/(\epsilon^2)$ is local, its maximal ideal being the one generated by ε. Local rings are so called because they contain information about the neighborhood of the point represented by their single maximal ideal relative to some ambient structure such as a manifold, variety, scheme, etc. Considering the pairs $(R, \mathfrak{m}_R)$ where $\mathfrak{m}_R$ denotes the maximal ideal of the local ring R, the correct morphisms that reflect the geometrical content are the ring morphisms $f : R \to S$ for which $f(\mathfrak{m}_R) \subset \mathfrak{m}_S$. Such maps are called *local*. The main geometrical example of a local ring is the ring $\mathcal{O}_x$ of germs of respectively smooth, holomorphic, etc., functions (roughly speaking) in neighborhoods of the point x in respectively a k-manifold, (variety, scheme), complex manifold, etc. Then a local map $\mathcal{O}_x \to k[\epsilon]/(\epsilon^2)$ corresponds to the mapping of the single geometric object represented by the maximal ideal of $k[\epsilon]/(\epsilon^2)$ to the point x. (This can be interpreted roughly as "gluing" the singleton spectrum, i.e. the maximal ideal regarded as a single object, to the manifold, say, in question.) Then the result is: there is an isomorphism between the space of local maps $\mathcal{O}_x \to k[\epsilon]/(\epsilon^2)$ and the space of tangents at x. (Cf. [68], p. 200. The result depends upon the quotient field of $\mathcal{O}_x$ being k, as it clearly is for $k[\epsilon]/(\epsilon^2)$.) Thus, the ring $E(k) \cong k[\epsilon]/(\epsilon^2)$ — in other words the state space of a qubit or b-neuron, when $k = \mathbb{C}$ or $\mathbb{R}$ — determines, as its singleton spectrum, the generic, free-floating tangent vector, which can be glued onto geometric objects of the right type. (These are generally locally ringed spaces, i.e. spaces supporting a prescribed sheaf of local rings.) In a sense the algebraic structure of this state space, of a b-neuron say, reflects the purely directional nature of the b-neuron's polarity.

Chapter 3

The Logic of Many Networks

We have so far considered the generalization of Boolean logic called orthologic (**OL**) as the appropriate logic under our doctrine of hidden variables. This was used to justify the move to full finite dimensional real Hilbert spaces as appropriate to a logical treatment of the state spaces of our network models. Now **OL** is modeled by *single* ortholattices. That is to say, in our cases, by the subspaces of a *single* finite dimensional Hilbert space. However, we have already seen how different networks may be combined and connected to each other and how this entails combinations of, and maps between, their respective state spaces. **OL** does not seem capable of reflecting such interactions. In fact there is an overarching or *external* logic that can deal with the case of many spaces, relative to which **OL** may be regarded as the logic *internal* to a single space. In this chapter we introduce this logic and a slight variant of it.

In section 3.1 we briefly discuss natural deduction systems, and introduce the idea of a Gentzen sequent calculus. We then briefly describe Girard's Linear Logic of which our logic will appear as a fragment. In section 3.2.3 we show that Girard's *of course* operator, endowing its operand with storage capability, may be interpreted in the category of finite dimensional vector spaces as the exterior algebra or Fermi-Dirac space, which we have already met in the previous chapter. This will have a bearing on all of our subsequent work. In section 3.3 we introduce a timing element into our sequent calculus. This will enable us to track structural changes which the dynamics of a single network is by its nature incapable of doing. The final section is a digression which may be skipped.

We remark that this logic could have been adduced without giving even the brief history of modern proof theoretic computational logic we have outlined at the beginning of this chapter. In doing so we would have missed or lost track of the computational and logical possibilities that are inherent to the sequent calculus. This being said, no advanced logical results are used or appealed to in this chapter or later.

3.1 Natural deduction

Historically, the external logic referred to above evolved for intrinsic logical reasons, and its importance to the field of computation emerged rather gradually. In section 3.1.1 we briefly sketch this development since although it developed independently of our applications of it, it is just this computational aspect that is our interest here. We contend that, roughly speaking, this logic is to our quantum-like networks as Boolean logic is to the standard classical networks of the McCulloch and Pitts type.

3.1.1 *A minimal system*

A *deductive system* at minimum comprises a specified set of expressions called *axioms* and a set of *inference* rules for manipulating or rewriting them. A *deduction* in such a system is a sequence of replacements of expressions starting from axioms, using the rules. Such deductions are generally viewed as tree-like, with the axioms forming the leaves and the concluding expression lying at the root. Thus a deduction is very much like a computer program, which proceeds in steps to reconfigure patterns of data.

We note the obvious but important fact that the conclusions at the roots of deduction trees are produced by an entirely *constructive* process. The deduction tree is not unique to its conclusion since many other deductions may produce the same conclusion. A given expression may have many deductions or indeed none at all. From a constructivist viewpoint we should associate expressions with their entire repertoire of deduction trees. This kind of association lies at the heart of Heyting's interpretation of intuitionist logic, which arises from a wholesale adherence to constructivist principles. Cf. [38] and [95].

Here we shall briefly discuss a minimal deductive system for *implicational intuitionistic propositional logic*. This treatment combines elements from the early chapters of both [38] and [95]. The basic object in this system is a deduction of a formula A, say, which, after [38] we denote by

$$\vdots \\ A \tag{3.1.1}$$

where the dots stand for subdeductions. The first rule of inference is that a single formula by itself is a deduction (of itself). Strictly speaking we should assert this axiom only for a set of *atomic* formulas, the result then following for all formulas. We will follow custom in this brief overview and later by omitting the complication of specifying the atoms.

There are two further rules of inference. One rule *introduces* the implication sign $\Rightarrow$ and the other rule *eliminates* it. To express these rules we need some notational preliminaries. Suppose A appears in a *single* top node or leaf of deduction whose conclusion is B. Then we may unambiguously write

$$A \\ \vdots \\ B \tag{3.1.2}$$

In this case the introduction rule posits the new deduction

$$A \\ \vdots \\ \frac{B}{A \Rightarrow B} \; \Rightarrow I \tag{3.1.3}$$

(Here, the $\Rightarrow I$ is used to label the rule *implication introduction* extending the tree. It is frequently dropped in the absence of ambiguity.) The *occurrence* of A is said to be *open* or *live* in (3.1.2) but considered to be *closed*, or *killed*, or *discharged* by the application of $\Rightarrow I$ in (3.1.3). A may appear as open in other places such as the ambient deductions and in this case we would wish to keep track of which open occurrences of A are being discharged by the $\Rightarrow I$ inference rule. This can be accomplished by labeling A and then invoking the label at the point of inference. Thus in place of (3.1.3) we now write

$$
\begin{array}{c}
A^u \\
\vdots \\
\dfrac{B}{A \Rightarrow B} \quad u, \Rightarrow I
\end{array}
\qquad (3.1.4)
$$

This *intuitionistic* version of implication may be interpreted in light of the so-called *Heyting paradigm* which gives a semantics for formal intuitionistic logic (denoted **IL**). Cf. [95], Section 2.5.1, p. 55, where the attribution also includes Brouwer and Kolmogorov. In this interpretation a formula is intuitionistically *valid* only if a deduction of it can be explicitly presented or constructed. The interpretation of $A \Rightarrow B$ in (3.1.4) then becomes: if a deduction of A^u can be constructed then a deduction of B can be constructed by means of the deduction above the inference line in (3.1.4). After this encapsulation of the whole process in the formula $A \Rightarrow B$, A^u is no longer needed and may be discharged, the deduction leading to it being in a sense discarded.

As noted, open occurrences of A may appear in several places in the deduction leading to B, and we may choose to discharge a subset of these at the point of inference. The deductions involving the occurrences of A in the chosen collection are then all discarded simultaneously at the point of inference. Members of such a collection may then be grouped under a single label since there is no need to distinguish among them. The notation for such a collection of open occurrences of A is $[A]^u$. There may be other collections of open occurrences of A that are not chosen for discharge at the point of inference: these remain open after the inference. The complete statement of the $\Rightarrow I$ rule now reads

$$
\begin{array}{c}
[A]^u \\
\vdots \\
\dfrac{B}{A \Rightarrow B} \quad u, \Rightarrow I
\end{array}
\qquad (3.1.5)
$$

The degenerate case of $[A]^u$ being empty is allowed. This case would still require a label at the inference line as in

$$
\dfrac{B}{A \Rightarrow B} \quad v
\qquad (3.1.6)
$$

The v here labels the empty collection of occurrences which is discharged at the inference. Among the difficulties here is that $[A]$ labels a *pattern* of occurrences of the formula A and is not strictly speaking a *set*. We will ignore this difficulty in this brief informal sketch.

The other rule of inference in this system is a rule for eliminating $\Rightarrow$. It is just *modus ponens* written as

$$
\dfrac{\begin{array}{cc} \vdots & \vdots \\ A & A \Rightarrow B \end{array}}{B} \quad \Rightarrow E
\qquad (3.1.7)
$$

Two deductions, of A and $A \Rightarrow B$, are combined to produce a new deduction with conclusion B.

A careful treatment of the labeling scheme and a consideration of the Heyting paradigm leads to a connection between this deductive system and the λ-calculus of Church. We briefly sketch this connection. The labels were introduced above to keep track of the flow of discharges or closings of collections of open formulas as the inference is enacted. In the Heyting interpretation of **IL** a formula is valid only if a deduction of it can be produced. Thus, a formula may be *identified* with its set of deductions. In more formal language, a formula determines a *type*, A say, and a label u, say, of A is considered to be a *variable* of type A. The standard notation for this is $u : A$. Formal definitions of types, terms, variables, etc., may be found in the references cited. (For our purposes the informal intuitive notion of a type being a special kind of set, while variables are elements of such sets, will suffice.) Note that the Heyting interpretation of $A \Rightarrow B$ is now that given a deduction of A one has a deduction of B: that is, there exists a map from the set of deductions of A to the set of deductions of B.

This new interpretation of labels as variables brings a new interpretation to the deductions above. Thus A^u can now be written $u : A$, and $[A]^u$ can be written $[u : A]$: all the deductions of A in the posited collection are discharged simultaneously at the inference. The conclusion in (3.1.5) now "depends" upon the deduction (i.e. u) of A. The deduction that produced B from u is labeled t. With this in mind we can now rewrite (3.1.5) as

$$
\begin{array}{c}
[u : A] \\
\vdots \\
\dfrac{t : B}{\lambda u.t : A \to B}
\end{array}
\qquad (3.1.8)
$$

Here the symbol λ serves to *bind* u within t. The type $A \to B$ is a "function" type which can be interpreted via the Heyting paradigm as the set of functions from the set of deductions of A to the set of deductions of B. The expression $\lambda u.t$ is the name of the function which produces t upon the "input" u. Note that this binding of u within t via the symbol λ in the expression $\lambda u.t$ recapitulates exactly the discharging of the associated formula's occurrences.

Similarly the inference (3.1.7), which eliminates $\Rightarrow$, may be rewritten in type theoretic terms as

$$
\dfrac{\begin{array}{cc} \vdots & \vdots \\ s : A & t : A \to B \end{array}}{ts : B}
\qquad (3.1.9)
$$

Using these translations of the inference rules any deduction may be used to generate a "λ-term" which completely describes or encapsulates the deduction. What emerges is a version of Church's λ-calculus.

This was posited independently by Church in the 1930s as a means of investigating the computational and logical possibilities of pure functionality. Today the theory goes by the name "simply typed λ-calculus." The observation by Curry in 1958 [23] that the translation given above induces a complete structural isomorphism between the minimal natural deduction system outlined above and the simply typed λ-calculus was considered an advance by computational logicians.

The computational resources of simply typed λ-calculus and other λ-calculi have been well studied. The isomorphism sketched above may be extended to one that obtains between the minimal intuitionistic implicational fragment with inference rules for conjunction ($\wedge$, not to be confused with the exterior product of the last chapter) and disjunction ($\vee$) appended, and an appropriately supplemented version of simply typed λ-calculus. The inference rules for conjunction to be added to the natural deduction system are three in number: one Introduction and two Eliminations, namely

$$\frac{A \quad B}{A \wedge B} \quad \wedge I \tag{3.1.10}$$

$$\frac{A \wedge B}{A} \quad \wedge 1E \tag{3.1.11}$$

$$\frac{A \wedge B}{B} \quad \wedge 2E \tag{3.1.12}$$

Disjunction in an intuitionistic system is independent of conjunction, since De Morgan duality does not obtain, and is generally contentious for reasons to appear. In our system there are two Introduction rules

$$\frac{A}{A \vee B} \quad \vee 1I \tag{3.1.13}$$

and

$$\frac{B}{A \vee B} \quad \vee 2I \tag{3.1.14}$$

and one problematical Elimination rule:

$$\frac{\begin{array}{ccc} [A] & [B] \\ \vdots & \vdots & \vdots \\ A \vee B & C & C \end{array}}{C} \quad \vee E \tag{3.1.15}$$

The problem here is the extraneous C, which introduces an uncontrollable element into the business of deriving general theorems about deductions: cf. [38], Chapter 10.

To extend the Curry isomorphism to this supplemented natural deduction system, we again appeal to the Heyting paradigm. In order for the conjunction $A \wedge B$ to be intuitionistically valid we must possess a deduction of A *and* a deduction of B, and know which deduction belongs to which formula. So we must possess an ordered pair of formulas. If a formula is identified with its set of deductions then the set of deductions of $A \wedge B$ should be identified with the Cartesian product of the set of deductions of A and the set of deductions of B. Thus the $\wedge$ of formulas should be associated with the product $\times$ of the corresponding types in the extended Curry correspondence. Similarly, $A \vee B$ is intuitionistically valid only if we have a deduction of A *or* a deduction of B and an indication of which one of these formulas has been deduced. The collection of such pairs constitutes the disjoint union (or direct sum in the category of sets) of the sets of deductions of the constituent formulas.

The Curry correspondence thus extended is part of W. A. Howard's contribution to the full isomorphism which now bears the name Curry-Howard. (The other part of Howard's contribution involves quantifiers, $\exists$, $\forall$, which we are ignoring here.) The importance to computational theory of isomorphisms of the Curry-Howard type is that, since formulas may be regarded as types in such isomorphisms, deductions may be concomitantly regarded as computations or programs, which transform types in a stepwise fashion. Reversing this perspective, such isomorphisms allow us to regard the apparently static program represented by a λ-term in a dynamical light, since such a term may be unfolded to reveal its underlying deductive structure, with its flow of openings and closings of assumptions. It is this aspect of the Curry-Howard isomorphism that has arguably had the most impact.

3.2 The Gentzen sequent calculus

The Gentzen sequent calculus was originally intended to be a *metacalculus* for handling deductions in a natural deduction system. In this paradigm the basic objects in such a calculus stand for the deductions in a natural deduction system and are treated wholesale as objects in their own right. It has been developed in various directions as a style of deductive reasoning without necessarily having an underlying deductive system in mind. In its original guise as a metacalculus for natural deduction, the sequent calculus delineates certain symmetries and structural aspects of the underlying system which remain hidden, or are less apparent, if one remains fixed at the natural deduction level. The organizing power of the style has had a major impact on proof theoretic aspects of deductive logic.

The basic object is the *sequent*:

$$\Gamma \vdash \Delta \qquad\qquad (3.2.1)$$

Here, Γ and Δ stand for (possibly empty) finite sequences of formulas. The informal reading of (3.2.1) is "$\bigwedge \Gamma \Rightarrow \bigvee \Delta$". This stands for a deduction in which the conclusion lies in Δ. (Note that this is consistent with the doctrine of hidden variables: the turnstile symbol, $\vdash$ in (3.2.1) "hides" the details of the posited deduction.) The interpretation of $\emptyset \vdash \Delta$ is then that it asserts the truth of $\bigvee \Delta$. The empty set is usually just omitted from the notation. Thus $\Gamma \vdash$ is interpreted as asserting the falsity of $\bigwedge \Gamma$.

In keeping with this paradigm, and noting that disjunction is disruptive in intuitionistic systems, sequents in which Δ consists of at most a single formula are termed "intuitionistic."

In Gentzen calculi the inference rules are generally divided into classes: structural rules, logical rules and an "identity" group. A deduction in such a system is called a *proof*.

As an example we shall briefly describe some of the rules for a non-intuitionistic minimal propositional sequent caclculus. As above, the horizontal line in a rule represents the inference of the sequent below it from the sequent(s) immediately above it. As before the appropriate label appears to the right of the inference line. Thus LE for left exchange, etc.

Structural Rules

EXCHANGE

$$\frac{\Gamma, A, B, \Gamma' \vdash \Delta}{\Gamma, B, A, \Gamma' \vdash \Delta} \text{ LE} \qquad \frac{\Gamma \vdash \Delta, A, B, \Delta'}{\Gamma \vdash \Delta, B, A, \Delta'} \text{ RE} \qquad (3.2.2)$$

WEAKENING

$$\frac{\Gamma \vdash \Delta}{A, \Gamma \vdash \Delta} \text{ LW} \qquad \frac{\Gamma \vdash \Delta}{\Gamma \vdash \Delta, A} \text{ RW} \qquad (3.2.3)$$

CONTRACTION

$$\frac{A, A, \Gamma \vdash \Delta}{A, \Gamma \vdash \Delta} \text{ LC} \qquad \frac{\Gamma \vdash \Delta, A, A}{\Gamma \vdash \Delta, A} \text{ RC} \qquad (3.2.4)$$

Here, WEAKENING corresponds to the possibility of introducing spurious collections of occurrences of the formula A, while contraction corresponds to the possibility of amalgamating certain collections of occurrences of A.

A revolution occurred in the late 1980s when Girard realized that sequent calculi could be given an *operational* interpretation. In this reading, formulas, considered as types *à la* Curry-Howard, are regarded as *resources* and $\Gamma \vdash \Delta$ has the informal reading as: "Use up Γ to produce Δ." Then LC (3.2.4) acquires the connotation that while two As are required to produce Δ we can get away with only one use of it to produce Δ. The "resource" A must then be *storable* and can be reused, by copying or cloning in some way. One might say that *A admits storage*. Clearly many real resources, like coins, do not have this convenient property: if an item requires two coins for its purchase, one will not suffice.

We shall briefly discuss some important aspects of the remaining two divisions of the sequent calculus inference rules.

The Identity Group

AXIOM

$$A \vdash A \quad \text{Ax} \qquad\qquad (3.2.5)$$

CUT

$$\frac{\Gamma \vdash \Delta, A \qquad A, \Gamma' \vdash \Delta'}{\Gamma, \Gamma' \vdash \Delta, \Delta'} \text{CUT} \qquad\qquad (3.2.6)$$

The terminology here seems to be due to Girard [38]. He notes that CUT is another way of expressing identity. The AXIOM asserts that *A* on the left of the turnstile is used to produce *A* on the right: *A* produces *A*. On the other hand CUT asserts that if *A* appears on the right of the first turnstile then it may be used again to the left of the second turnstile: again *A* must have "produced" *A* to effect the CUT. So CUT is a redundant expression of identity. In fact its analogue in natural deduction may be deduced from the other rules. Its use is akin to the use of a lemma in a mathematical proof, or the use of subroutine in a computer program. In these forms the cut rule would seem to be essential to both of these disciplines. It is however problematical from the point of proof theory itself since the coming and going of the possibly extraneous and uncontrollable *A*s greatly complicate the tree handling techniques used in the subject. It may therefore come as a surprise to learn that even in quite general Gentzen sequent calculi cuts can be removed from any proof, demonstrating their redundancy. That is to say, any proof involving uses of CUT may be recast without using CUT. This *CUT elimination* is the gist of Gentzen's justly famous "Hauptsatz": cf. the references already cited. It is a rigorous expression of Girard's subsequent insight that CUT is a redundant expression of identity. It will fail for our applications since we will be adding extra "non-logical" axioms.

Logical Rules

We omit the rules for $\vee$ and $\Rightarrow$ and display here only the rules for conjunction.

$$\frac{\Gamma, A_i \vdash \Delta}{\Gamma, A_0 \wedge A_1 \vdash \Delta} \; \mathrm{L}i\wedge, i = 0, 1 \qquad \frac{\Gamma \vdash A, \Delta \quad \Gamma' \vdash B, \Delta'}{\Gamma, \Gamma' \vdash A \wedge B, \Delta, \Delta'} \; \mathrm{R}\wedge \qquad (3.2.7)$$

For an intuitionistic system, all of these logical rules, except those for $\vee$, are restricted merely by allowing at most one formula on the right of the turnstiles. For the $\vee$ rules (not shown) special considerations are called for.

For an intuitionistic Gentzen sequent calculus it is generally possible to produce a natural deduction system that might be presumed to underlie it. This is done by judiciously assigning terms to sequents and then regarding these as λ-like terms for underlying deductions. Interested readers may refer to the works cited, in particular [1].

3.2.1 *Linear Logic and the minimal system GN*

Taking these lessons into account, Girard effected his revolution by positing a vast generalization of deductive logic, replacing the usual connectives and inference rules of the Gentzen calculus with new ones based on these new principles. He called the resulting calculus Linear Logic (**LL**). The repercussions were wide and deep and the literature has become oceanic. References include, among many many others: [1, 5, 7, 38, 94, 95]. Since we will be applying only a small fragment of this logic, we will not go into a discussion of the entire scheme, but only the parts relevant to our application.

We note in particular via the coin example above that what we have called storage capability is not a given: that is to say, not all resources (or types) A are subject to the Contraction rule as in LC above (3.2.4). Girard introduced a modal-like operator !( ), which he called *of course*, as in !A, which renders A storage capable: that is to say !A is the storage capable version of A. This operator gets its own set of inference rules. We note also that WEAKENING may not obtain, since the inclusion of extraneous resources may interfere with the resources already present. The ! operator applies in that case as well to effect WEAKENING.

The set of rules we have found appropriate to our application turns out to be equivalent to a fragment of a system called *intuitionistic linear logic*, or **ILL**, itself a fragment of **LL**, which we have dubbed **GN**. It is given below. (A not dissimilar system of logic applied to the field of molecular biology was put forward independently in [14] and developed further in [86].)

GN
Structural Rules
EXCHANGE

$$\frac{\Gamma, A, B, \Gamma' \vdash D}{\Gamma, B, A, \Gamma' \vdash D}\text{LE} \qquad (3.2.8)$$

WEAKENING

$$\frac{\Gamma \vdash D}{\Gamma, !A \vdash D}\text{LW} \qquad (3.2.9)$$

CONTRACTION

$$\frac{!A, !A, \Gamma \vdash D}{!A, \Gamma \vdash D}\text{LC} \qquad (3.2.10)$$

The Identity Group
AXIOM

$$A \vdash A \quad \text{Ax} \qquad (3.2.11)$$

CUT

$$\frac{\Gamma \vdash A \quad A, \Gamma' \vdash D}{\Gamma, \Gamma' \vdash D}\text{CUT} \qquad (3.2.12)$$

Multiplicative Logical Rules
CONJUNCTIVE (MULTIPLICATIVE) CONNECTIVE

$$\frac{\Gamma, A, B \vdash D}{\Gamma, A \otimes B \vdash D} \text{L} \otimes \qquad (3.2.13)$$

$$\frac{\Gamma \vdash A \quad \Gamma' \vdash B}{\Gamma, \Gamma' \vdash A \otimes B} \text{R} \otimes \qquad (3.2.14)$$

Additive Logical Rules
DISJUNCTIVE (ADDITIVE) CONNECTIVE

$$\frac{\Gamma \vdash A}{\Gamma \vdash A \oplus B} \text{R} \oplus_1 \qquad (3.2.15)$$

$$\frac{\Gamma \vdash B}{\Gamma \vdash A \oplus B} \text{R} \oplus_2 \qquad (3.2.16)$$

$$\frac{\Gamma, A \vdash D \quad \Gamma, B \vdash D}{\Gamma, A \oplus B \vdash D} \text{L} \oplus \qquad (3.2.17)$$

$$!$$

$$\frac{\Gamma, A \vdash D}{\Gamma, !A \vdash D} \text{L}! \qquad (3.2.18)$$

$$\frac{!\Gamma \vdash A}{!\Gamma \vdash !A} \text{R}! \qquad (3.2.19)$$

Here capital Greeks stand for finite sequences of formulas including possibly the empty one, and D stands for either a single formula or no formula, i.e. the empty sequence, and when it appears in the form $\otimes D$, the $\otimes$ symbol is presumed to be absent when D is empty. If Γ denotes the sequence $A_1, A_2, \ldots, A_n$ then $!\Gamma$ will denote the sequence $!A_1, !A_2, \ldots, !A_n$. The Girard *of course* exponential operator ! is sometimes pronounced "bang." The sign $\otimes$ should strictly speaking just be regarded as an abstract symbol, though of course we shall soon interpret it as usual. In applications, one adds "non-logical" axioms to the **GN** rules, for instance to depict a brain as a family of linked networks (see next paragraph), and other sorts of given sequents. The resulting family of proofs or deductions is known by logicians as a **GN**-*theory*.

The logical rules for the analogue of disjunction, which is denoted $\oplus$, will be eventually interpreted (cf. section 3.2.2 below) as the external direct sum in a category of vector spaces. As mentioned in section 2.2.4, this combination of state spaces does not reflect a discernible operation upon the underlying networks unless they are disjoint or the shared nodes are multimodal. So this will limit our applications to network state spaces. There is another objection to the use of $\oplus$ in application to networks of our type. Namely, it is promiscuous as to the pairings it allows. Thus it allows pairings between disparate networks that may have nothing in common and such profligacy is not likely to reflect the reality of our networks *in vivo*. Its use must be reined in. Likewise the $\otimes$ connective is apparently similarly promiscuous. In fact these two profligacies are related by the identity equation (B.1.8) and we will address these issues briefly in sections 5.6.1 and 5.6.2 and more comprehensively in Chapter 6.

(The set of axioms for full linear logic, also known as *classical linear logic*, **CLL**, is to be found in [7] and our axioms above correspond to those in the latter work, given the intuitionistic embargo on more than one formula in the succedents, as follows: Section 4.4.2, p. 85 *et seq.*: (id) $\leftrightarrow$ Ax; (exc, r) $\leftrightarrow$ LE; (cut) $\leftrightarrow$ Cut; $(\otimes, \text{r}) \leftrightarrow$ R$\otimes$; $(\otimes, \text{l}) \leftrightarrow$ L$\otimes$; $(\oplus,\text{r},1) \leftrightarrow$ R$\oplus_1$; $(\oplus,\text{r},2) \leftrightarrow$ R$\oplus_2$, $(\oplus, \text{l}) \leftrightarrow$ L$\oplus$. Section 5.5.2, p. 115: (weak-l) $\leftrightarrow$ LW; (contr-l) $\leftrightarrow$ LC; (!, l) $\leftrightarrow$ L!; (!, r) $\leftrightarrow$ R!.)

Note that our calculus lacks a negation. In the standard interpretation of full **CLL** in categories of vector spaces, this negation, of a type A, say, is realized as the appropriate dual space $[\![A]\!]^*$ of linear functionals on the vector space $[\![A]\!]$ which interprets A: cf. section 3.2.2. The reason for its omission here is that there is no apparent interpretation of the dual space of the state space of a network in terms of network attributes. We should note here yet another cross-disciplinary notational misappropriation. The notation that has become standard for negation in the **LL** literature is a superscripted $\perp$, as in $A^\perp$, though, as noted, it is interpreted as the dual space of the space interpreting A. This is in conflict with the usual notations of linear algebra where $\perp$ denotes the orthogonal complement of its argument. To confuse matters further, the orthogonal complement is actually the negation in the internal **OL** of closed subspaces of a Hilbert space. As far as possible we have adhered to the linear algebra notational conventions in this work where they apply. Note also that although we are denied the use of negation in the logic, the dual is available in the interpreting category and may be used there in intermediate steps which do not violate any logical conclusions. Moreover, in the categories of Hilbert spaces we shall be using as interpreting categories, the appropriate dual spaces may be identified with the spaces themselves, via the Riesz-Fréchet theorem. (Ours is not the only fragment of **LL** to omit negation. There are other well known logics that omit both negation and disjunction, namely the ones known as *intuitionistic multiplicative exponential linear logic* or **IMELL** and **MILL**, which is the multiplicative version of **ILL**, which also lacks the ! operator as well as negation and disjunction. In particular, a non-intuitionistic variant of multiplicative exponential linear logic is used in [86] in application to molecular biology very much in the spirit of our attempts in this work.)

Our rules here are provisional and are the minimal set we have found it possible to get away with.

This being said, it turns out that an intuitionistic fragment of **OL** may be translated into the sequent calculus **GN** in such a way that sequents in the former translate to sequents in the latter. Interested readers may refer to the digressive last section of this chapter for a discussion.

We shall cut through the rigorous formal procedures usually imposed by logicians (*pace* them) and apply this system informally. Namely, we shall regard the formulas (also known as *types*) $A, B, \ldots$, as standing for finite dimensional real Hilbert spaces (formerly written $\mathfrak{H}_A, \mathfrak{H}_B, \ldots$) read as representing the state spaces of unicameral networks $\mathcal{N}_A, \mathcal{N}_B, \ldots$ (This is no restriction on finite dimensional vector spaces since any such space is the state space of at least one network, namely the degenerate cluster with the same number of nodes as the dimension of the space). Sequents $A \vdash B$ will then represent linear maps of A into B. There are special ones that arise by synaptically connecting $\mathcal{N}_A$ to $\mathcal{N}_B$ as in the previous chapter. Such connecting sequents may be composed via the CUT rule to produce connections of arbitrary length. This type of map is like a *wiring diagram* of a connection of $\mathcal{N}_A$ to $\mathcal{N}_B$ describing a potential (i.e. *possible*) situation. It is important to note that other sorts of maps of state spaces must also be allowed, and are treated as sequents on the same logical footing as these linking sequents. CUT will then also apply to them. (Thus a sequent resembles, but is more general than, a synaptic-like wiring connection between one layer of a standard neural network and the next.)

In this interpretation of formulas or types as vector spaces and sequents as maps thereof, we interpret the empty set of formulas as the vector space $\mathbb{R}$, and sequences of types (appearing to the left of the turnstile) as the tensor product of their members, which is consistent with the rule L⊗. Note that **GN** allows only single formulas, or none, to the right of the turnstile. As we have indicated this is of some computational import: namely it renders the system *intuitionistic* or *constructive*.

We may motivate the non-! rules by thinking of the types as standing for state spaces of networks of the kind described in the previous chapter. We will address the interpretation of the ! operator separately in section 3.2.3 below. A sequent of the form $A \vdash D$ will indicate that $\mathcal{N}_A$ *affects* $\mathcal{N}_B$. A sequent of the form $A, B \vdash D$, can then be read as: both $\mathcal{N}_A$ and $\mathcal{N}_B$ affect $\mathcal{N}_D$, and we shall write $\mathcal{N}_A$ *with* $\mathcal{N}_B$ affect $\mathcal{N}_D$. We shall extend the subscript notation to sequences Γ of types so that $\mathcal{N}_\Gamma$ denotes the substrate-connected set of networks associated with the elements of Γ. Then

LE: Clearly, the order of listed types in the antecedent does not matter.

LW: The ordinary logical meaning of WEAKENING is that any hypothesis may be conjoined, or ANDed, to a valid inference without vitiating the validity of the inference. However, in the paradigm in which the types stand for evanescent resources, such as our networks, the intrusion of an arbitrary network into the axonal/synaptic connection expressed by $\Gamma \vdash D$ may interfere with such a connection, by blocking or otherwise changing the axonal terminals of the Γ network, and so we must abandon WEAKENING as a general rule. However, it can be brought back for networks which have been in a sense isolated or insulated via the application of the ! operator.

LC: As noted, CONTRACTION is associated with the notion of *storage*. For a general network this would seem to require at least some form of manipulation or copying of the network itself, which may cause damage unless it is insulated or protected in some way, so CONTRACTION had better not be presumed to hold for an arbitrary network. It turns out that the same operator ! applied to a general network affords this type of protection. Moreover, storage is accomplished by a process of quantum-like entanglement: cf. section 3.2.4 below.

AXIOM: We take this $A \vdash A$ to be induced by the identity affect on $\mathcal{N}_A$ which has no effect.

CUT: This is just a reflection of the fact that affects may be composed. If $\mathcal{N}_\Gamma$ affects $\mathcal{N}_A$, and $\mathcal{N}_A$ with $\mathcal{N}_{\Gamma'}$ affect $\mathcal{N}_D$, then $\mathcal{N}_\Gamma$ with $\mathcal{N}_{\Gamma'}$ affect $\mathcal{N}_D$ (with $\mathcal{N}_\Gamma$ acting via $\mathcal{N}_A$).

L⊗: If $\mathcal{N}_A$ with $\mathcal{N}_B$ affect $\mathcal{N}_D$ then the substrate-connected network will also affect $\mathcal{N}_D$.

R⊗: The connections of $\mathcal{N}_\Gamma$ to $\mathcal{N}_A$ and $\mathcal{N}_{\Gamma'}$ to $\mathcal{N}_B$ signified by the turnstiles remain intact when $\mathcal{N}_A$ and $\mathcal{N}_B$ are substrate connected. The conclusion follows.

⊕ rules: These are all obvious from the definition given in section 2.2.2.

L!: If $\mathcal{N}_\Gamma$ with $\mathcal{N}_A$ affect $\mathcal{N}_D$ then the same should be true *a fortiori* when $\mathcal{N}_A$ is replaced by its storage capable version.

R!: If the storage capable version of $\mathcal{N}_\Gamma$ affects $\mathcal{N}_A$ then, since the former may be brought out of storage repeatedly, it should also affect the repeatable (i.e. storage capable) version of $\mathcal{N}_A$.

By way of examples, here are some simple proofs in this calculus. We shall say A is *logically equivalent to* B, and write $A \rightleftharpoons B$, if $A \vdash B$ and $B \vdash A$. The standard notation for this is $\equiv$. This is rather misleading since it conveys the idea of isomorphism. This is of course generally a much weaker relation than isomorphism in categories in which the logic may be interpreted, and in particular the one of interest to us here, namely the category of finite dimensional real Hilbert spaces, if we take the symbols for the connectives in the logic to be their linear counterparts (section 3.2.2). That is to say, logical equivalence does not generally imply isomorphism in the interpreting category, though sometimes isomorphism does obtain, not necessarily via the interpreted maps. In fact, all the logical equivalences proved below, except the third, correspond to actual isomorphisms in the category of finite dimensional vector spaces.

Proposition 3.2.1.

1.
$$A \otimes B \rightleftharpoons B \otimes A \tag{3.2.20}$$

2.
$$A \oplus B \rightleftharpoons B \oplus A \tag{3.2.21}$$

3.
$$A \oplus A \rightleftharpoons A \tag{3.2.22}$$

4.
$$\frac{\Gamma, A \vdash D \quad \Gamma', B \vdash D'}{\Gamma, \Gamma', A \otimes B \vdash D \otimes D'} \tag{3.2.23}$$

5.
$$!A \vdash \overbrace{!A \otimes \cdots \otimes !A}^{n} \quad for \ n = 0, 1, \ldots \tag{3.2.24}$$

6.
$$(A \otimes B) \otimes C \rightleftharpoons A \otimes (B \otimes C). \tag{3.2.25}$$

Proof.
1.
$$\frac{\dfrac{\dfrac{A \vdash A \quad B \vdash B}{A, B \vdash A \otimes B} \text{R}\otimes}{\dfrac{B, A \vdash A \otimes B}{B \otimes A \vdash A \otimes B.} \text{L}\otimes} \text{LE}}{} \tag{3.2.26}$$

Reversing the rôles of A and B yields the converse implication.

2.

$$\frac{B \vdash B}{B \vdash A \oplus B} \text{R}\oplus_2 \qquad \frac{A \vdash A}{A \vdash A \oplus B} \text{R}\oplus_1 \qquad (3.2.27)$$

Now apply L$\oplus$ to these last concluding sequents to infer $B \oplus A \vdash A \oplus B$. Reversing the rôles of A and B yields the converse implication.

3.

$$\frac{A \vdash A}{A \vdash A \oplus A} \text{R}\oplus_1 \qquad (3.2.28)$$

and

$$\frac{A \vdash A \qquad A \vdash A}{A \oplus A \vdash A} \text{L}\oplus \qquad (3.2.29)$$

4.

$$\frac{\dfrac{\Gamma, A \vdash D \quad \Gamma', B \vdash D'}{\Gamma, A, \Gamma', B \vdash D \otimes D'} \text{R}\otimes}{\dfrac{\Gamma, \Gamma', A, B \vdash D \otimes D'}{\Gamma, \Gamma', A \otimes B \vdash D \otimes D'} \text{L}\otimes} \text{LE} \qquad (3.2.30)$$

5. For $n = 0$ apply LW with Γ and D empty. The $n = 1$ case follows from Ax. Then

$$\frac{\dfrac{!A \vdash !A \quad !A \vdash !A}{!A, !A \vdash !A \otimes !A} \text{R}\otimes}{!A \vdash !A \otimes !A} \text{LC} \qquad (3.2.31)$$

Repeat this with the leftmost $!A \vdash !A$ in the top line replaced with $!A \vdash !A \otimes !A$ to obtain the case of $n = 3$. And so on.

6.

$$\frac{\dfrac{\dfrac{A \vdash A \quad B \vdash B}{A, B \vdash A \otimes B} \text{R}\otimes \quad C \vdash C}{A, B, C \vdash (A \otimes B) \otimes C} \text{R}\otimes}{\dfrac{A, (B \otimes C) \vdash (A \otimes B) \otimes C}{A \otimes (B \otimes C) \vdash (A \otimes B) \otimes C} \text{L}\otimes} \text{L}\otimes \qquad (3.2.32)$$

Similarly, we have

$$
\frac{
  \dfrac{
    \dfrac{
      \dfrac{
        \dfrac{
          \dfrac{
            \dfrac{B \vdash B \quad C \vdash C}{A \vdash A \qquad B, C \vdash B \otimes C} \text{R}\otimes
          }{A, B, C \vdash A \otimes (B \otimes C)} \text{R}\otimes
        }{C, A, B \vdash A \otimes (B \otimes C)} \text{LE,LE}
      }{C, A \otimes B \vdash A \otimes (B \otimes C)} \text{L}\otimes
    }{A \otimes B, C \vdash A \otimes (B \otimes C)} \text{LE}
  }{(A \otimes B) \otimes C \vdash A \otimes (B \otimes C)} \text{L}\otimes
}{} \qquad (3.2.33)
$$

This proves the proposition.   □

The fifth sequent in the statement expresses the fact that a given $!A$ "reproduces" itself any number of times.

We note also the following logical import of some of the rules.

LE: Clearly, this allows any permutation of formulas in the antecedent of a sequent.

L⊗: *Both A and B together must affect D* in order for $A \otimes B$ also to affect D. So $\otimes$ is interpreted as a *parallel* conjunction.

R⊗: In this case the two antecedent sequents are completely independent and share no "resources" until the parallel conjunction $\otimes$ is applied in the conclusion.

⊕: It is apparent that *either* hypothesis in the right rules is sufficient to ensure the $\oplus$ conclusion. Since this conclusion involves a *choice*, this connective is interpreted as a disjunction.

3.2.2 *Interpretation of the calculus in the category of finite dimensional vector spaces*

Informally, an *interpretation* of a logical calculus of our type in a category is a mapping of types of the calculus to objects in the category and of logical connectives to connectives in the category in such a manner that the logical inferences hold in the category. That is to say, the interpretation should preserve the inferences of the logic. The conventions of formal logic entail a notation for the interpreted formula, such as for instance $[\![A]\!]$ for the interpretation of the formula A. Since our treatment will be informal here, we will not make this distinction but simply identify a formula with its interpretation: indeed, we have already done so in the last section.

Proper logicians use such interpretations mainly for reasons of proving consistency and other logical concerns. Our use of it will be for the rather different and mundane purpose of realizing logical outcomes in terms of our applications to network state spaces. As such, this interpretation of logical outcomes in the family of state spaces of our networks is, where it is possible, our most important step. We shall pursue logical procedures while keeping in mind the interpretations in terms of the state spaces of our networks. There are two problems with this. The first, of our own making, concerns the lack of distinction we have made between a formula and its interpretation as a Hilbert space. To make any sort of notational distinction, such as the standard use of the bracket notation mentioned above, would lead to prolific notational clutter and would tend to obfuscate rather than clarify. On the other hand, by neglecting to notate the distinction, confusion is risked. We hope to avoid this by making the network interpretation our general operating procedure. The other problem involves our desire to apply the logic to actual networks. For one thing, we access our networks through their spaces of states and and not every logical operation that can be done to or with such spaces reflects discernible, concrete operations upon the corresponding underlying networks, at least when these are considered as classical constructs. The $\oplus$ combinator is a case in point. Another such example is the ! operator. As we shall soon see, although $!A$ has a realistic though non-classical network interpretation in case A is interpreted as the state space of a unicameral cluster — namely, as the space of firing patterns of the corresponding b-network — $!!A$ does not: there are more logical possibilities available than are allowed to the notional underlying networks whose state spaces are the interpretations. Rather than going to great and pedantic lengths to avoid doubling the ! operator via the logic itself, even if this were possible, we shall exercise caution in deploying it in our applications.

Let us denote the category of finite dimensional k vector spaces by $\mathscr{V}_{F,k}$. To carry out the interpretation of **GN** in $\mathscr{V}_{F,\mathbb{R}}$ we need to specify, for each atomic **GN** formula, a corresponding object of $\mathscr{V}_{F,\mathbb{R}}$. Supposing this to have been done (and it will be done later) we then obtain for each **GN** formula A, not involving the operator !, an object in $\mathscr{V}_{F,\mathbb{R}}$ merely by interpreting the occurrences of $\otimes$ and $\oplus$ as carrying their usual meanings in $\mathscr{V}_{F,\mathbb{R}}$. For such formulas, **GN** sequents $A \vdash B$ are then interpreted inductively as elements of $\text{Hom}(A, B)$ according to the interpretations specified for the inference rules. For general sequents, we replace each comma in a non-empty sequence Γ by $\otimes$, so that Γ is interpreted as $\bigotimes \Gamma := \bigotimes_i A_i$ if $\Gamma = A_1, \ldots, A_n$, and we replace each empty sequence by $\mathbb{R}$. Thus for instance $A \vdash A$ (Ax) shall be interpreted as (or by) the identity map $1_A \in \text{Hom}(A, A)$. The other rules not involving ! hold in the category $\mathscr{V}_{F,\mathbb{R}}$ and linear maps may be built up which interpret **GN** proofs in an obvious manner. Thus for the case of CUT, since Γ, etc., is going to be interpreted by the single space $\bigotimes \Gamma$, etc., we may consider Γ, Γ', to be replaced by single spaces. Then, given interpretations $f : \Gamma \to A$, $g : A \otimes \Gamma' \to D$, the composition

$$\Gamma \otimes \Gamma' \xrightarrow{f \otimes 1_{\Gamma'}} A \otimes \Gamma' \xrightarrow{g} D \qquad (3.2.34)$$

is indeed the interpretation in $\mathscr{V}_{F,\mathbb{R}}$ of the sequent $\Gamma, \Gamma' \vdash D$ given by the CUT rule applied to the antecedent sequents, so we conclude that this rule may be derived in the interpreting category. The other rules not involving ! can be treated similarly, using the properties of the connectives in $\mathscr{V}_{F,\mathbb{R}}$.

A model for the ! operator in $\mathscr{V}_{F,\mathbb{R}}$ will be treated separately in section 3.2.3.

The upshot is that a *proof* of a sequent $\Gamma \vdash D$, say, reviving the standard notation, gives rise to a linear map $[\![\Gamma]\!] \to [\![D]\!]$, such that if another equivalent proof of $\Gamma \vdash D$ is used, the same linear map is produced. Here, the notion of equivalence of proofs needs to be defined, but it is not necessary for us to do so here.

As noted, more identities hold in $\mathscr{V}_{F,\mathbb{R}}$ than may be deduced from **GN**, and since this is our category of interest we shall not hesitate to use them when needed.

3.2.3 A model for the ! operator in the category $\mathscr{V}_{F,k}$

It turns out that the exterior algebra $E(\ )$ precisely satisfies the logical rules for the *of course* operator ! in $\mathscr{V}_{F,k}$ and it seems to be a remarkable and happy circumstance that the functor !$(\ )$ should simultaneously induce storability while also describing a space of firing patterns in case A labels the set of output nodes of a b-network: i.e. firing patterns are storable commodities in this sense. That is to say, if A represents the state space of the set of output nodes of a b-network ($\mathfrak{H}_A$ in our old notation), then we can take !A (or $E(\mathfrak{H}_A)$ in our old notation) to represent the storage capable version of A. At the same time, !A *also* represents the space of *firing patterns* of the b-network whose space of output nodes is A (or $\mathfrak{H}_A$ in our old notation). Thus firing patterns are storable commodities. (This does not exhaust the convergence of interpretations meeting in the exterior algebra. By the Plücker embedding (section B.1.4) each firing pattern determines a unique subspace of A itself and therefore these patterns reflect the logic of the space A. This will emerge as an important aspect of the behavior of such networks: cf. Chapter 5).

To demonstrate this we return to a discussion of the exterior algebra $E(V)$ of a finite dimensional k vector space V. This object has additional elements of structure we shall briefly describe. First consider the vector space maps given by $V \to V \oplus V$, $v \mapsto (v,v)$ and $V \to \{0\}$, $v \mapsto 0$. As in section B.1.3 these respectively induce algebra maps

- $\psi_V : E(V) \to E(V \oplus V) \cong E(V) \otimes E(V)$ called the *coproduct*, or *comultiplication*, which is an algebra map given on elements $v \in V \subset E(V)$ by

$$\psi_V(v) = 1 \otimes v + v \otimes 1. \tag{3.2.35}$$

This coproduct makes $E(V)$ a *coalgebra*. Together with the algebra structures — the exterior algebra structure of $E(V)$ and the graded algebra structure $E(V) \otimes E(V)$ — these maps give $E(V)$ the structure of a graded *Hopf algebra* in which the product, coproduct, unit, and counit intertwine in certain ways that need not concern us yet;

- $c_V : E(V) \to E(\{0\}) = k$, called the *counit*, given by the projection of $E(V)$ onto its first component.

The resulting structure has many interesting properties, but the existence of these maps, and the additional fact that $E(V^*) \cong E(V)^*$, where W^* denotes the space of functionals on W, will suffice for our purposes here, namely to show how $E(V)$ provides a model for $!V$ in our case of $k = \mathbb{R}$.

There are four rules to consider.

LW: The interpretation of a sequent $\Gamma \vdash D$ is as a linear map of real vector spaces $\bigotimes \Gamma \to D$. Compose this map with the map $1 \otimes c_A : \bigotimes \Gamma \otimes E(A) \to \bigotimes \Gamma \otimes \mathbb{R} \cong \bigotimes \Gamma$ to obtain a map $\bigotimes \Gamma \otimes E(A) \to D$. This is the required interpretation of $\Gamma, !A \vdash D$.

LC: A similar argument using $\psi_A : E(A) \to E(A) \otimes E(A)$. We will argue later (in section 3.3) that this process is physically implementable at the network level and requires a certain time, labeled τ_c^A, to implement.

L!: A similar argument using the projection $E(A) \to A$. This does not reflect a physical change at the same level of resolution but rather a change in that level of resolution.

R!: It suffices to show this for a single formula since, if $\Gamma = A_1, \ldots, A_n$, $!\Gamma$ is interpreted as $\bigotimes (!\Gamma) = !A_1 \otimes \cdots \otimes !A_n = !(A_1 \oplus \cdots \oplus A_n)$, a now familiar identity. Then $!\Gamma \vdash A$ is interpreted as a map $E(\Gamma) \to A$. Dualizing, we get a map $A^* \to E(\Gamma)^* = E(\Gamma^*)$. From the UMP of $E(\ )$ this map lifts to a map $E(A^*) \to E(\Gamma^*)$. Dualizing again, we obtain a map $E(\Gamma) \to E(A)$. This is the required interpretation of $!\Gamma \vdash !A$. This is also just a change in resolution.

3.2.4   *The logical interpretation of storage*

Returning to the rule LC, we note that in the basic case of an atomic formula a, which, according to our earlier conventions, will be interpreted by a one-dimensional space isomorphic with $\mathbb{R}$, we have $!a$ interpreted as $E(\mathbb{R}) = \mathbb{R} \oplus \mathbb{R}$, where the first $\mathbb{R}$ is generated by the unit 1 in the algebra, so that $\psi_{\mathbb{R}}(1) = 1 \otimes 1$, and if x denotes the generator of the second component we have

$$\psi_{\mathbb{R}}(x) = 1 \otimes x + x \otimes 1. \qquad (3.2.36)$$

As noted in Chapter 2, in quantum computing parlance the state 1, representing the vacuum state, would be written $|0\rangle$ and the state x, the state of single occupancy, would be written $|1\rangle$, so that the last equation becomes

$$\psi_{\mathbb{R}}(|1\rangle) = |0\rangle \otimes |1\rangle + |1\rangle \otimes |0\rangle. \qquad (3.2.37)$$

This is the map implementing the duplication operation in the rule LC:

$$\frac{!a, !a, \Gamma \vdash D}{!a, \Gamma \vdash D}. \qquad (3.2.38)$$

Moreover, $\psi_{\mathbb{R}}(|1\rangle)$ is a *maximally entangled* state [69]. In actual quantum mechanics no information about the internal constitution of the state — its *preparation* — can be obtained by separately performing local experiments upon the component subsystems. Thus quantum and quantum-like duplication or storage is implemented by quantum or quantum-like entanglement. Moreover, in the actual quantum case, the conclusions of the no-cloning theorem — which asserts that no quantum state can be copied *per se* — are avoided [69]. This is basic to the discipline known as quantum computing, and it is likely to persist in our case of quantum-like behavior.

As we have noted, the rule of CONTRACTION is generally regarded as expressing storability of the type $!A$: if two instances of $!A$ are required, $!A, !A, \Gamma \vdash D$, then only one is needed, thus we infer $!A, \Gamma \vdash D$. The type $!A$ is implicitly "stored" (somewhere) and can be reused. We have dubbed $!A$ *storage capable*. This is interesting when interpreted in the category of finite dimensional real vector spaces if we take, as above, $!A := E(A)$. For then, with $A \cong \mathbb{R}^n \cong \bigoplus_{i=1}^n \mathbb{R}$, we have as above

$$!A \cong E(\bigoplus_{i=1}^n \mathbb{R}) \cong \bigotimes_{i=1}^n E(\mathbb{R}) \qquad (3.2.39)$$

and thus the *smallest* storage capable unit is realized as $E(\mathbb{R})$, the real version of the *qubit* of quantum computation (and the state space of a single b-neuron). Moreover, every storage capable type is then necessarily interpreted as a tensor product of these, i.e. a qubit *register*, in quantum computing parlance. (Note that we did not start with the classical bit and then quantize it: the classical bit arises instead intrinsically and for logical reasons as the classical degeneration of the quantum-like atom of storability, namely the qubit.) In view of the above, we may now interpret sequents of the form $!A \vdash !B$ as maps of state spaces, or firing patterns, of b-networks $\mathcal{N}_A^b$ and $\mathcal{N}_B^b$ implementing connections $\mathcal{N}_A^b \to \mathcal{N}_B^b$ which may or may not be the result of synaptic-like communications. In this way, we can take into account general types of connection involving other means of signalling such as chemical fluxes or other mechanisms promoting substrate connections. A general map interpreting such a sequent must be of the form:

$$e_{i_1}^A \wedge \ldots \wedge e_{i_p}^A \mapsto \sum_{k \geqslant 0} \sum_{j_1 < \ldots < j_k} \alpha_{j_1 \ldots j_k}^{i_1 \ldots i_p} e_{j_1}^B \wedge \ldots \wedge e_{j_k}^B \qquad (3.2.40)$$

for all $i_1 < \ldots < i_p$, with some real functions $\alpha_{j_1 \ldots j_k}^{i_1 \ldots i_p}$ depending on time. Here we note that for the single basis elements e_k^A such an assignment followed by projection upon the $k = 1$ component of the summation on the right yields the effect of a synaptic-like connection of $\mathcal{N}_A^b$ to $\mathcal{N}_B^b$. The higher-grade terms in the sum must represent non-synaptic kinds of maps of firing patterns of $\mathcal{N}_A^b$ to firing patterns of $\mathcal{N}_B^b$.

We describe the move from the unicameral situation to the more highly resolved bicameral case in a little more detail. For network $\mathcal{N}_A$ with nodes n_i^A we denoted the unicameral state space by A, with basis $\{e_i^A\}$. In moving to the bicameral version of the network we replace each node by a b-neuron, namely a pair of nodes $(n_i^{A^I}, n_i^A)$, where the original node n_i^A is now regarded as the output chamber of a b-neuron to which an input chamber $n_i^{A^I}$ has been appended. The state space A is replaced by $!A \cong {!}\mathbb{R}e_1^A \otimes \cdots \otimes {!}\mathbb{R}e_n^A$. We may promote a unicameral sequent $A \vdash B$ to its bicameral form by applying L! and R! to obtain the sequent $!A \vdash !B$, which may be checked to interpret the result of applying the functor $E(\ )$. This maps an OFF state of the left hand side (or upstream) network to the OFF state of the right hand (or downstream) network with a certain probability depending upon the coefficients, and firing patterns on the left to firing patterns on the right, again with certain probabilities. In general both sides consist of superpositions of firing patterns.

Our basic sequent $!A \vdash !B$ represents a connection going from the b-network $\mathcal{N}_A^b$ to the b-network $\mathcal{N}_B^b$ and manifests as a linear map from the exterior algebra $!A$ to the exterior algebra $!B$. We have identified maps of the space of states A to the space of states B with direct axonal *synaptic* connections from the b-neurons in $\mathcal{N}_A$ to certain b-neurons in $\mathcal{N}_B$: for it seems unlikely that a substrate or chemical influence emanating from highly specific neurons in $\mathcal{N}_A^b$ would target highly specific neurons in $\mathcal{N}_B^b$. (It is more likely that groups of A-neurons could influence groups of B-neurons in this manner.) Such linear maps $f : A \to B$ extend canonically to algebra maps $E(f) : !A \to !B$ so, as a matter of terminology, we have dubbed the corresponding connection *synaptic-like*.

3.2.5 *Modes of connection I*

A purely synaptic connection which is *many to one*, i.e. which assigns some of the states e_i^A to one state $e_{j_0}^B$, exactly emulates a *neural projection* from those A-neurons *converging* from n_i^A onto the B-neuron $n_{j_0}^B$. We will always include the conditional in this term so as not to confuse it with the mathematical term for a projection *operator*, a confusion that may threaten us. (It is to be noted that such a purely synaptic convergent neural projection, f say, cannot be extended to higher grades of basic firing patterns by the canonical process since $E(f)$ vanishes on them.)

A general connection $!A \to !B$ is determined by an assignment of each basic A-firing pattern to a general B-firing pattern, i.e. a superposition of basic B-firing patterns. This allows for representations of more general convergent neural projections: i.e. from groups of unitized co-spiking collections of A-neurons, whose states are the basic A-firing patterns, to a single B-neuron or to a group of unitized co-spiking B-neurons, whose state is represented by a single basic B-firing pattern. A general connection, in which a group or groups of unitized co-spiking neurons influence another such group, might be implemented by means of synaptic connections or by extra-synaptic volumetric substrate connections or signaling modes involving the influence of electromagnetic or chemical fluxes, such as neurotransmitters, hormones, peptides, etc., or through the medium of interneurons, or combinations of all of these. Our formalism allows for a plethora of such modes of connectivity, but cannot yet distinguish between them.

In a little more detail: for singleton firing patterns e_i, e_j, say, the state written $e_i \wedge e_j$ entails a directed connection between the corresponding b-neurons in one direction or the other, because $e_i \wedge e_j = -e_j \wedge e_i$, while disallowing self-connection, because $e_i \wedge e_i = 0$. The $\otimes$ connection allows self-connection, because for any non-zero ξ, $\xi \otimes \xi \neq 0$ but is unreliable if excitatory since blowup might occur.

The basic linear map (B.1.5):

$$\bigwedge^m V \otimes \bigwedge^n W \longrightarrow \bigwedge^{m+n} (V \oplus W) \tag{3.2.41}$$

given by $(v_1 \wedge \ldots \wedge v_m) \otimes (w_1 \wedge \ldots \wedge w_n) \mapsto v_1 \wedge \ldots \wedge v_m \wedge w_1 \wedge \ldots \wedge w_n$, which replaces the $\otimes$ connective in certain contexts with the $\wedge$ connective, may not always be physically implementable in networks of actual neurons. And may not be implemented even when it is possible. For instance, we postulated in [85] that it may fail in certain regions of the brains of schizophrenia victims.

Although we did not postulate specific reasons for this failure we noted that the $\otimes$ connective is likely only to operate reliably in the presence of inhibitory influences, since otherwise neuronal self-connection via $\otimes$ is possible and if these were excitatory then blowup might occur. Hence a deficiency of (inhibitory) GABA might, in such cases as schizophrenia, prohibit this connective from functioning. Such a deficiency has been hypothesized for schizophrenia. On the other hand, too much excitatory substrate connectivity might be expected to disable any tensor-like cooperation since the neurons involved are likely to self-connect and become thereby disabled.

So too much glutamate, for instance, may also be involved. Since the map above underlies the basic identity (B.1.6) we find its application denied in certain operations involving memory formation and maintenance (Chapter 4) and in other operations involving cognitive brain regions in the case of schizophrenia patients. We will argue that such failures would lead to problems with context formation, memory formation and retrieval and that the failure of ⊗ might also explain certain language deficits observed in schizophrenia patients (section 6.5.3).

3.3 Timing sequents

The narration so far refers to a necessarily static situation since time has not been introduced into the formalism. Thus, certain questions regarding the states of the b-neuron nodes — such as: when the output node is occupied; at what point of the passage through it of the action potential is it? — must remain unanswerable until a temporal scheme is imposed, a task to which we now turn. In so doing we will depart from standard logic. First we review some standard aspects of actual quantum theory.

Temporal evolution in elementary non-relativistic quantum mechanics is pictured in two ways. As noted earlier, the action of time upon states is as an operator, $U(t)$, say, necessarily unitary since normalized states are assumed to develop into normalized states. A state ξ develops in time $t \in \mathbb{R}$ to

$$\xi_S(t) = U(t)\xi. \tag{3.3.1}$$

Under mild constraints, a theorem of Stone and von Neumann applies to yield the representation

$$U(t) = \exp(-iHt), \tag{3.3.2}$$

where H is an Hermitian operator, and this leads immediately to the Schrödinger equation

$$\frac{\partial}{\partial t}\xi_S(t) = -iH\xi_S(t). \tag{3.3.3}$$

Here H is identified with the (time independent) Hamiltonian of the system in units in which $\hbar = 1$.

This arrangement, in which the states evolve in time whereas the operators (such as H in the time independent case) do not, has come to be called the *Schrödinger picture*.

The time dependence of the states in the Schrödinger picture may be transferred to the operators upon the system, a move which has certain advantages. Thus for any operator $\mathcal{O}_S$, the time dependence may be transferred to it by the *ansatz*

$$\mathcal{O}_H(t) = \exp(-itH)\mathcal{O}_S \exp(itH). \qquad (3.3.4)$$

We note that the states $\xi_H(t) := \exp(iHt)\xi_S(t)$ are constant in time and that the time variation of the operators is simply expressed by

$$\frac{\partial}{\partial t}\mathcal{O}_H(t) = i[\mathcal{O}_H(t), H]. \qquad (3.3.5)$$

All the time dependence has been merely shifted to the operators, relative to the rotated states $\xi_H(t)$. This *picture* is named for Heisenberg. It is completely equivalent (i.e. unitarily equivalent) to the Schrödinger picture and cannot take into account any independent variation in time of the states and the operators.

Let us now return to the **GN** calculus as realized in the category $\mathscr{V}_{F,\mathbb{R}}$ (section 3.2.2). A *proof* consists of a set of chains of inferences which generally has a tree-like form, with the axioms as the leaves and the concluding sequent as the root, if the proof terminates. At each level of, or horizontal slice through, such a proof, the ensemble of sequents found there may be held to represent simultaneous processes among the resources involved, and may be characterized by their time development operators at that moment. Thus a sequent $\Gamma_i \vdash D_i$, $i = 1, \ldots, n$, in such a slice corresponding to a time t might be interpreted as a Heisenberg picture operator $\mathcal{O}_{H,i}(t)$ at that time t and note that the temporal development of the states ξ_S in the Schrödinger picture for that process are independent of the temporal development (in t) of the operators. This is because the vertical (i.e. logical) progress signifies a possible *structural* change of the physical circumstances, not the change induced by any dynamics inherent to a component system. Such a slice represents an ensemble of n possibly different operators $\mathcal{O}_{H,i}(t)$ at the time t. As such, it corresponds to an ensemble of *parallel* Schrödinger pictures associated with that slice or level of the proof tree. (The non-relativistic physical picture is as follows. For each $t \in \mathbb{R}$ we have a family of n systems, say, in which n in general will depend on t, each with its own Hamiltonian $H_{i,t}$, say, for $i = 1, \ldots, n$. This arrangement constitutes a certain bundle of Hilbert spaces over $\mathbb{R}$. In the relativistic case $\mathbb{R}$ would be replaced by a spacetime manifold and we would have a quantum field theory.) The upshot, to put it briefly, is that each level of a proof in the sequent calculus may be read horizontally as a sequence of possible simultaneous snapshots in the Schrödinger picture, the corresponding set of Hamiltonian-like operators $H_{i,t}$ remaining fixed for that slice. Read vertically, each horizontal slice represents a different set of $H_{j,t}$ and so such a reading corresponds to a series of timed *snapshot albums* in a Heisenberg picture, though now we are no longer condemned to join these two pictures at the hip, since the vertical change is structural (or external) while the horizontal ones are dynamical (or internal). They may proceed independently. Such independent variation is to be expected in so volatile an environment as a brain network *in vivo*, in which the network itself is subject to changes of gross topology as well as to simultaneous changes in its microscopic electrochemical properties.

Since we have to hand a sequent calculus of this kind, and little else in the way of detailed knowledge of the underlying mechanics of a network of this type, we will attempt this vertical Heisenberg picture album-like reading of certain proofs to derive results pertaining to context and memory formation and retrieval expressed in the language of the sequent calculus. This is the business of subsequent chapters. In this we shall depart from the standard development of sequent calculi.

Since we are interested only in applying this logic to our b-networks, we shall henceforth informally restrict our formulas, and the timing of sequents relating them, to those applicable to b-networks, so far as this is possible, since we have not addressed the dynamics of general unicameral networks. Among other things, this is to say we shall avoid occurrences of such non-network-related formulas as $!!A$. (As noted, this rule could no doubt be formalized but we shall not need to do so in this work.)

Thus, with this restriction, a basic sequent $A \vdash B$, of the type considered above, at time t, which we shall write $A \vdash_t B$ where t is a real number residing in some subinterval of the non-negative real numbers, represents in a sense a wiring-*cum*-firing *diagram* at time t. It represents a certain state of affairs relating to how the network $\mathcal{N}_A$ is connected to the network $\mathcal{N}_B$ at time t, which also includes information concerning the possible firing activity of $\mathcal{N}_A$'s nodes relative to $\mathcal{N}_B$'s nodes at time t. As such it represents an *eventuality*, or possible outcome at time t, rather than an event. At a particular time t, a collection of such networks, such as those comprising a brain for instance, may be describable in terms of these basic irreducible one-step or singly linked sequent types, taken to be non-logical *axioms* added to the logic, together with the whole panoply of relations deducible from them via the rules of **GN**, all at time t. The use of the CUT rule produces chains of these basic maps so that in general, if two basic sequents $A \vdash_t B$, $B \vdash_t C$ obtain at t then so does the sequent $A \vdash C$, which represents a wiring diagram from $\mathcal{N}_A$ to $\mathcal{N}_C$ which also obtains at time t (and is subject to change at later times) and so we write $A \vdash_t C$. (Of course, there may be other sequents/wiring diagrams $A \vdash_t C$ at time t). In these cases the map depicted in the concluding sequent in (3.2.12) would involve products of the relevant adjacency matrices at t and compositions of the functions implicit in the scaling factors.

Sequents of the form $!A \vdash_t B$ are now interpreted as maps of the *firing patterns* (of A) at the time t which is to say, insofar as the sequent represents a wiring diagram, it concerns only those wirings relevant to (all possible) subsets of firing b-neurons. This is what ultimately underlies the truth of Tsien's postulate (cf. Chapter 5). In the case of bicameral networks (interpretations involving banged types) we shall allow non-logical axioms in the form of sequents that do not necessarily arise from synaptic connections: these will model chemical or other indirect or substrate connective possibilities.

Certain inferences of the logic may themselves take time to implement, specifically if they can be judged to be *physically implementable*.

We revisit the axioms to judge which of them may be deemed to be physically implementable and therefore might entail an interval of time to effect the transition from antecedent sequent to the inferred succedent sequent.

LE: This takes no time, since it represents merely a reordering of a list. So the timed rule LE_t leaves the timed $\vdash_t$ in the antecedent unchanged in the succedent.

LW: Referring to the LW rule suppose we have a connection $\mathcal{N}_{\otimes\Gamma} \to \mathcal{N}_D$. Then we also have such a connection from the nodes of $\mathcal{N}_{\otimes\Gamma}$ paired via $\otimes$ with the vacua or input nodes of the b-network $\mathcal{N}_A^b$, since these pairs are assigned to the same nodes in $\mathcal{N}_D$ that the nodes in $\mathcal{N}_{\otimes\Gamma}$ are assigned to, by the unitizing nature of the $\otimes$-product. This is seen to implement the rule. It clearly takes no time to implement this step since the map in the succedent is implicit in the map in the antecedent.

LC: We shall argue that the fundamental operation of CONTRAC-TION, associated with a notion of storage, is in fact physically implementable. Suppose $\mathcal{N}_A^b$ is a given b-network. We consider the entangled state $1 \otimes e_i^A + e_i^A \otimes 1 \in !A \otimes !A$. It represents the state in which the ith b-neuron has its input node and output node unitized (thus co-acting), while at the same time the b-neuron itself is ON. (The two possible states reflecting this, namely $1 \otimes e_i^A$ and $e_i^A \otimes 1$, must be superposed.) This can happen only when the b-neuron is firing *and* the entire cell membrane is at the same potential (section 2.3). We have assumed that this occurs at the initiation of the firing, after which the action potential moves down the axon and effectively the two nodes are disentangled, the output node being temporarily occupied (ON), before the refractory period is entered. That is, leaving the b-neuron in the state e_i^A. The disentangling process will thus last about as long as the duration of the action potential's spike, which we shall denote by $\tau_{c,i}^A(> 0)$. Now consider the interpretation of a sequent $!A, !A, \Gamma \vdash_t D$, implemented as a map $f\colon !A{\otimes}!A \otimes \bigotimes \Gamma \to D$, as a wiring diagram. As such it specifies that at the time t, each state of the form $\psi_A(e_i^A) \otimes \gamma$, with $\psi_A(e_i^A) = 1 \otimes e_i^A + e_i^A \otimes 1$ and $\gamma \in \bigotimes \Gamma$ should be mapped (via the wiring) onto a state $f(\psi_A(e_i^A) \otimes \gamma)$ of the D network. However, at time $t + \tau_{c,i}^A$ a new diagram needs to be specified for the ith b-neuron since its output node would have been disentangled from its input node. So this new map must go from $!A \otimes \bigotimes \Gamma$ to D and is exactly $e_i^A \otimes \gamma \mapsto f(\psi_A(e_i^A) \otimes \gamma)$. Note that this is exactly the restriction to the basic states of the intermediate map implementing LC (cf. section 3.2.3). We have conducted the argument for a singleton firing pattern but it applies, *mutatis mutandis*, to a general basic firing pattern of the form $e_{i_1}^A \wedge \ldots \wedge e_{i_k}^A$ since the corresponding set of b-neurons are unitized and co-spiking. Thus the state corresponding to the entangled state $\psi_A(e_i^A)$ is in this case $\psi_A(e_{i_1}^A) \wedge \ldots \wedge \psi_A(e_{i_k}^A)$ since each component b-neuron attains the various states simultaneously and must be combined as fermionic states, i.e. via the $\wedge$ product. We shall assume that this duration is the same for all basic firing patterns in a network and is a characteristic of that network. That is to say, the cells that are considered to be part of the *same* network are not too different from each other (cf. Chapter 6). Consequently the timed LC rule is:

$$\frac{!A, !A, \Gamma \vdash_t D}{!A, \Gamma \vdash_{t+\tau_c^A} D}\text{LC}_t \qquad\qquad (3.3.6)$$

where we have written τ_c^A for the assumed common value of $\tau_{c,i}^A$. Of course, the network dynamics as dictated by the appropriate p-Hamiltonians will be operating inside the networks during this time, but are independent of the structural rearrangements of the type described in this rule for sequents. The p-Hamiltonian dynamics addresses only the changes in the states of occupation of the networks, not the structural aspects entailed by a vertical traversal of a sequent proof.

As it stands, this value τ_c^A may itself depend on t so we shall take it to be the average such time, pending a finer analysis. In humans the duration of an action potential is approximately 5 ms after initiation, so the average interval between this and the attainment of the rest state is $\tau_c^A \approx 3$ ms. In fact the precise value of this duration will makes little material difference to any of our conclusions, as long as it is non-zero.

It is known that the sort of synchronization exhibited by networks of b-neurons is an important component of brain behavior for the processes we wish to discuss [98]. We note that this rule of contraction may *fail* in nature: for instance if there is a pathology involving the interneurons or a deficit in the supply of appropriate neurotransmitter, or some other condition that thwarts the formation of tensor products of networks. Thus the use of this rule must be guarded and used with these provisos.

We may note that in this context, *synchronicity promotes storage.*

CUT: The non-timed nature of the CUT rule has been discussed above.

L⊗, R⊗: The rules involving ⊗ present a timing problem. This combinator is supposed to implement a substrate connection, via multiple possible processes, such as flows of neuromodulators, diffusion of various kinds, interneuron involvement, etc. Thus some time may be expected to elapse across the inference line in the application of these rules. This time would depend on many unknowable factors, and timing the ⊗ rules would introduce an uncontrollable complexity which is work for the future. Consequently we shall assume that this operation is a reflection of a presence in the networks of a preexisting matrix or substrate, and suppose that connection thereby effected takes a negligible time to effect, if any. That is to say, we shall assume, in the normal run, that the substrate connections are already in place and operational when the wiring diagrams or maps represented by the sequents involving the tensor products are posited. This assumption may have to be revised in future, but we shall adopt it here. (Here again, as in the previous rule, the formation of tensor products may itself be thwarted by abnormalities in the substrate, deficits of appropriate neurotransmitters, etc.)

⊕-rules: These involve the formation of general superpositions, and as such will be assumed to be instantaneous or to take a negligible period of time to effect. As in the last rule this assumption may have to be revised in future, but we adopt it here.

L!, R!: The axiom L! also does not take any time since the projection map $!A \to A$ implementing it may be viewed in two ways. First it may be viewed as the result of a measurement act yielding the singleton firing patterns of a b-network and this takes no objective time. Secondly, the map $!A \to A$ may also be viewed as just a change in resolution from finer (bicamerality) to coarser (unicamerality): this is just a change in perspective and should take no objective time at all. R! is just the opposite change of perspective, so also should take no objective time.

Thus, the only logical rule we shall deem at this point to require timing to effect is the fundamental storage operation of CONTRACTION.

We note that non-logical axioms added later may also be subject to timings which would have to be ascribed when they are introduced. We shall avoid this issue in the present work.

The picture that emerges of such a timed calculus requires, for each time t (or very short time interval), a family of non-logical axioms describing the basic building blocks of (in our case) an ensemble of networks at that time, together with all the sequents generated by these axioms and those rules of **GN** not involving any time-sensitive rules. The conclusions of the proofs involving one or other of the time-sensitive rules will then belong to another family of sequents holding at a later time, in case these rules do not fail. We shall make the assumption in the applications to follow that none of the rules fail initially, which we take to be the state of affairs in a normally functioning biological network.

In all the discussions to follow, we may take the t labels to represent a finite number of points of a finite, sufficiently fine partition of a closed subinterval of $\mathbb{R}$, so that we are confronted with a large but finite total number of added non-logical axioms.

(It may be noted that such timed sequents are interpreted as linear maps between finite dimensional vector spaces, such maps being indexed by t, meant to be interpreted as the time variable.

As such, with t taken be a continuous parameter, they may perhaps be interpreted as describing the behavior of the analogue devices now going by the general term *memristor*, which are being used to implement new kinds of neuromorphic neural net-like computers, in which they play the rôle of synapses. Please see the following chapter for more on this.)

3.4 Digression: A translation theorem

We support the claim that **GN** is an "externalization" of **OL** by showing that an intuitionistic fragment of **OL** may be translated into formulas of **GN** so that sequents in the former translate to sequents in the latter. (Here we ignore timing.)

This fragment of **OL** must be intuitionistic (because **GN** is) and we must also omit the axioms involving the complement/negation in **OL** (because **GN** lacks one). We shall denote this fragment of **OL** by **IOL⁻**, the superscript denoting the absence of the complement. Since the intuitionistic rules of **OL** *only* involve the complement, **IOL⁻** is simply **OL** with the rules involving the complement omitted, and the ones derived in **OL** through the application of De Morgan's Law added. That is:

AXIOMS

IO1. $\alpha \vdash \alpha$
IO2. $\alpha \sqcap \beta \vdash \alpha$
IO3. $\alpha \sqcap \beta \vdash \beta$
IO4. $\alpha \vdash \alpha \sqcup \beta$
IO5. $\beta \vdash \alpha \sqcup \beta$

INFERENCE RULES

IO6. $\dfrac{\alpha \vdash \beta \quad \beta \vdash \gamma}{\alpha \vdash \gamma}$

IO7. $\dfrac{\alpha \vdash \beta \quad \alpha \vdash \gamma}{\alpha \vdash \beta \sqcap \gamma}$

IO8. $\dfrac{\beta \vdash \alpha \quad \gamma \vdash \alpha}{\beta \sqcup \gamma \vdash \alpha}$

As with **OL** a string $s_1; s_2; \ldots; s_n$ of sequents is called a *proof* of its last member s_n if each s_i is either an axiom or follows from some preceding sequent through the use of one of the rules of inference. If there exists a proof of a sequent $\alpha \vdash \beta$ then we write

$$\alpha \vdash_{\mathbf{I}} \beta \qquad (3.4.1)$$

and say the β is *deducible from α in* **IOL**$^-$.

The translation may be derived by a careful consideration of a quantum-like Heyting paradigm applied to **GN** which we shall omit. The upshot is as follows. Assuming a common set of atoms for both logics, if α denotes an **IOL**$^-$ formula, its translation into **GN** is denoted α^g. Then the translation rules are as follows.

TRANSLATION RULES

T1. If α is an atom, $\alpha^g := \alpha$;
T2. $(\alpha \sqcap \beta)^g :=\, !\alpha^g \otimes !\beta^g$;
T3. $(\alpha \sqcup \beta)^g :=\, !\alpha^g \oplus !\beta^g$.

Recalling that the turnstiles in **GN** are unadorned (when timing is ignored), we have the following theorem whose proof may be skipped.

Theorem 3.4.1.

If $\alpha \vdash_I \beta$ then $!\alpha^g \vdash \beta^g$.

Proof. The proof is by induction on the length of a deduction. That is, induction on the number of steps n in a deduction $s_1; s_2; \ldots; s_n$ of the **IOL**$^-$ sequent s_n. Since deductions of length $n = 1$ are just the axioms of **IOL**$^-$, we must prove the theorem for these first.

Proof for IO1. For any α,

$$\frac{\alpha^g \vdash \alpha^g}{!\alpha^g \vdash \alpha^g}\, \text{L!} \qquad (3.4.2)$$

Proof for IO2. For any α, β

$$\frac{\dfrac{!\alpha^g \vdash \alpha^g}{!\alpha^g, !\beta^g \vdash \alpha^g} \text{ LW and above}}{\dfrac{!\alpha^g \otimes !\beta^g \vdash \alpha^g}{!(!\alpha^g \otimes !\beta^g) \vdash \alpha^g} \text{ L!}} \text{ L}\otimes \qquad (3.4.3)$$

the conclusion being $!(\alpha \sqcap \beta)^g \vdash \alpha^g$.

Proof for IO3. Similar to the proof for IO2.

Proof for IO4.

$$\frac{!\alpha^g \vdash !\alpha^g}{!\alpha^g \vdash !\alpha^g \oplus !\beta^g} \text{ R}\oplus_1 \qquad (3.4.4)$$

which is $!\alpha^g \vdash (\alpha \sqcup \beta)^g$.

Proof for IO5. Similar to the proof for IO4, using $\text{R}\oplus_2$.

The inductive hypothesis is now that the theorem holds for the last sequent in all $\mathbf{IOL}^-$ deductions of length less than n. So consider the deduction $s_1; \ldots ; s_n$ of length n. If s_n is an axiom, the theorem is proved. If s_n is not an axiom, then it must have followed from an inference applied to preceding sequents, each of which having been the result of a shorter deduction. The theorem holds for each of these preceding sequents by the induction hypothesis, so we consider each inference rule in turn.

IO6. If we suppose that s_n is of the form $\alpha \vdash_{\mathbf{I}} \gamma$ and follows via IO6 from preceding deductions $\alpha \vdash_{\mathbf{I}} \beta$ and $\beta \vdash_{\mathbf{I}} \gamma$, then, since $!\alpha^g \vdash \beta^g$ and $!\beta^g \vdash \gamma^g$, it follows from R! that $!\alpha^g \vdash !\beta^g$ and then from Cut that $!\alpha^g \vdash \gamma^g$ so the theorem follows for this s_n.

IO7. If s_n is of the form $\alpha \vdash_{\mathbf{I}} \beta \sqcap \gamma$ and follows via IO7 from preceding deductions $\alpha \vdash_{\mathbf{I}} \beta$ and $\alpha \vdash_{\mathbf{I}} \gamma$ then it follows as above that $!\alpha^g \vdash !\beta^g$ and $!\alpha^g \vdash !\gamma^g$, so

$$\frac{\dfrac{!\alpha^g \vdash !\beta^g \qquad !\alpha^g \vdash !\gamma^g}{!\alpha^g, !\alpha^g \vdash !\beta^g \otimes !\gamma^g} \text{ R}\otimes}{!\alpha^g \vdash !\beta^g \otimes !\gamma^g} \text{ LC} \qquad (3.4.5)$$

which is $!\alpha^g \vdash (\beta \sqcap \gamma)^g$, so the theorem holds for this s_n.

IO8. If s_n is of the form $\beta \sqcup \gamma \vdash_{\mathbf{I}} \alpha$, then as before we have

$$\dfrac{\dfrac{!\beta^g \vdash \alpha^g \quad !\gamma^g \vdash \alpha^g}{!\beta^g \oplus !\gamma^g \vdash \alpha^g}\,\text{L}\oplus}{!(!\beta^g \oplus !\gamma^g) \vdash \alpha^g}\,\text{L!} \tag{3.4.6}$$

which is $!(\beta \sqcup \gamma)^g \vdash \alpha^g$ and so proves the theorem. $\qquad\square$

Part II

Applications

Chapter 4

Memory-like Processes

The coins of our realm emerged in Chapter 2 as the so-called *firing patterns*. These are the states of possible subsets of co-operating or co-spiking b-neurons of a given b-network, along with their possible superpositions. The eigenstates of the p-Hamiltonian of the b-network were invested with cognitive significance since they are more stable than other states. Of course, a judgement as to what exactly this cognitive significance *is*, its cognitive semantics as it were, is beyond the remit of our model. Perhaps they are (possibly short-lived) memoranda. In Chapter 3 we were pleasantly surprised to find that these firing patterns are storage capable commodities in the sense of our sequent calculus **GN**. We pointed out in that chapter that these two attributes are independent of each other, eigenspectra being aspects of the internal dynamics of individual b-networks, while the sequent calculus refers to structural issues: how these networks combine and interact. In what follows we will exploit both the internal dynamics and the external structural logic, separately and in tandem, to derive results for our networks that seem to be analogous to actual memory processes in brains.

(If eigenstates are of cognitive significance then presumably networks with more of them, *ceteris paribus*, would be of selective advantage and therefore more frequently found in living organisms. This hypothesis was tested for the 13 simple graph theoretic motifs involving three b-neurons in [84]. It was found that the spectral "profiles" of these networks did indeed roughly follow their observed frequency in macaque brains, as reported by Sporns [90].)

In section 4.1 we prove two linked theorems that establish a version of *long term potentiation* entirely within our formalism and explain in section 4.1.2 how it is currently believed this consequence of our axioms has been implemented by evolution. An apparently precise implementation of the phenomenon known as *Hebbian learning* follows as a mathematical consequence, as shown in section 4.1.3. In section 4.2.1 we discuss context and memory formation and exhibit a fragment of proof in **GN** that seems to implement the cued pattern completion process famously described by Proust on tasting a tea-soaked fragment of madeleine. Although crude, we claim this to be an example of a "neural computation."

4.1   The fragility of connections

Let us consider a general sequent $A \vdash_t B$ supposed to obtain at the time t which realizes a map of state spaces of networks. As a map from the (real) finite dimensional Hilbert space A into the other such space B it is an element in the Hilbert space $\mathrm{Hom}(A, B) \cong A^* \otimes B \cong A \otimes B$ since as real Hilbert spaces $A^* \cong A$ by the Riesz-Fréchet theorem. As such a quantum-like state, it, *se ipse*, is subject to inherent quantum-like fluctuations. Thus, if a particular such normalized state $\varphi_t \in A \otimes B$ (representing a map in $\mathrm{Hom}(A, B)$ interpreting the sequent above) obtains at instant t we may have at some later instant t', $\|\varphi_{t'}\| \neq 1$ since the dynamics is not orthogonal. Hence the pure state φ_t evolves into a non-pure, or mixed, state which may be a superposition and is therefore subject to decoherence or other forms of decay. This models the fragile stochastic nature of such connections *in vivo*.

(This fragility is inherent in the quantum-like nature of the model. The toy examples in section 2.4.3 are cases in point.)

In view of the stochastic behavior of sequent connections, we now specify a constraint upon timed sequents which in a sense protects action potentials in progress from stochastic variability. Namely

Definition 4.1.1. For $\varepsilon > 0$ we shall call a sequent $A \vdash B$ *ε-persistent* at t_0 if $A \vdash_{t_0} B$ implies $A \vdash_t B$ for all $t \in [t_0, t_0 + \varepsilon]$.

4.1.1 *The onset of Long Term Potentiation (LTP)*

We may now prove the following.

Theorem 4.1.1. *Suppose* $!A \vdash_s !B$ *for some* $s \in \mathbb{R}$ *and that there is a sequent* $!B \otimes !B \vdash_t !B$ *which holds for all* t *in some interval* $[s, t_F]$. *Then*

$$!A \vdash_{s+n\tau_c^A} !B \qquad (4.1.1)$$

for $n = 0$ *and for integer* $n \geqslant 1$ *as long as* $s + n\tau_c^A \leqslant t_F$.

Proof. For s in the hypotheses we have

$$\frac{\dfrac{!A \vdash_s !B \quad !A \vdash_s !B}{!A, !A \vdash_s !B \otimes !B} \, \text{R}\otimes_s}{!A \vdash_{s+\tau_c^A} !B \otimes !B} \, \text{LC}_s \qquad (4.1.2)$$

If $s + \tau_c^A > t_F$ the process stops at this point leaving us with $!A \vdash_s !B$ and proving the theorem for the $n = 0$ case.

If $s + \tau_c^A \leqslant t_F$ then the concluding sequent above may be CUT with the sequent $!B \otimes !B \vdash_{s+\tau_c^A} !B$ of the theorem to yield

$$!A \vdash_{s+\tau_c^A} !B. \qquad (4.1.3)$$

Since $s + \tau_c^A \leqslant t_F$ we may repeat the above sequent proof with s replaced by $s + \tau_c^A$ to obtain

$$!A \vdash_{s+2\tau_c^A} !B \otimes !B. \qquad (4.1.4)$$

If $s + 2\tau_c^A > t_F$ the process stops, proving the theorem for the $n = 1$ case.

If $s + 2\tau_c^A \leqslant t_F$ then this sequent may be CUT with the sequent in $!B \otimes !B \vdash_{s+2\tau_c^A} !B$ to yield

$$!A \vdash_{s+2\tau_c^A} !B. \qquad (4.1.5)$$

This process can continue until $s + (n+1)\tau_c^A > t_F$ leaving us with $!A \vdash_{s+n\tau_c^A} !B$. This proves the theorem. $\qquad \square$

Note that after time $t = t_F$ this intermittent implementation of the sequent $!A \vdash_t !B$ can no longer be asserted. The connection it reflects might be broken.

Theorem 4.1.2. *Suppose $!A \vdash_{t_0} !B$ for some t_0 and that the following hold:*

1. *There is a sequent $!B \otimes !B \vdash_t !B$ which holds for all $t \in [t_0, t_F]$.*
2. *The sequent $!A \vdash !B$ is τ_c^A-persistent at t_0.*

Then

$$!A \vdash_t !B \text{ for all } t \in [t_0, t_F]. \tag{4.1.6}$$

Proof. Note that if $t_F \leqslant t_0 + \tau_c^A$ then the conclusion of the theorem holds trivially by (2). Thus we assume that $t_0 + \tau_c^A < t_F$.

Consider the set $[t_0 + \tau_c^A, t_0 + 2\tau_c^A] \cap [t_0, t_F]$. This is not empty since $t_0 + \tau_c^A < t_F$. Choose any $t' \in [t_0 + \tau_c^A, t_0 + 2\tau_c^A] \cap [t_0, t_F]$. Then

$$t_0 + \tau_c^A \leqslant t' \leqslant t_0 + 2\tau_c^A \text{ and } t' \leqslant t_F \text{ so}$$
$$t_0 \leqslant t' - \tau_c^A \leqslant t_0 + \tau_c^A < t_F \text{ and } t' \leqslant t_F.$$

Then $s := t' - \tau_c^A$ satisfies

$$!A \vdash_s !B \tag{4.1.7}$$

by hypothesis (2) of the theorem and $!B \otimes !B \vdash_t !B$ holds for all $t \in [s, t_F]$ by hypothesis (1) of the theorem. Then by Theorem 4.1.1 we conclude that

$$!A \vdash_{s + n\tau_c^A} !B \tag{4.1.8}$$

for $n = 0$, and for $n \geqslant 1$ as long as $s + n\tau_c^A \leqslant t_F$. Since $s + \tau_c^A = t' \leqslant t_F$ this holds for $n = 1$ giving $!A \vdash_{t'} !B$. Since t' was chosen arbitrarily, we have shown that

$$!A \vdash_t !B \text{ for all } t \in [t_0 + \tau_c^A, t_0 + 2\tau_c^A] \cap [t_0, t_F]. \tag{4.1.9}$$

Now consider the set $[t_0 + 2\tau_c^A, t_0 + 3\tau_c^A] \cap [t_0, t_F]$. If this set is empty then $t_F < t_0 + 2\tau_c^A$, so that $t_0 + \tau_c^A < t_F < t_0 + 2\tau_c^A$ (recall the second sentence of this proof) and therefore $!A \vdash_t !B$ holds for all $t \in [t_0, t_F]$ since it holds for all $t \in [t_0, t_0 + \tau_c^A]$ by (2) and for all

$t \in [t_0 + \tau_c^A, t_0 + 2\tau_c^A] \cap [t_0, t_F]$ by the argument just given. So the theorem is proved in this case.

So suppose $[t_0 + 2\tau_c^A, t_0 + 3\tau_c^A] \cap [t_0, t_F]$ is not empty. Then $t_F \geqslant t_0 + 2\tau_c^A$. If $t_F = t_0 + 2\tau_c^A$, then $[t_0, t_F] = [t_0, t_0 + 2\tau_c^A]$ and the theorem is true in view of (2) and the argument above, namely (4.1.9).

So suppose $t_F > t_0 + 2\tau_c^A$, and choose any $t'' \in [t_0 + 2\tau_c^A, t_0 + 3\tau_c^A] \cap [t_0, t_F]$. Then

$$t_0 + 2\tau_c^A \leqslant t'' \leqslant t_0 + 3\tau_c^A \text{ and } t'' \leqslant t_F \text{ so}$$
$$t_0 + \tau_c^A \leqslant t'' - \tau_c^A \leqslant t_0 + 2\tau_c^A < t_F \text{ and } t'' \leqslant t_F.$$

Then $s := t'' - \tau_c^A \in [t_0 + \tau_c^A, t_0 + 2\tau_c^A] \cap [t_0, t_F]$ so that

$$!A \vdash_s !B \tag{4.1.10}$$

by (4.1.9) and $!B \otimes !B \vdash_t !B$ holds for all $t \in [s, t_F]$ by (1).
Then by Theorem 4.1.1 we conclude that

$$!A \vdash_{s+n\tau_c^A} !B \tag{4.1.11}$$

for $n = 0$, and for $n \geqslant 1$ as long as $s + n\tau_c^A \leqslant t_F$.

Since $s + \tau_c^A = t'' \leqslant t_F$ this holds for $n = 1$ giving $!A \vdash_{t''} !B$. Since t'' was chosen arbitrarily, we have shown that

$$!A \vdash_t !B \text{ for all } t \in [t_0 + 2\tau_c^A, t_0 + 3\tau_c^A] \cap [t_0, t_F]. \tag{4.1.12}$$

Now consider the set $[t_0 + 3\tau_c^A, t_0 + 4\tau_c^A] \cap [t_0, t_F]$. If this is empty then $t_F < t_0 + 3\tau_c^A$ and the theorem has been proved in the previous steps. If it is not empty we can proceed as before, adding further intervals of length τ_c^A — or part of one should the value t_F be encountered in it — to the range of validity of the sequent $!A \vdash_t !B$. If the value t_F is encountered within one of these intervals as eventually it must, then the theorem is proved and the next set of the form $[t_0 + k\tau_c^A, t_0 + (k+1)\tau_c^A] \cap [t_0, t_F]$ will be empty, whereupon the process stops. This proves the assertion. $\qquad\square$

In the application to come, we shall take the sequent in (3) to be interpreted as the "consolidation" map

$$\mu :!B \otimes !B \rightarrow !B \qquad (4.1.13)$$

given by $\xi_1 \otimes \xi_2 \mapsto \xi_1 \wedge \xi_2$ discussed in section 2.5.1 as possibly signifying an oriented flux of possibly excitatory influences upon the B b-network, implementing the exterior multiplication map. Such a flux or other agency could stop.

4.1.2 *Early LTP*

An earlier caveat applies in the case of applications of these theorems, which we shall call the *LTP theorems*, to our networks. Namely some care is required in applying them to tensor product types of the form $!A \otimes !B$ since unless $A \oplus B$ represents the state space of an underlying unicameral network, the tensor product may not be a banged type and the theorems may not apply.

Let us consider the case in which $\mathcal{N}_A^b$ and $\mathcal{N}_B^b$ each consist of single b-neurons, n_a and n_b, so that the sequent involved is $!a \vdash_{t_0} !b$. Remarkably, evolution seems to have produced molecular mechanisms to exactly implement the hyptheses of Theorem 4.1.2 according to current theory [16, 22, 89, 92]. Namely, the excitatory flux of glutamate implementing the sequent $!b \otimes !b \vdash_t !b$ could be realized by the firing of n_a. Then the NMDA receptor on the n_b input node will act if n_b *also fires within the time between t_0 and $t_0 + \tau_c^a$*. This NMDA receptor performs an ANDing operation upon these two events. Namely the simultaneous arrival of glutamate at the n_b node, and the change in the membrane potential there as a consequence of n_b's firing, causes the NMDA receptor to eject the Mg^{2+} ion blocking a relevant ion channel and allows a flux of Ca^{2+} ions into the n_b soma while at the same time backpropagating exosomes (and/or nitrous oxide) to the upstream presynaptic n_a, reconfiguring its electrochemical structure. This latter event has the effect of maintaining the $!a \vdash !b$ connection against stochastic fluctuations and effecting what we have dubbed persistence.

Thus we have the almost simultaneous firing of n_a and n_b implementing the continued validity of the sequent $!a \vdash !b$ via Theorem 4.1.2 and this continues as long as such firings are repeated. This is believed to constitute the initial phase of the process called *long term potentiation*, or LTP, involved in memory formation, and to implement the Hebbian hypothesis that synapses of co-firing neurons are strengthened upon repeated firing (cf. section 4.1.3 for more on this).

The picture that emerges from the theorem seems to give a fairly exact outline of the initial phase of the LTP process, called *early* LTP, which follows, literally, as a logical consequence of the axioms of our model. Moreover, most of the logical procedures are essentially choice free, such as the basic notion of storage and storability involving the rule of CONTRACTION and the interpretation of the operator $!(\ )$ in the category of finite dimensional Hilbert spaces.

As long as this process of almost joint firing of the two neurons continues, the conclusion of the theorem will guarantee the maintenance of the connection between them, but as soon as this stops (so that hypothesis 1 may not apply) the theorem may fail. In the case of actual neurons the next step, which is beyond the reach of our theorem, is believed to entail the intervention of the Cam II kinase process consequent upon the continued Ca^{2+} inflow to the n_b neuron, which inserts AMPAR-containing vesicles into the postsynaptic membrane and unsilences the previously silent connection even after step (2) of the theorem fails to hold. This is *late* LTP. (There is little doubt that dopamine plays a significant rôle in keeping this process going. We barely touch on the dopamine system in this work, which seems to be highly complex despite the apparent Boolean nature of most dopaminergic neurons. Cf. section 2.3.)

We note that our theorems are neutral as far as the cause of excitation upon the postsynaptic neuron is concerned. Thus it is possible that glutamate may spill over from extrasynaptic sources and this would have the effect of prolonging the connection without the need for the calmodulin kinase process. Similarly, one may envisage other processes entailing protein synthesis that could implement the hypotheses of the theorems and prolong the connection without

entering upon the kinase reaction. Thus we could sometimes observe
LTP occurring without the late LTP stage happening. There seems
to be some evidence for this phenomenon [73].

4.1.3 *Extrasynaptic connections and Hebb redux*

In this section we shall carry out some of the computations involved
in implementing the scheme dictated by Theorem 4.1.2, or for
simplicity, by its discrete time version Theorem 4.1.1. We will find
that synaptic connections alone are insufficient to initiate the onset
of LTP: an extrasynaptic component is required. When such a
component is included the Hebbian hypothesis may be verified in
the sense that a measure of synaptic strength is seen to change
monotonically with time. Both of these conclusions are consistent
with current knowledge.

Let us consider first a purely synaptic connection between the
b-neurons considered above, at time t_0. Such a connection is given
by assignments of the form:

$$f(|\mathbf{0}^a\rangle) = 0 \tag{4.1.14}$$

$$f(|\mathbf{1}^a\rangle) = c|\mathbf{0}^b\rangle. \tag{4.1.15}$$

Then from the theorems above (and section 3.2.4) after time τ_c^a
the map interpreting $!a \vdash_{t_0+\tau_c^a} !b$, with $\mu : !b \otimes !b \to !b$ denoting the
algebra (i.e. exterior) product is given (from section 3.2.3) by the
composition

$$(\mu \circ (f \otimes f) \circ \psi_a)(|\mathbf{0}^a\rangle) = (\mu \circ (f \otimes f))(|\mathbf{0}^a\rangle \otimes |\mathbf{0}^a\rangle)$$
$$= 0 \tag{4.1.16}$$
$$(\mu \circ (f \otimes f) \circ \psi_a)(|\mathbf{1}^a\rangle) = (\mu \circ (f \otimes f))((|\mathbf{0}^a\rangle \otimes (|\mathbf{1}^a\rangle + (|\mathbf{1}^a\rangle \otimes |\mathbf{0}^a\rangle))$$
$$= 0. \tag{4.1.17}$$

So in this case the LTP process fails at the word go: some extrasynaptic communication is required. (There seems to be some experimental
support for this kind of necessary cooperation between synaptic and
modularity behavior under these Hebbian conditions [53].)

At this point it is worth digressing to note that maps of the form
$\mu \circ (f \otimes g) \circ \psi_a$ may be realized as *convolutions* of the maps f and

g and may be written $f * g$ without abusing the notation: cf. section B.2. Note that convolutional structures have been found useful in the pursuit of AI implementations [4].

Now let us now suppose there is in addition a non-synaptic component via some other mechanism or mechanisms such as a flux of molecules, ions, neurotransmitters, hormones, exosomes, intramembrane allosteric signalling, etc., diffusing or directed from the whole soma of the upstream b-neuron n_a. Since the receptors, etc., of the downstream n_b are supposed to be localized to its input node, we postulate a connection interpreting the sequent $!a \vdash_{t_0} !b$ of a form which complements (i.e. adds to) a purely synaptic connection, namely, in its most general form

$$g(|\mathbf{0}^a\rangle) = c_1|\mathbf{0}^b\rangle \tag{4.1.18}$$

$$g(|\mathbf{1}^a\rangle) = c_2|\mathbf{0}^b\rangle. \tag{4.1.19}$$

Here it is likely that the scaling factors c_i change in time but we shall assume for simplicity, without sacrificing much generality, that they may be considered constant over short periods. Then we find that the sequent $!a \vdash_{t_0 + \tau_c^a} !b$ is interpreted by

$$\begin{aligned}
(g * g)(|\mathbf{0}^a\rangle) &= (\mu \circ (g \otimes g))(|\mathbf{0}^a\rangle \otimes |\mathbf{0}^a\rangle) \\
&= \mu(c_1^2|\mathbf{0}^b\rangle \otimes |\mathbf{0}^b\rangle) \\
&= c_1^2|\mathbf{0}^b\rangle
\end{aligned} \tag{4.1.20}$$

recalling that $|\mathbf{0}^b\rangle = 1 \in \mathbb{R}$, and

$$\begin{aligned}
(g * g)(|\mathbf{1}^a\rangle) &= (\mu \circ (g \otimes g))(|\mathbf{0}^a\rangle \otimes |\mathbf{1}^a\rangle + |\mathbf{1}^a\rangle \otimes |\mathbf{0}^a\rangle) \\
&= \mu(c_1|\mathbf{0}^b\rangle \otimes c_2|\mathbf{0}^b\rangle + c_2|\mathbf{0}^b\rangle \otimes c_1|\mathbf{0}^b\rangle) \\
&= 2c_1 c_2|\mathbf{0}^b\rangle.
\end{aligned} \tag{4.1.21}$$

Let us denote convolutional powers of g, say, by $g^{*n} := \overbrace{g * g * \ldots * g}^{n}$. Then we have as above, after another interval τ_c^a:

$$g^{*4}|\mathbf{0}^a\rangle = (g * g)^{*2}|\mathbf{0}^a\rangle = (\mu \circ (g^{*2} \otimes g^{*2}))(|\mathbf{0}^a\rangle \otimes |\mathbf{0}^a\rangle) \tag{4.1.22}$$

$$= c_1^4|\mathbf{0}^b\rangle \tag{4.1.23}$$

$$g^{*4}|1^a\rangle = (g * g)^{*2}|1^a\rangle = (\mu \circ (g^{*2} \otimes g^{*2}))(|0^a\rangle \otimes |1^a\rangle + |1^a\rangle \otimes |0^a\rangle)$$
$$(4.1.24)$$

$$= \mu(c_1^2|0^b\rangle \otimes 2c_1c_2|0^b\rangle + 2c_1c_2|0^b\rangle \otimes c_1^2|0^b\rangle)$$
$$(4.1.25)$$

$$= 2^2 c_1^3 c_2 |0^b\rangle \qquad (4.1.26)$$

from above. After n such intervals, and a little algebra, we obtain

$$g^{*2^n}|0^a\rangle = c_1^{2^n}|0^b\rangle \qquad (4.1.27)$$

$$g^{*2^n}|1^a\rangle = 2^n c_1^{2^n - 1} c_2 |0^b\rangle. \qquad (4.1.28)$$

After n such intervals the map interpreting the connection diagram described by the sequent $!a \vdash_{t_0 + n\tau_c^a} !b$ may be written in vector form as

$$(\lambda, \nu) \mapsto (c_1^{2^n} \lambda + 2^n c_1^{2^n - 1} c_2 \nu, 0). \qquad (4.1.29)$$

Thus, the two scaling factors, after the interval $n\tau_c^a$ has passed after l_0, are $c_1^{2^n}$ applied to the first value and $2^n c_1^{2^n - 1} c_2$ applied to the second value. The results are to be added and deposited in the input node of the n_b b-neuron. These factors are a measure of the strength of the connection. Thus, if a scaling factor is greater than one, it will have the effect of inducing an amplification of the downstream signal. Let us assume that the scaling factor c_1 is amplifying in this sense, i.e. $c_1 > 1$, and $c_2 \geqslant 0$. Then we have that both scaling factors are greater than one so the net scaling effect after the interval $n\tau_c^a$ has passed is amplifying. Moreover, these scaling factors increase hyperexponentially with time (and the repeated almost joint firing of the two b-neurons which implements the hypotheses of the theorems) showing that the connection strengthens in time until the t_F cutoff, bearing out the Hebbian hypothesis. We note the fact that the hyperexponential growth of the scaling factors with time depends only upon the non-synaptic component c_1 since the synaptic-like factor c_2 does not exponentiate with time. This suggests a possible mechanism to evade an exponential explosion. Namely, since the synaptic contribution c_2 does not itself grow exponentially, if the number of synaptic contributions could increase during memory formation, either by somehow un-silencing extant silent synapses, or by growing new ones, the contribution of each would be shared among the target input nodes and the c_2 scaling factors would effectively decrease: an increase in synaptic density would dampen the per synapse strength and counter hyperexponential explosion. A similar ramification of modulatory sources or tissue could similarly vitiate the growth of the other scaling factors. Such plasticity, if real, may be amenable to experiment. (In this connection, very recent work, at the time of this writing, namely [27], may be relevant.) On the other hand, if c_1 could become < 1 we would have an exponential decrease in the scaling factors, which entails increasing inhibition. Other values of c_i would modulate the scaling factors differently of course.

This would affect similarly any other presynaptic b-neurons communicating with a postsynaptic one, and one may envisage chains of such neurons forming pathways or networks with the connections strengthening as nearly coincident firings are repeated. The formation of such pathways or networks is what is termed *Hebbian learning*.

We note that this sort of phenomenon would apply to any link between b-neurons of this type assuming the conditions of the theorem(s) hold. Moreover, it is a necessary consequence of our few axioms.

(Such calculations can be done for general connections of the form $!A \vdash !B$, assuming the conditions hold, but they get much more complicated for general firing patterns, not least because one must use the graded multiplication to expand the convolutions. Thus, for instance, switching to the other notation for the ON b-neurons, we have

$$\psi(e_i^A \wedge e_j^A) = \psi(e_i^A)\psi(e_j^A) \qquad (4.1.30)$$
$$= (1 \otimes e_i^A + e_i^A \otimes 1)(1 \otimes e_j^A + e_j^A \otimes 1) \qquad (4.1.31)$$

where juxtaposition here means graded multiplication

$$= (1 \otimes e_i^A)(1 \otimes e_j^A) + (e_i^A \otimes 1)(1 \otimes e_j^A)$$
$$+ (1 \otimes e_i^A)(e_j^A \otimes 1) + (e_i^A \otimes 1)(e_j^A \otimes 1) \qquad (4.1.32)$$

$$= (-1)^{1.0}(1 \otimes e_i^A \wedge e_j^A) + (-1)^{0.0}(e_i^A \otimes e_j^A)$$
$$+ (-1)^{1.1}(e_j^A \otimes e_i^A) + (-1)^{0.1}(e_i^A \wedge e_j^A \otimes 1) \quad (4.1.33)$$
$$= 1 \otimes e_i^A \wedge e_j^A + e_i^A \otimes e_j^A - e_j^A \otimes e_i^A + e_i^A \wedge e_j^A \otimes 1. \quad (4.1.34)$$

This having been said, as time passes, the result will again be an exponentiation of the convolution with itself of the function interpreting the sequent.)

Such links between b-neurons may now efficiently convey firing patterns between b-networks if, like eigenstates, they remain stable for long enough and which cause the source network to fulfill the condition of the LTP theorems.

4.1.4 *Modes of connection II: White matter tracts*

In vertebrate brains there are analogues of electronic *buses* or *data highways* in the jargon of computer engineering. These are the *white matter tracts* which are masses of densely packed bundles of fast-acting myelinated axons which connect certain brain regions and play an essential rôle in brain connectivity. These tracts generally consist of elongated masses of parallelly oriented axons, often including isolated interstitial interneuron-like cells and other bodies. The detailed anatomy of these tracts has been difficult to determine since most of the more intimate methods have been invasive, making dissection difficult. However, their importance in determining cognitive function has been well established (cf. [33] for one example among many). Much of the postulated connectivity we will be invoking, particularly in Chapters 5 and 6, may be attributed to their presence, promoting perhaps massive axonal/synaptic projectivity.

Such white matter tracts fit into our model in the usual way as maps interpreting the usual sort of sequents for b-networks, of the form $!A \vdash !B$, say. However, in this case we shall specialize them to be purely synaptic: that is to say, they will be of the $\wedge$-preserving (i.e. algebra map) form $E(f): !A \to !B$ where as usual $f: A \to B$ maps the underlying A state space to the B state space. Thus,

$$E(f)(e_{i_1}^A \wedge \cdots \wedge e_{i_k}^A) = f(e_{i_1}^A) \wedge \cdots \wedge f(e_{i_k}^A). \quad (4.1.35)$$

We note in this case that basic firing pattern *grade*, or *occupation number*, is preserved by such maps. Thus synaptic neural convergence is defeated since any many-to-one f would produce a zero on the right. Thus the basic firing patterns of the standard form using the $\wedge$ connective would be transferred intact and in parallel fashion, regardless of the physical length of the tract, as long as f is one-to-one. If f is not one-to-one, reflecting certain synaptic convergences, then the firing patterns involved will never arrive at the target network. (Moreover, since the internal dynamics of the network formed by amalgamating the A network, the tract itself and the B network into a single network, is always of the form H in equation (2.4.17), the time evolution operator of the entire network, namely $\exp(-tH)$, is also an algebra map, so will likewise not change the grade of a basic firing patterns, not interfere with parallelism, and maintain the integrity of the transmitted firing pattern. This is assuming there is no external interference.)

For the case of basic A-firing patterns in which the replacement $\otimes \mapsto \wedge$ has not taken place we would have, for example, assignments of the form

$$e_{i_1}^A \wedge \cdots \wedge e_{i_l}^A \otimes e_{i_m}^A \wedge \cdots \wedge e_{i_k}^A$$

$$\mapsto E(f)(e_{i_1}^A \wedge \cdots \wedge e_{i_l}^A) \otimes E(f)(e_{i_m}^A \wedge \cdots \wedge e_{i_k}^A) \qquad (4.1.36)$$

$$= f(e_{i_1}^A) \wedge \cdots \wedge f(e_{i_l}^A) \otimes f(e_{i_m}^A) \wedge \cdots \wedge f(e_{i_k}^A) \qquad (4.1.37)$$

because the multiplexing of $E(f) : !A \to !B$ is $E(f) \otimes E(f) : !A \otimes !A \to !B \otimes !B$ (cf. the rules of **GN**). Although grade is maintained through the tract, possible convergences are risked, and more likely, if any of the basis elements on one side of the $\otimes$ in the left hand side of equation (4.1.36) are matched by at least one on the other side. Moreover, if there are such matches, the firing pattern in equation (4.1.37) would be prone to similar dangers of self-connection (via a substrate) and blowup that $\otimes$ concatenations are prone to. In an environment in which firing patterns are rife with $\otimes$ connectives, such tracts would not be reliable, unless there was enough of an inhibitory influence at the terminus.

We conclude:

- White matter tracts, modeled as above, will reliably and exactly transmit normal firing patterns over arbitrary distances and are likely therefore to be selected by evolution where they are advantageous;
- White matter tracts, modeled as above, will safely, if not necessarily exactly, transmit firing patterns involving the $\otimes$ connective if there are sufficient inhibitory influences at the terminus, and are therefore likely to be selected if they are advantageous in regions of high occurrence of this connective.

As noted, if tracts of the kind discussed in the second point above were selected as postulated, then whatever advantages are bestowed by them would be vitiated in victims of schizophrenia.

These considerations will be of importance to us later (sections 6.5.1–6.5.3).

4.2   Memory and retrieval

It is believed that generally permanent or long term memories are established via the Cam II kinase mechanism, which fixes the pattern in the appropriate synaptic complex, though such LTMs are not exclusively effected by this mechanism [73]. (Our logic is not yet flexible enough to cope computationally with this step, though the techniques developed in [14, 86] seem to offer promise in this direction.) In this and the remaining sections of this chapter we shall discuss the possible memory processes that may obtain once such memory traces are established.

4.2.1   *Context formation and pattern completion*

The notion of *context* is discussed in [17], Chapter 7. The version of quantum theory used there is similar to ours, being adopted from the standard physical model, and stays mainly in the "single space" modality. To make the connection we return to our earlier treatment of **OL** in Chapter 1. Namely, we consider the logic of a single space, corresponding to the ortholattice of *propositions* in a

Kripke model for **OL**, i.e. a proximity space, exemplified in this case by the ortholattice of closed subspaces of a Hilbert space. In this view, the Kripkean propositions are regarded as representing *concepts*, or *perceptions*, and it is illuminating to contemplate their structure in the language of proximity spaces, divorced from the distraction of the linear structure, which is in a sense incidental. Namely, a *concept* associates states according to the transitive closure of the proximity, or *similarity*, relation. For example, in a putative space of *tastes* the concept of *confection* might collect states associated with sweetness and other attributes. Then, if a taste state x is such that every taste state similar to x is also similar to one recognized as confectionary, x should itself be recognized as confectionary. That is to say, with C representing the set of states representing the concept *confection*, we have:

$$\text{if } x \text{ is such that for each } y \approx x \text{ there exists a } z \in C$$
$$\text{such that } y \approx z, \text{ then } x \in C.$$

In a proximity space this is equivalent to $C = C^{\perp\perp}$ or C being a proposition. In the proximity space based on a Hilbert space this is equivalent to C being a closed subspace. Thus, by enforcing transitive closure via similarity in this way, states are corralled into the "conceptual space" *confection*. (See [36, 100].)

Now, vectors in a Hilbert space may be resolved along many choices of basis elements. In [17] each basis choice for a given *concept*, i.e. closed subspace and therefore a Hilbert space in its own right, is identified as a *context* for that concept. A change of context for a concept is then tantamount to an orthogonal transformation of the representing space or subspace. For such a space A, let us represent a given context by the basis $\{\xi_1, \ldots, \xi_n\}$. Then we may write $A = \mathbb{R}\xi_1 \oplus \cdots \oplus \mathbb{R}\xi_n$ so that $!A = !\mathbb{R}\xi_1 \otimes \cdots \otimes !\mathbb{R}\xi_n$ which is of the form $!A_1 \otimes \cdots \otimes !A_n$ which is the interpretation of $!\Gamma := !A_1, \ldots, !A_n$. Thus our use of the term *context* is a storage capable (i.e. banged) version of its use in [17], generalized to the logic of many spaces.

A *context* is thus identified in general as a resolution of a "conceptual" space into a direct sum of subconcepts/subspaces.

A different resolving set of subconcepts/subspaces affords a different context in which to view or apprehend the same space of concepts.

We note that the presence of many tensor products entails much possible entanglement among states sharing a context, a point emphasized in [17].

We shall first outline a proof fragment in the sequent logic which implements the *pattern completion* function of memory *in abstracto*, and then specialize to possible specific functional processes. (*Pattern completion* refers to the process by which the Proustian taste of a madeleine brings into consciousness a flood of associated memories. It exemplifies the general notion of *cued recall*.) As in the consolidation condition of the LTP theorems, to effect a model of memory, we may add the non-logical consolidation axiom,

$$!M \otimes !M \vdash_t !M \qquad\qquad (4.2.1)$$

where some care must be taken over the range of t and where we noted that the interpreted map is just multiplication in the exterior algebra $!M$. This may easily be generalized to a sequent $\bigotimes^k !M \vdash_t !M$ interpreted for each k as the map sending firing patterns $\xi_1 \otimes \cdots \otimes \xi_k$ of the corresponding bicameral network to their exterior product $\xi_1 \wedge \cdots \wedge \xi_k$, which is another firing pattern of the same network. Let us assume that the status of the M b-network, as a long term memory "module" (or LTM), is established by the interval for which the sequent $\bigotimes^k !M \vdash_t !M$ is maintained. For the sake of our argument here, let us take this as unlimited, holding effectively for all t.

We assume a finite collection of persistent sequents of the form $!A_i \vdash_{t_i} !M$, with $i = 1, \ldots, n$ and $0 \leqslant t_0 \leqslant t_1 \leqslant \cdots \leqslant t_n$, and the cutoff time t_F being assumed large enough. The intuition here is of incoming firing patterns of networks $\mathcal{N}^b_{A_i}$, representing memory fragments, or sense impressions, mutually associated in some way, possibly by time or space or emotional state, having been consigned to, or encoded into, a "memory module", namely the network $\mathcal{N}^b_M$. (Thus the A_i networks comprise memory engrams or sets of them or contain them or sets of them.) Then in view of the persistence of these sequents and the LTP theorems, eventually a time t' will come when we will have, for $t \geqslant t'$:

$$!A_1 \vdash_t !M \qquad !A_2 \vdash_t !M \quad \ldots \quad !A_n \vdash_t !M. \qquad (4.2.2)$$

Repeated applications of $R\otimes_t$ then yields the contextual sequent

$$!A_1, \ldots, !A_n \vdash_t \overset{n}{\bigotimes} !M \qquad (4.2.3)$$

for $t \geqslant t'$.

For t so large that the last sequent has been reached by applications of $R\otimes$, the order of such applications does not affect the ordering of the sequence to the left of the turnstile, since this sequence may be arbitrarily permuted (LE), and the bracketing entailed on the right of the turnstile may be removed in light of Proposition 3.2.1, part 6. We shall here assume this to have been done and delay the problem of real-time selection via $R\otimes$ until later.

From the discussion above, we have $\bigotimes^n !M \vdash_t !M$. Then CUTing this with the last sequent yields

$$!A_1, \ldots, !A_n \vdash_t !M \qquad (4.2.4)$$

for any time t later than that specified above. That is to say, firing patterns of the sequence of types on the left of the turnstile are eventually "stored" — or accreted by consolidation via $\wedge$ as in the LTP theorems — in the one on the right. Thus $\mathcal{N}_M^b$ functions like a long term memory module (LTM). This is *context formation*. Now let us *access* or retrieve one of these stored types, $!A_{i_0}$ say, at some time t'' say, later than the last t. That is to say, it is "fed out" to an output network with state space $!D_{i_0}$, a *working memory* module perhaps, by some external prompt or cue, such as the Proustian taste of a madeleine, at time t''. This is expressed by the sequent

$$!A_{i_0} \vdash_{t''} !D_{i_0} \qquad (4.2.5)$$

persistence being assumed. Then for any $s \geqslant t''$ we have

$$\frac{\dfrac{!A_1, \ldots, !A_n \vdash_s !M \qquad !A_{i_0} \vdash_s !D_{i_0}}{!A_1, \ldots, !A_{i_0}, !A_{i_0}, \ldots, !A_n \vdash_s !M \otimes !D_{i_0}} R\otimes_s}{!A_1, \ldots, !A_{i_0}, \ldots, !A_n \vdash_{s+\tau_c^{A_{i_0}}} !M \otimes !D_{i_0}.} LC_s \qquad (4.2.6)$$

(Here we have elided the various permutations needed to apply LC_s and then return the antecedent sequence back to the order shown.)

Thus, for $s \geqslant t'' + \tau_c^{A_{i_0}}$ we have

$$!A_1, \ldots, !A_n \vdash_s !M \otimes !D_{i_0}. \tag{4.2.7}$$

Now, it is easy to show using first the axiom $!M \vdash !M$, and then LW, that for any t, $!M \otimes !D_{i_0} \vdash_t !M$, and similarly $!M \otimes !D_{i_0} \vdash_t !D_{i_0}$, so for s large enough for (4.2.6) to be valid we may CUT the sequent (4.2.7) with both of these last two sequents to obtain

$$!A_1, \ldots, !A_n \vdash_s !M \tag{4.2.8}$$

and

$$!A_1, \ldots, !A_n \vdash_s !D_{i_0} \tag{4.2.9}$$

which should be compared with (4.2.4) and (4.2.5).

To summarize, we have encoded into the formalism:

- the sequential long term storage of a set of network inputs into a single memory module: (4.2.2);
- the retrieval later of one of these stored inputs by feeding it out to a (perhaps *working*) memory module $!D_{i_0}$: (4.2.5);
- the applications of LC_t in the above represent a memory step, in which Hebbian-like changes in the synaptic strengths are simulated, according to our earlier LTP results.

The formalism then delivered the sequent (4.2.7) at a later time, which has the following interpretation. Retrieving one type from storage has the effect of a little later retrieving all of them willy-nilly: the network state space $!D_{i_0}$ now appears on the right of the turnstile. Moreover, the circumstances of the types returned to the LTM module might now have been affected by interference and entanglement effects produced by the presence of $!D_{i_0}$. At the same time or later the formalism produces the sequents (4.2.8) and (4.2.9). This shows that the memory fragments have been restored to the LTM by being copied back into it, and possibly changed by that process, while being available to working memory. The LTM is now ready to repeat this process if another memory fragment is accessed. It would seem that the order of access now must count, since the different order of processes may alter circumstances. This simulates *contextual* memory retrieval [88]. Note that each time a type is retrieved, the whole collection of them is re-established in the LTM so retrieval has a reinforcing effect. This mimics exactly the operation of memory retrieval in actual brains. Evoking one memory evokes associated memories. Moreover, the stored memories may be affected by this act of retrieval, though this retrieval reinforces their continued presence.

We remark that as an algorithm, and indeed as for any such algorithm or proof in the system, no central executive or master control "unit" is required.

As we have indicated, a description of the pattern completion phenomenon at the end of the Overture to Proust's *À la recherche du temps perdu (In search of lost time)* has become the *locus classicus* of such descriptions. The taste by the narrator of a fragment of madeleine dissolved in tea provokes an involuntary flood of associated childhood memories:

> "... so now all the flowers in our garden and those of Mr. Swann's park, and the water lilies of the Vivonne, and the good people of the village and their little houses and the church and all of Combray and its environs, all this takes form and substance, emerging, city and gardens, from my cup of tea." (From the end of the Overture to *Combray*, private translation by Haifa Nouaime.)

Informally and ignoring the timings: the inputs to the storage capable (i.e. banged) memory "module" M are the As, one at a time, as in reality. For instance, if $M_{\text{childhood}}$ is the LTM associated with the "childhood" *Gestalt*, then A_{tea} could label the network enabled by the first taste of Aunt Léonie's tea-soaked fragment of madeleine. Its affect is then "stored" in $M_{\text{childhood}}$ via $!A_{\text{tea}} \vdash !M_{\text{childhood}}$, and so on. In reality there may be thousands of these. Then, one of these experiences ($A_{\text{madeleine}}$, say) is recalled (into the STM $!D_{\text{madeleine}}$) perhaps by tasting one again. This is expressed by $!A_{\text{madeleine}} \vdash !D_{\text{madeleine}}$. Then the formalism yields the sequent (4.2.7): $!A_{\text{Aunt Léonie}}, !A_{\text{waterlilies}}, \ldots, !A_{\text{tea}}, \ldots, !A_{\text{madeleine}} \vdash !M_{\text{childhood}} \otimes !D_{\text{madeleine}}$. This says: *all* the associated memories are recalled into $!D_{\text{madeleine}}$, and when put back into the memory module for "childhood" they may be changed by the presence of the $\otimes !D_{\text{madeleine}}$ on the right hand side via possible entanglement. Note that $!A_{\text{madeleine}}$, etc., may also have been stored elsewhere, such as in the LTM for *confection*, $!M_{\text{confection}}$, and therefore memories of other kinds of confection, etc., may be brought forth, resulting in an experience of great complexity.

In the following sections we shall look in further detail at the retrieval process of the type invoked above in (4.2.5) and postulate brain regions which may accommodate the networks labeled above.

4.2.2   *Gestalts I*

(*Pace* German speakers!)

The proof segment above illustrates the need to delimit the sets of sequents available in the pattern completion process. The idea is that they should all have common features or attributes that mutually associate them. Thus we have alluded to the *Gestalt* comprising the networks accommodating the memory fragments associated with Proust's narrator's childhood. If *any* memory fragments — not just the ones associated with *childhood* — were allowed on the left of the turnstiles in (4.2.2) the effect would be an overload of disassociated memories. Another application would be to sets of neurons "tuned" to classes of experiences such as fear, taste, etc., as in the next chapter. Although such sets and their common features are difficult to precisely specify in general, we shall refer to them as *Gestalts*. Thus there would be a childhood Gestalt, a fear Gestalt, a taste Gestalt, etc. We will return to this issue in Chapter 6.

4.2.3   *Retrieval*

In this section we shall make a first pass at addressing the notion of *retrieval* of stored memoranda as put forward in the previous section, returning to it later (Chapter 6). In ordinary computing parlance, this operation is dual to the operation of storage. In our setup it has a somewhat different connotation. Thus, in ordinary quantum mechanics, "retrieval" would correspond to a measurement operation, and in the standard Copenhagen interpretation this amounts to a projection upon an eigenvector or eigenspace of the Hamiltonian. And so in this interpretation there arises in some minds the "measurement problem" since a projection operator is not unitary so cannot describe an ordinary quantum mechanical process, and consequently apparently lies outside the main purview of quantum dynamics itself, and would seem to represent an unwarranted intrusion of the classical macrocosm. This schism is known as the Copenhagen Cut (unrelated to the logical CUT). It does not arise in our quantum-like setup since we did not require the analogous real property of orthogonality of our p-Hamiltonian, which is indeed not orthogonal in general. (In other minds the projection postulate of ordinary quantum mechanics is not a problem: for them the collapse of the wave function represents merely a change in the viewpoint of the observer or state of the *epi*system, and is anyway not part of the dynamics of the system.)

If we regard a retrieval operation in our context as being similarly implemented by a projection upon an eigenspace, then such projections, being linear maps, may be included as non-logical axioms in our basic logical setup. In this sense the operations of storage and retrieval, both being implemented via sequents in the calculus, are *unified*. (In this sense the unitary store models discussed in [54] are supported by our model.) Moreover, as such our retrieval processes are subject to the possibility of satisfying the conditions of the LTP theorems, which would prolong the retrieval process and at the same time involve the consolidation process. This may help explain the observation that retrieving memories seems to reinforce them. In our context, this process of projection, onto an eigenvector say, reflects the collapse of all the possible choices of firing patterns at a particular moment onto one that is of cognitive significance, such as a memory trace. (How exactly such patterns are encoded cognitively is not addressed in our setup as yet.) A projection (operator) onto an eigenspace is the generalization of this in which eigenvectors may be superpositions.

In neural applications the notion of storage entails temporal stability or maintenance of cognitively relevant firing patterns. These will be associated in various ways with the eigenstates of the p-Hamiltonians discussed above.

(Before proceeding, we note the following proviso. A sequent involving b-networks of the form $!A \vdash !B$ itself defines a new network, namely the one formed by keeping the two networks involved in the sequent and including the connecting links between them. This gives rise to a new p-Hamiltonian appropriate to the new system. We shall assume that the dynamics of the two original networks are negligibly impacted by the addition of the links, whose net effect may be regarded as a small additive perturbation of the individual network dynamics. The inclusion of additional links manifests as additional exchange terms added to the p-Hamiltonian: we are assuming that their addition does not impinge forcibly on the separate eigenstates of the original networks. A rigorous investigation along these lines, using methods from many-body physics, could be carried out but we shall overlook this here. For the purposes of this work we shall understand the eigenstates involved as being appropriate to the respective original networks.)

Our standing hypothesis is that the semi-stable eigenstates of the p-Hamiltonian of a network have cognitive significance. In the present context these will be supposed to implement memory traces or *memoranda*. The *experience* of a memory *is* the firing of the neurons involved in the relevant firing pattern. As we have argued, these may have various degrees of stability, depending for instance on the degree of degeneracy, or other factors involved in the decoherence of the states, such as noise, or other forms of breakdown. (These may explain what differences there are between long term and short term memories.) In fact, the topological symmetry of a network may enhance its stability and therefore its value as an efficient receptacle for memoranda, as we have seen in section 2.4.4.

Now we note that the image under a linear map of an eigenstate inherits its stability, etc. For, if ξ_λ denotes an eigenstate of the p-Hamiltonian for an A-network, and, with our usual notations, $f_t : E(\mathfrak{H}_A) \rightarrow E(\mathfrak{H}_B)$ denotes the interpretation of the sequent

$!A \vdash_t !B$ at time t, then

$$\xi_\lambda(t_0 + t) = \exp(-\lambda t)\xi_\lambda(t_0) \qquad (4.2.10)$$

so that

$$f_{t_0+t}(\xi_\lambda(t_0 + t)) = \exp(-\lambda t)f_{t_0+t}(\xi_\lambda(t_0)). \qquad (4.2.11)$$

Thus $f_{t_0+t}(\xi_\lambda(t_0 + t))$ determines the same state in the space of the B-network's firing patterns as does $f_{t_0+t}(\xi_\lambda(t_0))$, regardless of the local dynamics of the B-network. (The only requirement is that f_t be linear, though in our biological context such maps will in general also be smooth on their supports.)

Now we may parse the retrieval step (4.2.5) in more detail. We suppose retrieval via (operator) projection has been cued by the incoming stimulus corresponding to the firing of the "stored" or already established A_{i_0}-network. (For example the taste of the madeleine.) This causes a collapse or (operator) projection onto an eigenspace of the space $E(\mathfrak{H}_{A_{i_0}})$. We denote this eigenspace by $\mathfrak{H}_{i_0}$. It will be spanned by linear combinations of the basis elements of $E(\mathfrak{H}_{A_{i_0}})$ which are tuples of wedged-together basic states $e_j^{A_{i_0}}$s. Let $\mathscr{N}_{S_{i_0}}^b$ denote the subnetwork of the A_{i_0}-network consisting of those b-neurons whose output nodes are the ones appearing in the $\mathfrak{H}_{i_0}$, together with whatever connections run between them. Then $\mathfrak{H}_{i_0}$ may be considered to be a subspace of the space of firing patterns of the S_{i_0} network, namely $E(\mathfrak{H}_{S_{i_0}})$. Then we have a composition

$$E(\mathfrak{H}_{A_{i_0}}) \xrightarrow{p_{t_0}} \mathfrak{H}_{i_0} \hookrightarrow E(\mathfrak{H}_{S_{i_0}}) \qquad (4.2.12)$$

in which the first map is the projection at time t_0 and the second map is inclusion. It is the subnetwork $\mathscr{N}_{S_{i_0}}^b$ that we assume may be connected to a working memory "module" $\mathscr{N}_{D_{i_0}}^b$ via a further connection which interprets as

$$f : E(\mathfrak{H}_{S_{i_0}}) \to E((\mathfrak{H}_{D_{i_0}}), \qquad (4.2.13)$$

say, giving the composition

$$g_{t_0} : E(\mathfrak{H}_{A_{i_0}}) \xrightarrow{p_{t_0}} \mathfrak{H}_{i_0} \hookrightarrow E(\mathfrak{H}_{S_{i_0}}) \xrightarrow{f} E(\mathfrak{H}_{D_{i_0}}). \qquad (4.2.14)$$

This interprets a sequent $!A_{i_0} \vdash_{t_0} !D_{i_0}$ which has an interesting property. Namely, for any $\xi \in E(\mathfrak{H}_{A_{i_0}})$ we have

$$g_{t_0+\tau}(\xi) = \exp(-\lambda\tau)g_{t_0}(\xi) \qquad (4.2.15)$$

for small τ, since $p_{t_0}(\xi)$ is in the eigenspace, where λ denotes the corresponding eigenvalue. The map (4.2.15) interprets a sequent $!A_{i_0} \vdash_{t_0+\tau} !D_{i_0}$. Thus, if $\tau_c^{A_{i_0}}$ is small enough, we would have $\tau_c^{A_{i_0}}$-persistence at t_0 and therefore, by Theorem 4.1.2, we would have $!A_{i_0} \vdash_t !D_{i_0}$ for all $t \geqslant t_0$ until a corresponding t_F. This possibly lengthy lifetime of the connection seems appropriate to a working memory. In general, the lifetime of the retrieved memorandum may be modulated in various ways. Note the possibly interesting circumstance that the network outside the subnetwork $\mathcal{N}_{S_{i_0}}^b$ is excluded, or "pruned," from the LTP process.

As noted above, in general the image of an eigenstate of the antecedent network in a sequent in a sense inherits its protected properties within the target network without being necessarily an eigenstate of the target network. However, as noted, it is subject to other sources of decay and may be fragile, or *short term*. On the other hand, if this image does happen to land up inside an eigenspace of the target network, it may gain extra stability and be of *longer term*. In this respect, there is no difference between LTM and STM other than that accounted for by the environment of the receiving network: the representation and/or coding aspects will be the same. As noted, this is in agreement with the conclusions of [54]. Moreover, if memory maintenance bestows a selective advantage, one may expect unusual conditions in the memory storing networks (of the neocortex, say) which encourage the formation of highly degenerate eigenspaces — that is to say eigenspaces of high dimension — such as those networks exhibiting high degrees of symmetry. In this connection see [42, 77]. Such high degeneracy could also explain how the many different inputs via the associated A_i-networks in (4.2.8) could be accommodated into the working memory D_{i_0}-subnetwork. The same remarks apply to the networks involved in the considerations in the next chapter, namely Tsien's FCMs, which do seem to exhibit a high degree of connective symmetry.

To continue to parse the program segment in section 4.2.1, we may now conjecture that the A_i-networks reside in the hippocampus or elsewhere in the medial temporal lobe, where types of incoming stimuli are usually temporarily stored as short term memory traces, before migrating out to the neocortex or parcellations of it, where we conjecture that the M-network or collection of networks reside, which store longer term memory traces. It seems that this division of labor between hippocampus and neocortex is not cut and dried, since each may harbor memory traces of either type.

If retrieved memories are experienced when semi-stable, and therefore evanescent, eigenstates (i.e. firing patterns) actually fire, where are the "programs" for them stored or *written*? It would seem in our model that they lie jointly in the synaptic and extrasynaptic connections between b-neurons or networks of b-neurons, and in the satisfaction of conditions such as the ones in the hypotheses of LTP theorems above. That is to say, in the disposition of the synapses and the nature of the signaling modes they conduct (persistence and consolidation, for example). This idea seems to be confirmed in recent work on larval zebrafish [27]. Pathways or networks of such connected b-neurons would then constitute what are usually referred to as *engrams*. We will return to this idea in section 5.6.

4.3 Conclusions

We have used our model to derive some interesting properties of memory-like systems and have shown:

- that the model implements non-synaptic neuromodulatory connections in addition to the synaptic ones;
- that an essentially choice-free computation (in memory) arises in the model that reproduces the neurological process known as early long term potentiation, and supports the Hebbian hypothesis that applies in this context;
- that Hebbian change in synaptic strength requires extrasynaptic input (bearing out the finding in [53]);
- that a version of the mathematical procedure known as convolution emerges spontaneously during the course of this computation;

- that memory "instructions" or "programs" to reconstruct a memory reside mainly in the synaptic complexes of engrams and their substrate connections. These are analogous to the "hard wiring" of primitive computers, except that the wiring is not hard, since synaptic wiring is plastic, and the substrate connection is fluid, in a manner of speaking;
- that the model realizes computationally such phenomena as global synchronization, storage and retrieval mechanisms;
- that as far as **GN** is concerned, no central executive or control unit is required to perform computations.

In particular, the phenomena of neuromodulation, computation-in-memory and convolution are all of interest in the study of neuromorphic artificial intelligence.

It is further to be remarked that, having set up the machinery, we have merely interpreted a very simple fragment of **GN** in the category of finite dimensional real Hilbert spaces representing the concise dynamics of our finite quasispin systems. This has led directly, in a virtually choice-free manner, to a computational description of certain experimentally verified brain processes namely: LTP and Hebbian learning; the insufficiency of purely synaptic connectivity therein; and changes in synaptic density during memory formation (at least in the brains of larval zebrafish). Further applications will follow.

Chapter 5

Tsien's Theory of Connectivity

J. Z. Tsien formulated his "theory of connectivity" to explain certain cognitive processes which were revealed in a long series of careful *in vivo* recording experiments upon rodent brains [60, 96, 97, 101]. Tsien's "power-of-two" law has also been found serendipitously (for the case of three inputs) in a more recent study of macaque brains [67]. There is evidence for it in humans as well [34]: cf. section 5.5. Roughly speaking, the idea is that certain assemblages of neurons become specialized over the course of natural selection to respond to stimuli of particular significance to the survival of the organism. Moreover, the memories induced upon the introduction of these stimuli follow a particular pattern: namely, as new types of stimuli are introduced, all possible combinations of them are registered and stored for future use. This immediately implies a pre-existing hierarchy in which all combinations, from the most general in a category to the individual stimuli in the category, may be quickly apprehended and processed. The primitive selective advantage of such a system is apparent: a general DANGER! response to a frightening experience of some kind is clearly more survivable than a time-consuming pattern-matching search through a set of memories of other frightening experiences to decide the lethality of the one we are confronting. (Cf. the pattern completion mechanism discussed in section 4.2.) As a side effect, such an arrangement can contribute to higher cognitive functions, as seems to be the case in humans.

This phenomenon arises spontaneously in our model, being a function of the nature of bicameral neurons and their multiplicity of possible synaptic and substrate connectivities. Bicameral networks, apart from being "tuned" to some input modality, require no special hardwiring other than a plurality of projection axons.

In fact, fear-inducing stimuli in mice were the first ones investigated. These are more likely to have been selected in the course of evolution, and moreover would induce stronger responses upon exposure than other types of stimuli. For instance, mice and other rodents seem to have evolved to be extremely fearful of a puff of air directed at their backs. Such an event presumably presages the imminent attack from the air of a bird of prey such as a hawk or an owl. Such puffs, as well as drops, as in a rapidly descending elevator, and shakes, as in an earthquake, were used in some of Tsien *et al.*'s first experiments. Other types of stimuli, such as tastes and social cues, were also tested in these experiments with the power-of-two law verified in each case.

A group of such specialized neural assemblages, each tuned to a specific stimulus type, such as a particular fear-inducing event, or a particular taste, etc., is called by Tsien *et al.* [101] a *functional connectivity motif*, or FCM. Thus a particular Gestalt, such as the *fear* Gestalt, may be processed by an FCM comprising a collection of b-neuron assemblages each tuned to a particular kind of fear-inducing event, such as *puff*, *quake*, *drop*, etc. Such an FCM, responding to a class of stimuli such as fear-inducing events, contain smaller groups of cells called by Tsien *projection-cell cliques* or just *cliques*, which are assemblages of principal projection neurons whose disposition within the ambient assemblage is observed to follow a certain pattern in the recording experiments. Namely, as noted above, they are disposed in such a way that all possible combinations are processed and stored.

CAVEAT!
The terms *motif* and *clique* are not used here in their technical graph theoretic senses.

This notion of FCM is necessarily a bit vague since a cluster of such FCMs might itself constitute an FCM, owing to the resolutional self-symmetry of the pattern that emerges, as we shall see. We shall endeavor at all times to make it clear which collections we are considering to be FCMs. Roughly speaking, FCMs correspond to special cases of what we have informally dubbed *Gestalts*. Here they will generally be thought to model most closely assemblies or networks of pyramidal cells typically residing in the hippocampal CA1 region and amygdala or principal excitatory neurons in cortical layers L2–L6, with axonal projections into other areas. (A reference that might be relevant to Tsien's conjectures concerning the size and nature of such FCMs is [43]).

We shall examine such scenarios derived entirely within our model and then compare our results to the experimental findings. First, in section 5.1, we shall give an algorithmic account of the course of a Tsien *et al.* recording experiment in terms of our sequent calculus, and then, in sections 5.2 to 5.4.2, we give the corresponding physics-like account, and seek to explain at least in part, the experimental findings of Tsien *et al.* on the basis of our model.

In section 5.5 we attempt to compute the expected time course for certain types of stimuli and apply it to the experimental findings so far available for the power-of-two law in human language processes [34].

Section 5.6 takes a few steps towards the important subject of *forgetting* in the context of our model. In particular section 5.6.2 raises an important issue having to do with our notion of Gestalt, which will have a bearing upon the work of the following chapter.

In the final section we summarize some of the results of the chapter and posit some possible hypotheses concerning short term memory loss and its associated neuronal environment.

5.1   The view from GN

We shall in this section apply our logical formalism to model the experimental setup used by Tsien *et al.* to confirm his power-of-two law in organizing cell assemblies and show how some of these findings follow from our assumptions. We shall assume that our networks satisfy the conditions laid out in the discussion in the last chapter wherever necessary. (The model to keep in mind being sets of pyramidal or other excitatory principal cells, residing in certain areas such as the hippocampal CA1 region, amygdala or in cortical layers L2–L6 with multiple axonal projections into other areas.)

Here we shall attempt to simulate the general experimental setup. We assume that some b-network, $\mathcal{N}_{A_1}^b$ say, established by evolution, say, will be activated by the first stimulus in the chosen Gestalt (fearful event-1, i.e. earthquake; taste-1, i.e. candy, etc.) to which it is "tuned," at time t_1 say. This is described as in section 4.2.1 by an "activation" sequent of the form:

$$!A_1 \vdash_{t_1} !D_1 \tag{5.1.1}$$

with the target D_1 network being a dedicated working memory module, which we may assume to be a short term memory module (or STM) residing perhaps in the hippocampal region. Its "short term" attribute will be interpreted as the time the connection exists, which may be short. We suppose that such STMs are "hardwired" or connected for a sufficiently long time to an LTM M b-network via the sequent $!D_1 \vdash_t !M$, where, as before, we assume that the M-consolidation map is essentially permanent. Let us suppose that the sequent $!A_1 \vdash_{t_1} !D_1$ is τ_c^A-persistent at t_1. Then for $t \in [t_1, t_1 + \tau_c^A]$ we have

$$\frac{!A_1 \vdash_t !D_1 \quad !D_1 \vdash_t !M}{!A_1 \vdash_t !M} \text{CUT} \tag{5.1.2}$$

so that the sequent $!A_1 \vdash_t !M$ is also τ_c^A-persistent at t_1 and therefore persists for as long as the M b-network consolidation sequent lasts, by the second LTP theorem. This starts the establishment of A_1 in LTM, which may not get as far as permanent status, and may remain relatively short term.

So we have

$$!A_1 \vdash_t !M \quad \text{for all } t \geq t_1 \text{ as long as the conditions of the}$$
$$\text{second LTP theorem hold.} \tag{5.1.3}$$

In other words, some time after the first activation, $!A_1$ is mapped into LTM.

Now suppose that the second stimulus (fear-2 or taste-2, etc.) is experienced by a second network and its feed-out working memory, both indexed by 2:

$$!A_2 \vdash_{t_2} !D_2 \qquad (5.1.4)$$

where t_2 is some later time. Then, again we have

$!A_2 \vdash_t !M$ for all $t \geqslant t_2$ as long as the conditions of the
second LTP theorem hold.   (5.1.5)

Then, with (5.1.3) we have

$$\frac{\dfrac{!A_1 \vdash_{t_2} !M \qquad !A_2 \vdash_{t_2} !D_2}{!A_1, !A_2 \vdash_{t_2} !M \otimes !D_2} \text{R} \otimes}{!A_1 \otimes !A_2 \vdash_{t_2} !M \otimes !D_2.} \text{L} \otimes \qquad (5.1.6)$$

As before we then have

$$!A_1 \otimes !A_2 \vdash_{t_2} !M \qquad (5.1.7)$$

and

$$!A_1 \otimes !A_2 \vdash_{t_2} !D_2 \qquad (5.1.8)$$

and at later times in view of the second LTP theorem. Equation (5.1.8) shows that the activation of $!A_2$ entails the *reactivation* of $!A_1$ (which will fire into the D_2 network but that is still an activation).

Here it is important to note that, as in the actual experiments, we are assuming that these stimuli are each occurring for the first time. The A-networks may be in place but have not been activated until the appropriate stimulus arrives. This is why only the previously activated networks, presumed to belong to the same Gestalt, and no others, are selected by the R$\otimes$ rule. This delays until the next chapter the problem of how such selections may be made among Gestalts.

So now we have from the above, L$\otimes$ and Cutting with the appropriate M consolidation map

$!A_1, !A_2 \vdash_t !M$ for all $t \geqslant t_2$ as long as the conditions of the
second LTP theorem hold.   (5.1.9)

This pattern is repeated with each new stimulus so that eventually we obtain the sequents

$$!A_1, \ldots, !A_n \vdash_t !M \quad \text{for all } t \geqslant t_n \text{ as long as the conditions of the}$$

second LTP theorem hold. (5.1.10)

It is important to notice that if we had repeated the first stimulus instead of inducing the second, we would have had instead of (5.1.9):

$$\frac{\dfrac{!A_1 \vdash_t !M \qquad !A_1 \vdash_t !M}{!A_1, !A_1 \vdash_t !M \otimes !M} \text{LC}_t}{!A_1 \vdash_{t+\tau_c^1} !M \otimes !M} \quad (5.1.11)$$

for all $t > t_2$ as long as the conditions of the second LTP theorem hold, from which follows as usual that $!A_1 \vdash_t !M$. Similarly, $!A_1 \vdash_t !D_1$ for $t \in [t_2, t_2 + \tau_c^A]$. That is to say, we return to the initial pair of sequents (though with different interpreting maps). This clearly also applies to any repeat further along in time. So we note that:

- The activation by the last stimulus reactivates the prior networks (though with a different interpreted Hilbert space map);
- The repeat of any stimulus does not activate any new network.

To see how Tsien's power-of-two law now arises, it is illuminating to take the simplest possible candidates for the "tuned" networks we have labeled A_i above. Namely, let us take them to be *single* b-neurons, now labeled a_i and interpreted as the space $\mathbb{R}e_i$ where e_i denotes the basic state representing the output node. Then we have the following interpretations, in the category of real Hilbert spaces, of the left hand sides of the turnstiles in the sequents (5.1.9) and (5.1.10):

$$!a_1 \otimes !a_2 = !(a_1 \oplus a_2) \quad (5.1.12)$$

$$= \mathbb{R} \oplus (\mathbb{R}e_1 \oplus \mathbb{R}e_2) \oplus \bigwedge^2 (\mathbb{R}e_1 \oplus \mathbb{R}e_2) \quad (5.1.13)$$

$$= \mathbb{R} \oplus (\mathbb{R}e_1 \oplus \mathbb{R}e_2) \oplus \mathbb{R}(e_1 \wedge e_2) \quad (5.1.14)$$

$$= \mathbb{R} \oplus (a_1 \oplus a_2) \oplus (a_1 \wedge a_2). \quad (5.1.15)$$

Each of these summand spaces is mapped/synaptically projected into the LTM state space and the last expression depicts the possible firing patterns after the second stimulus has been applied. The first factor $\mathbb{R}$ denotes the "vacuum" or state of no firing patterns.

If either a_1 or a_2 fires then the co-activated pair (or *clique*) of cells represented by $a_1 \wedge a_2$ must fire as a unit, presumably axonally projecting as a pair of cells into some other brain region ($!D_1$ or $!D_2$ then $!M$). This then is the "general" response to any of the repertoire of stimuli (startles, or tastes, etc.), which so far consists only of the first two experiences. An obvious barcode-like diagram (as in [101] for instance) which depicts the firing activity of the two neurons after the second stimulus would look like this:

$$a_1 \; \blacksquare\square$$
$$a_2 \; \square\blacksquare \qquad (5.1.16)$$
$$a_1 \wedge a_2 \; \blacksquare\blacksquare$$

where the columns correspond to the neurons so far activated listed in temporal order from left to right, and the rows represent the firing patterns of the set of cells, two in this instance, yielding $2^2 - 1 = 3$ possible co-firings (or firing patterns) of the constituent subsets (or cliques) which is the number of rows. To reiterate, the first column shows possible activity because the second stimulus reinstates/reactivates the first associated neuron.

Adding a third stimulus and attendant neuron yields

$$!a_1 \otimes !a_2 \otimes !a_3 = !(a_1 \oplus a_2 \oplus a_3) \qquad (5.1.17)$$

$$= \mathbb{R} \oplus (a_1 \oplus a_2 \oplus a_3) \oplus \bigwedge^2 (a_1 \oplus a_2 \oplus a_3)$$

$$\oplus \bigwedge^3 (a_1 \oplus a_2 \oplus a_3) \qquad (5.1.18)$$

$$= \mathbb{R} \oplus (a_1 \oplus a_2 \oplus a_3) \oplus (a_1 \wedge a_2 \oplus a_2 \wedge a_3 \oplus a_1 \wedge a_3)$$

$$\oplus (a_1 \wedge a_2 \wedge a_3). \qquad (5.1.19)$$

The same phenomenon occurs with the new stimulus leading to reactivations of the earlier firing patterns. We note again the emergence of the power-of-two law: there are $2^3 - 1 = 7$ cliques or firing patterns. Again every possible combination of such patterns is accommodated and mapped into LTM, with $a_1 \wedge a_2 \wedge a_3$ covering the general case of this class of stimulus, and the others becoming more specialized as we lower the grade of firing pattern (i.e. as we go up the rows of the barcode). Again the barcode is of the expected form after the third stimulus.

In the case of four inputs, there are $2^4 - 1 = 15$ cliques (or number of rows):

$$
\begin{array}{rl}
a_1 & \blacksquare\square\square\square \\
a_2 & \square\blacksquare\square\square \\
a_3 & \square\square\blacksquare\square \\
a_4 & \square\square\square\blacksquare \\
a_1 \wedge a_2 & \blacksquare\blacksquare\square\square \\
a_1 \wedge a_3 & \blacksquare\square\blacksquare\square \\
a_1 \wedge a_4 & \blacksquare\square\square\blacksquare \\
a_2 \wedge a_3 & \square\blacksquare\blacksquare\square \\
a_2 \wedge a_4 & \square\blacksquare\square\blacksquare \\
a_3 \wedge a_4 & \square\square\blacksquare\blacksquare \\
a_1 \wedge a_2 \wedge a_3 & \blacksquare\blacksquare\blacksquare\square \\
a_1 \wedge a_2 \wedge a_4 & \blacksquare\blacksquare\square\blacksquare \\
a_1 \wedge a_3 \wedge a_4 & \blacksquare\square\blacksquare\blacksquare \\
a_2 \wedge a_3 \wedge a_4 & \square\blacksquare\blacksquare\blacksquare \\
a_1 \wedge a_2 \wedge a_3 \wedge a_4 & \blacksquare\blacksquare\blacksquare\blacksquare
\end{array}
\qquad (5.1.20)
$$

which may be compared to Figure 1B of [101].

In this simplified case of one cell per stimulus each barcode is contained as a sub-barcode of the next one. As each new single cell is added the reactivations of the previous ones maintain the same pattern within the new barcode.

It is a easy to see that this pattern is maintained for *any* finite collection of networks here again labeled A_i. Namely, we have the following simple identity:

Lemma 5.1.1. *Suppose* $G := \{G^{(n)}\}_{n \geqslant 0}$ *is a positively graded vector space with grades starting from 0. Put* $G_+ := \bigoplus_{n=1} G^{(n)}$. *Thus for* example, $(!A)_+ = A \oplus \bigwedge^2 A \oplus \cdots$. Then, as vector spaces,

$$(!A_1 \otimes !A_2)_+ \cong [(!A_1)_+ \otimes (!A_2)_+] \oplus (!A_1)_+ \oplus (!A_2)_+ \qquad (5.1.21)$$

Proof.

$$
\begin{aligned}
!A_1 \otimes !A_2 :={} & (\mathbb{R} \oplus (!A_1)_+) \otimes (\mathbb{R} \oplus (!A_2)_+) && (5.1.22) \\
\cong{} & [\mathbb{R} \otimes \mathbb{R}] \oplus [(!A_1)_+ \otimes \mathbb{R}] \oplus [\mathbb{R} \otimes (!A_2)_+] \\
& \oplus [(!A_1)_+ \otimes (!A_2)_+] && (5.1.23) \\
\cong{} & \mathbb{R} \oplus (!A_1)_+ \oplus (!A_2)_+ \oplus [(!A_1)_+ \otimes (!A_2)_+]. && (5.1.24)
\end{aligned}
$$

The result follows immediately. $\qquad\qquad\qquad\qquad\qquad\square$

Of course the analogous statement is true for any similarly graded vector space but in this case it holds also at the "cellular" level of individual elements of a single exterior algebra, as in equation (5.1.15).

We claim that this relation now generates barcodes of the type found by Tsien *et al.* To see this we consider the formal equation

$$(xy)_+ = x_+ y_+ + x_+ + y_+ \qquad (5.1.25)$$

noting its similarity to equation (5.1.24). Now substitute yz for y in the above equation and expand accordingly. We obtain:

$$
\begin{aligned}
(xyz)_+ ={} & x_+ (yz)_+ + x_+ + (yz)_+ && (5.1.26) \\
={} & x_+ [y_+ z_+ + y_+ + z_+] + x_+ + [y_+ z_+ + y_+ + z_+] && (5.1.27) \\
={} & x_+ y_+ z_+ + x_+ y_+ + y_+ z_+ + x_+ z_+ + x_+ + y_+ + z_+. \\
& && (5.1.28)
\end{aligned}
$$

Note that with three inputs on the left, there are $2^3 - 1 = 7$ summands on the right. It is easy to prove that with n inputs on the left we get $2^n - 1$ summands on the right, every possible subcombination of the inputs appearing.

The direct sums of state spaces represent the independent operation of the subnetworks underlying the summands and so this decomposition generates barcodes representing every possible tensored combination of firing networks. As noted above, these combinations, corresponding to rows of the barcode, are states, or firing patterns, of what Tsien calls "cliques." (Later we shall refer to the cellular cliques used above as *microcliques* to distinguish them from the general cliques which will be supposed to model the states of ensembles of many cells.)

In the case of three inputs the barcode can be depicted as follows:

$$(!A_1)_+ \ \blacksquare\square\square$$
$$(!A_2)_+ \ \square\blacksquare\square$$
$$(!A_3)_+ \ \square\square\blacksquare$$
$$(!A_1)_+ \otimes (!A_2)_+ \ \blacksquare\blacksquare\square \qquad (5.1.29)$$
$$(!A_2)_+ \otimes (!A_3)_+ \ \square\blacksquare\blacksquare$$
$$(!A_1)_+ \otimes (!A_3)_+ \ \blacksquare\square\blacksquare$$
$$(!A_1)_+ \otimes (!A_2)_+ \otimes (!A_3)_+ \ \blacksquare\blacksquare\blacksquare$$

The interpretation of the sequent (5.1.10) for $n = 3$ maps each of the spaces of (non-null) firing patterns labelled by the rows in (5.1.29) into the state space of the network $\mathcal{N}_M^b$, implying a possible wiring or connectivity constraint on the networks involved to be discussed later.

As more stimuli are experienced, this pattern repeats, with the prior barcodes being incorporated into the current one. This seems to be borne out by the barcodes depicted for instance in [101]. Since we are allowing superpositions of states, current contributions may interfere with prior ones and complicate the picture as the experiment proceeds. This "power-of-two" rule — namely that the number of rows or co-spiking groups of tuned b-networks is $2^n - 1$ where n is the number of inputs — was postulated by Tsien as the basic wiring logic underlying the cell assembly-level computation involved in many brain cognitive functions, since such a structure is capable of covering all possible combinations for a given number of inputs, and might therefore be selectively advantageous. As noted earlier, the postulate has been supported by a series of careful large scale recording experiments with mice and hamsters (cf. the references above) and evidence for it has also been found in other species, including humans.

We note that since we are faced again with tensor products of the tuned b-networks labeled A_i, they may be tuned or modulated (via substrate or volumetric connections or signaling) so as to effect all-to-all graph theoretic clique-like symmetries, which would have the selective advantages indicated in section 2.4.4. There seems to be some experimental evidence for this phenomenon: cf. [43].

Moreover, the barcode-like structure has the fractal-like property of containing nested barcode-like substructures similar to the ambient structures, since each A_i may itself generally be decomposed as a direct sum of state spaces of sub-networks, and so on, until one gets down to the single b-neuron level. That is to say, each black square in the diagram above may be replaced by another entire barcode, and so on until the level of single b-neurons is reached.

5.1.1 *Digression: How many black squares?*

How many black squares are there in a barcode with n inputs/columns? Answer: $n2^{n-1}$. One way of seeing this is to note that

there is one choice for each singleton input: that is, $\binom{n}{1} \times 1$ black squares;

there are $\binom{n}{2}$ choices of doublets of inputs: that is $\binom{n}{2} \times 2$ black squares;

there are $\binom{n}{3}$ choices of triplets of inputs: that is $\binom{n}{3} \times 3$ black squares;

and so on.

So the total number of black squares is

$$\sum_{k=1}^{n} k \binom{n}{k} = n2^{n-1}. \tag{5.1.30}$$

This last identity can be seen either by manipulating the summands on the left or by expanding $(1 + x)^n$ by the Binomial Theorem, differentiating each side with respect to x and then putting $x = 1$.

5.2   The interaction Hamiltonian

In section 2.4.1 we pointed out that the p-Hamiltonian (or what now we shall just call the Hamiltonian), which is the infinitesimal form of the dynamical operator, for a network of b-neurons $\mathcal{N}_A^b$, with b-neurons labeled n_i^A, is of the form

$$H_{\text{Net}} := -\sum_{i,j>0} J_{ij} a_i^\dagger a_j \qquad (5.2.1)$$

where the J_{ij} reflect the (synaptic) network structure and the strength of the j to i connections, and the $a^\dagger$, a operators are the corresponding creation and annihilation operators on the space which is the exterior algebra (or Fermi-Dirac space) on $\mathbb{R}^n$, where n is the number of b-neurons in the network. If there are external influences then the simplest non-synaptic "interaction" Hamiltonian reflecting this will be, at least to a linear approximation, of the form

$$H_I := \sum_{i>0} (l_i a_i^\dagger + h_i a_i). \qquad (5.2.2)$$

The coefficients J_{ij}, l_i, h_i are generally functions of time, and can sometimes be zero, though we shall assume they are not all zero all the time. Neither of these operators need be orthogonal, and usually will not be. By "external" we mean only that the influence modeled comes from a source external to the network $\mathcal{N}_A^b$, hence the omission of the self-interaction term. This can be the outside world or another part of a brain. We assume here that equation (5.2.2) is a reasonable approximation to the circumstances of the typical Tsien *et al.* experiment. The l_i terms are excitatory upon the ith neuron, since $a_i^\dagger$ creates a state of occupation there (i.e. turns it ON), while the h_i terms are inhibitory since the a_i term annihilates a state of occupation there (i.e. turns it OFF).

(In the rest of this section we will continue to temporarily suppress the time dependence of the functions l_i and h_i, though this dependence should be borne in mind. Thus the results derived in the theorem below are to be interpreted as obtaining at any fixed time.)

The expression H_I has an interesting property which follows immediately from the anticommutation relations of the operators involved. Namely

Theorem 5.2.1.

$$H_I^2 = (\mathbf{l} \cdot \mathbf{h})I, \tag{5.2.3}$$

where $\mathbf{l} = (l_1, \ldots, l_n)$, $\mathbf{h} = (h_1, \ldots, h_n)$ *and* $\mathbf{l} \cdot \mathbf{h} := \sum l_i h_i$.

Proof. We recall the anticommutation relations:

$$\{a_i, a_j^\dagger\} = a_i a_j^\dagger + a_j^\dagger a_i = \delta_{ij} I, \tag{5.2.4}$$

$$\{a_i, a_j\} = 0, \tag{5.2.5}$$

$$\{a_i^\dagger, a_j^\dagger\} = 0, \quad \text{from which it follows that}$$

$$a_i^2 = (a_i^\dagger)^2 = 0. \tag{5.2.6}$$

Then we have

$$H_I^2 = \sum_{i,j} (l_i a_i^\dagger + h_i a_i)(l_j a_j^\dagger + h_j a_j) \tag{5.2.7}$$

$$= \sum_{i<j} l_i l_j \{a_i^\dagger, a_j^\dagger\} + \sum l_i h_j \{a_j, a_i^\dagger\} + \sum_{i<j} h_i h_j \{a_i, a_j\} \tag{5.2.8}$$

$$= \left(\sum l_i h_i \right) I \tag{5.2.9}$$

$$:= (\mathbf{l} \cdot \mathbf{h})I, \tag{5.2.10}$$

as required. $\qquad\square$

We note that this immediately specifies the (real) eigenvalues of H_I: they are $(\mathbf{l} \cdot \mathbf{h})^{\frac{1}{2}} (= \pm\sqrt{\mathbf{l} \cdot \mathbf{h}})$ if $\mathbf{l} \cdot \mathbf{h} \geqslant 0$. Thus, in the absence of a network structure other than the degenerate one, the system will have no eigenvectors other than the zero vector unless $\mathbf{l} \cdot \mathbf{h} > 0$. In other words it will be unstable if $\mathbf{l} \cdot \mathbf{h} \leqslant 0$, but please see below.

The time evolution operator $T_I(t) := e^{-tH_I(t)}$, where t is the time measured from some point chosen as $t = 0$, then assumes an interesting form as follows, using the last equation:

$$T(t) = e^{-tH_I(t)} \tag{5.2.11}$$

$$= I - tH_I(t) + \frac{t^2}{2!}H_I(t)^2 - \frac{t^3}{3!}H_I(t)^3 + \cdots \tag{5.2.12}$$

$$= I - tH_I(t) + \frac{t^2}{2!}(\mathbf{l} \cdot \mathbf{h})I - \frac{t^3}{3!}(\mathbf{l} \cdot \mathbf{h})H_I(t) + \cdots \tag{5.2.13}$$

Putting $\kappa := \mathbf{l} \cdot \mathbf{h}$ this is soon seen to give:

$$T(t) = \omega(\kappa, t)I + \sigma(\kappa, t)H_I(t) \tag{5.2.14}$$

where

$$\omega(\kappa, t) := \sum_{j=0} \frac{t^{2j}}{(2j)!}\kappa^j; \tag{5.2.15}$$

$$\sigma(\kappa, t) := -\sum_{j=0} \frac{t^{2j+1}}{(2j+1)!}\kappa^j. \tag{5.2.16}$$

Both of these series clearly converge for all values of t and κ and are uniformly convergent in any closed t-interval.

There is a case to be made that the excitatory coefficients l_i should all be less than or equal to zero (because $T(t) := \exp(-tH_I)$) while the inhibitory coefficients h_i should all be greater than or equal to zero. Then $\kappa \leqslant 0$. The $\kappa = 0$ case is immediate from equation (5.2.13). In the case of $\kappa < 0$ let us put $\kappa = -\varrho$ where ϱ is positive. Then it is easy to see that

$$\omega(-\varrho, t) = \cos(\varrho^{\frac{1}{2}}t), \tag{5.2.17}$$

$$\sigma(-\varrho, t) = -\varrho^{-\frac{1}{2}}\sin(\varrho^{\frac{1}{2}}t) \tag{5.2.18}$$

so that from equation (5.2.14)

$$T(t) = \cos(\varrho^{\frac{1}{2}}t)I - \varrho^{-\frac{1}{2}}\sin(\varrho^{\frac{1}{2}}t)H_I(t). \tag{5.2.19}$$

(Note that $T(t)$ remains the same whichever sign is chosen for $\varrho^{\frac{1}{2}}$.) So in this rather realistic case, we get wave-like forms for the coefficients of the time evolution operator, even though we have not invoked any complex exponentials. However, these coefficients are not necessarily simply periodic in t since ϱ is itself dependent on t: here, as earlier, we have suppressed the time dependence of κ, and shall also do so later for typographical reasons, but it should be borne in mind. This is a case in which there are no real eigenvalues so the system is unstable (in the absence of internal network dynamics). However, the instability in this case manifests as oscillatory behavior rather than catastrophic blowup or death.

In what follows, and again for typographical reasons, we shall generally drop the κ dependence from the notation of these functions.

We note that equation (5.2.14) is exact, and not an approximation to an exponential, which it superficially resembles.

5.3 Inside the FCM for a single input

We model first the case of a single stimulus acting upon a brain region such as a part of the hippocampal CA1 region of a mouse that has, as noted, presumably been selected by evolution to respond to the fear stimuli induced by: air puffs, earthquakes, drops, etc. We assume that a single stimulus may involve neural inputs from many sources which may affect each relevant neuron or group of neurons separately. For instance, in the case of a taste, there may be inputs from different clusters of taste buds over the time it takes for the taste to register, not to mention olfactory contributions. Thus we assume for instance that the fear region or FCM can be modeled by an N-fold b-network $\mathscr{N}_A^b$, say, where N is likely to be astronomically large, with interaction Hamiltonian given as in equation (5.2.2). Thus, $\mathscr{N}_A^b$ denotes the b-network presumed to underlie the fear-detecting network. We assume it may be decomposed into disjoint subnetworks $\mathscr{N}_{A_{\text{puff}}}^b, \mathscr{N}_{A_{\text{quake}}}^b, \ldots$ Then the fear Gestalt or FCM, or that part of it we are concerned with, is associated with the set $\{!A_{\text{puff}}, !A_{\text{quake}}, \ldots\}$ of state spaces. (This is a simplification, to put it mildly. The neurological fear system in vertebrates is known to be highly complex, involving many brain circuits such as those in the amygdala, thalamus, hippocampus, etc. Cf. [3, 102].) We have simplified matters further by assuming disjointness of the basic assemblages: this obviates the need to initially consider overlaps containing possible multimodal b-neurons. Their possible presence will be considered later.

It will be convenient in this thought experiment to label with a superscript the input interaction Hamiltonians according to the order of the stimuli applied. In this section we consider only the first such, *puff*, say, with interaction Hamiltonian:

$$H_I^{(1)} := \sum_{i=1}^{m_1} (l_i^{(1)}(t)a_i^\dagger + h_i^{(1)}(t)a_i) \tag{5.3.1}$$

where we have explicitly annotated the time dependence of the coefficients.

Here we have assumed that there is a population of m_1 neurons labeled arbitrarily $\{n_1^A, \ldots, n_{m_1}^A\}$ within the network $\mathcal{N}_A^b$ that may respond to the *puff* stimulus while others do not. It will prove convenient to view this subnetwork as a separate network and label the states corresponding to the neurons listed above accordingly, namely as $\{e_1^{(1)}, \ldots, e_{m_1}^{(1)}\}$. We may adopt our old notation in which e_i represents the ON state of the b-neuron n_i^A, now dropping the A superscript. Then we have stipulated that the *puff*-sensitive neurons should be listed first. That is to say, $e_i^{(1)} = e_i$ with $i = 1, \ldots, m_1$. Later we shall stipulate that the second stimulus-sensitive cluster of neurons shall be listed next. We have temporarily suppressed the influence of the network dynamics, equation (5.2.1), which considerably complicates the picture, returning to a discussion of it later. Consequently we may initially regard our network to be degenerate, namely a cluster of b-neurons, not connected synaptically. In this case of a single stimulus, the FCM involved consists of this single cluster.

The various components of this stimulus, namely the coefficients in equation (5.3.1), may change in the time it takes for the stimulus to register. Thus we are confronted with the difficulties of a time dependent Hamiltonian. We shall approximate the time development of the system by taking discrete time steps.

Suppose the system starts, at $t = 0$, from "rest," which we interpret as being the vacuum or OFF state denoted by e_0. (Here we use the notation for the basis of the exterior algebra, where the

$e_i^{(1)}$, $i = 1, \ldots, m_1$, label the basis of the state space of the *puff*-sensitive cluster.) We note that the vacuum state does not require the total absence of activity but only that it be below the neuronal threshold level: thus it conforms to the idea of "baseline" activity. Then from equation (5.2.14) and dropping the κ reference, we have at time δt_1, the state

$$T_1(\delta t_1)e_0 := (\omega(\delta t_1)I + \sigma(\delta t_1)H_I^{(1)}(\delta t_1))e_0 \tag{5.3.2}$$

$$= (\omega(\delta t_1)I + \sigma(\delta t_1)\sum_{i=1}^{m_1}(l_i^{(1)}(\delta t_1)a_i^\dagger + h_i^{(1)}(\delta t_1)a_i))e_0 \tag{5.3.3}$$

$$= \omega(\delta t_1)e_0 + \sigma(\delta t_1)\sum l_i^{(1)}(\delta t_1)e_i^{(1)}. \tag{5.3.4}$$

So at the end of the first interval the system is in a superposition of the state of being OFF and the states of the individual *puff* neurons being ON. This means that the system will be in the OFF state with a certain probability and that each individual neuron will be ON with a certain probability, such probabilities being dependent upon the respective coefficients when the state is normalized.

After another interval δt_2, and some laborious but simple algebra, the system will be in the state

$$T_1(\delta t_2)T_1(\delta t_1)e_0 = \left[\omega(\delta t_2)I + \sigma(\delta t_2)\sum(l_i^{(1)}(\delta t_2)a_i^\dagger + h_i^{(1)}(\delta t_2)a_i)\right]$$
$$\times (\omega(\delta t_1)e_0 + \sigma(\delta t_1)\sum l_i^{(1)}(\delta t_1)e_i^{(1)}) \tag{5.3.5}$$
$$= \left[\omega(\delta t_2)\omega(\delta t_1) + \sigma(\delta t_2)\sigma(\delta t_1)\sum l_i^{(1)}(\delta t_1)h_i^{(1)}(\delta t_2)\right]$$
$$\times e_0 + \sum[\omega(\delta t_2)\sigma(\delta t_1)l_i^{(1)}(\delta t_1)$$
$$+ \omega(\delta t_1)\sigma(\delta t_2)l_i^{(1)}(\delta t_2)]e_i^{(1)}$$
$$- \sigma(\delta t_2)\sigma(\delta t_1)\sum l_i^{(1)}(\delta t_1)l_j^{(1)}(\delta t_2)e_i^{(1)} \wedge e_j^{(1)}. \tag{5.3.6}$$

After the second interval, the grade 2 *puff* firing patterns (i.e. the last summation term in equation (5.3.6)) have been been recruited to join the superposition, while the former summands (e_0 and a superposition of the $e_i^{(1)}$) are reactivated, although with different probabilities.

Note that we found the same phenomenon, namely the reactivation of prior firing patterns though with different probabilities, in the view from the logic: cf. the first bulleted remark following (5.1.11). (In the present section the A_is appearing in the **GN** proof segment refer to single cells.)

It is clear that as time goes on the repeated applications of the creation operators $a_i^\dagger$, for which the corresponding l-terms are not zero, recruit higher-grade basic firing patterns, while the annihilation operators a_i, for which the corresponding h-terms are not zero, lower the ranks of previous firing patterns, thereby adding them to the previous superposition. Moreover, the identity operator reproduces (thus maintains) the previous superposition. Thus, if enough time is allowed, and the l-coefficients are not all zero, a basic firing pattern of each available rank is added to the superposition which therefore appears with a certain probability.

Each set of b-neurons in a basic firing pattern operates independently, according to the mathematics of superposition. That is to say, each basic firing pattern, which is the state representing the simultaneous unitized co-action of only its component b-neurons and no others, belongs to an orthogonal subspace of the ambient Fermi-Dirac space, i.e. exterior algebra, and therefore has a generally different probability of manifesting in the above final superposition.

Moreover, the logic of the situation, as we have noted above, allows for the possible manifestation, or registration, of each basic firing pattern via another network it may be attached to through an axonal projection or otherwise. That is to say, such a target network may be a single neuron or a co-acting group of neurons which is the target for connections convergently projecting from the pyramidal cell assemblies whose states are described by the basic firing patterns in the regions of interest to us. As noted, these connections could be via synapses or other signaling modes, such as interneurons or volumetric substrate connections, or a combination of these. (Our formalism says nothing about the spatial location of such target cells, which may be physically near or inside the ambient network or cluster, but are not manifestly part of it.) These assumptions figure prominently in the Tsien scenario. In this case it will be the firing of these convergence targets that are recorded in a Tsien-like experiment.

Unlike the case of actual fermionic particles our firing pattern states refer not to physical particle state but rather to states of *affairs*: namely, to groups of neurons being ON or OFF. This introduces complications not present in the case of physical particles. For instance, suppose a group of co-spiking unitized b-neurons n_{i_1}, n_{i_2}, n_{i_3}, say, and only them, are firing or are ON at a particular moment. Then at that moment the system is in the state $e_{i_1} \wedge e_{i_2} \wedge e_{i_3}$, where this describes a unitized basic firing pattern which projects to a cellular ensemble which responds only to the simultaneous firing of the three neurons. (As we have noted there are chemical mechanisms such as those associated with NMDA receptors that perform such an ANDing operation.) Although the system is *not* in the state $e_{i_1} \wedge e_{i_3}$, say, the neurons n_{i_1} and n_{i_3} will neurally project via direct converging physical synaptic links or other pathways to the target neuron, or assemblage of neurons, which will respond only to the simultaneous firing of n_{i_1} and n_{i_3}: that is, the projective target that responds to the firing of the microclique corresponding to $e_{i_1} \wedge e_{i_3}$. Thus, a recording would in this case capture the effect of the system being effectively also in the state $e_{i_1} \wedge e_{i_3}$ at that moment. This would apply to all occurrences of $e_{i_1} \wedge e_{i_3}$ in any other basic firing pattern. And similarly for all the basic firing patterns that have similar convergent targets. All of these should be counted in any attempt to simulate such a recording experiment. (Here we are interpreting the response intensity as being tantamount to spike frequency: this is a crude approximation to the actual processing of the signals obtained in the Tsien experiments.)

Returning to our network $\mathcal{N}_A^b$, let us assume first that these circumstances are maximally and ideally fulfilled: namely, that *every* subset of *puff*-sensitive neurons can form a unitized co-acting assemblage describable by the corresponding firing pattern state, that each such assemblage convergently projects synaptically or by other means of communication onto a *distinct* target, and that the *l*-coefficients are all non-zero. Then each *puff* firing pattern could yield a different response signature and all occurrences of each firing pattern should count towards an accumulated recorded response over time for that firing pattern. This accumulated response would additively increase the recorded inherent response index over and above that given by the dynamics and described by the superposed final state shown above.

Although we do not have access to the details of the probabilities inherent in the superposition derived above, it is worth the attempt to gauge the effect of such accumulations in the idealized case of neglecting the differences between these probabilities. Consider the response of b-neuron n_1, say, in the case of 3 *puff* b-neurons. In the ideal case, over time, each firing pattern $e_1^{(1)}$, $e_2^{(1)}$, $e_3^{(1)}$, $e_1^{(1)} \wedge e_2^{(1)}$, $e_2^{(1)} \wedge e_3^{(1)}$, $e_1^{(1)} \wedge e_3^{(1)}$, $e_1^{(1)} \wedge e_2^{(1)} \wedge e_3^{(1)}$ may manifest and be recorded. In this example there are 4 occurrences of $e_1^{(1)}$ in the array of basic firing patterns. So this neuron might over the time of the recording fire 4 times, as the groups to which it belongs themselves fire together. That is, if some set of co-spiking unitized neurons containing n_1 fires at some moment then n_1 itself must be firing at that moment over and above what is specified by the dynamics giving rise to the equation (5.3.6) and its temporal descendants. And similarly for the other single firing b-neurons. Thus, as noted above, we would expect the response index of individual b-neurons to be enhanced or amplified by a factor of 4 beyond what might be expected to occur at a particular instant if it only fired once, given our idealization. Similarly there are 2 occurrences of each doublet, and 1 of the triplet.

This count is easy to generalize in the ideal case of full responsiveness. Let us consider as above a cluster or network of m_1 *puff*-sensitive b-neurons, and take the case of the number of occurrences in the basic firing patterns of the single b-neuron state $e_1^{(1)}$. Its next appearance in the hierarchy is in those doublets $e_1^{(1)} \wedge e_i^{(1)}$ in which it appears. Since there are $m_1 - 1$ e_is to choose from there are

$$\binom{m_1 - 1}{1} \tag{5.3.7}$$

possible choices of pairs $e_1^{(1)} \wedge e_i^{(1)}$. Going up to the next grade, there are

$$\binom{m_1 - 1}{2} \tag{5.3.8}$$

possible choices for triplets $e_1^{(1)} \wedge e_i^{(1)} \wedge e_j^{(1)}$. And similarly for the higher grades. Since this applies to all grade 1 firing patterns (i.e. each singleton $e_i^{(1)}$) we obtain for the number of occurrences $N_1^{(1)}$ of the grade 1 *puff* elements in the whole array of basic *puff* firing patterns:

$$N_1^{(1)} = \sum_{i=0} \binom{m_1 - 1}{i} = 2^{m_1 - 1} \tag{5.3.9}$$

where the first appearance of the first grade singlet itself accounts for the $i = 0$ term. A similar argument applies to the occurrences of the higher grades of basic firing patterns, so that the number of occurrences of a k-th grade basic element is

$$N_k^{(1)} = \sum_{i=0} \binom{m_1 - k}{i} = 2^{m_1 - k}. \tag{5.3.10}$$

Since we are assuming that each subset of b-neurons ($\{n_{i_1}^A, \ldots, n_{i_p}^A\}$, say) acts as a unit whose state is the corresponding basic firing pattern $e_{i_1}^{(1)} \wedge \cdots \wedge e_{i_p}^{(1)}$, with its response signature assumed to be independent, each such firing pattern acts as a neural unit via synaptic or other types of projective connection to registering ensembles. Then, modulo the absence so far in our account of a possible network structure and precise knowledge of the coefficient values in the superposition, not to mention our other idealizations and simplifications, we may expect a recording experiment over time to show the response index going from high to low, as we move from the individual b-neuron responses through the responses of higher grades, i.e. those involving the number of units of increasingly more cooperating b-neurons. This decrease would not be uniform as a function of the number of basic neuronal units since the states of the basic cooperating groups follow a binomial distribution. That is, for our original cluster or network of m_1 b-neurons:

there are $\binom{m_1}{1}$ neural units with response accumulation $N_1^{(1)} = 2^{m_1-1}$

(namely the singletons with states $e_i^{(1)}$);

there are $\binom{m_1}{2}$ neural units with response accumulation $N_2^{(1)} = 2^{m_1-2}$

(namely the doublets with states $e_i^{(1)} \wedge e_j^{(1)}$);

there are $\binom{m_1}{3}$ neural units with response accumulation $N_3^{(1)} = 2^{m_1-3}$

(namely the triplets with states $e_i^{(1)} \wedge e_j^{(1)} \wedge e_k^{(1)}$);

and so on.

The diagram in Figure 5.1 may clarify matters. The oval on the left depicts the FCM for a single stimulus, which in this simple

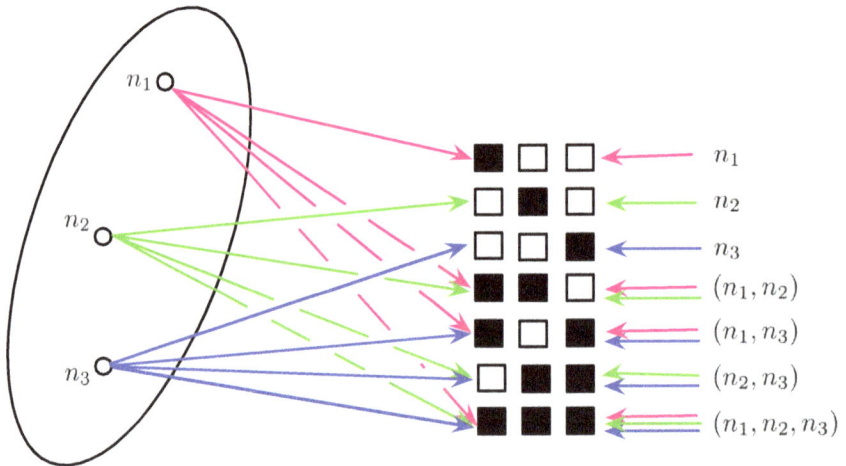

Figure 5.1   An FCM for one stimulus, comprising three b-neurons.

example comprises only three b-neurons n_1, n_2, n_3, presumed to be required for the single stimulus: that is to say, $m_1 = 3$ in the example of this section. Connections between these neurons are not shown. We denote by (n_i, n_j), (n_i, n_j, n_k), etc., the unitized *microcliques* comprising the unitized subsets of b-neurons whose firing patterns are represented by $e_i^{(1)} \wedge e_j^{(1)}$, $e_i^{(1)} \wedge e_j^{(1)} \wedge e_k^{(1)}$, respectively. The colored arrows depict the synaptic (or other kind of connected) projections, convergent in the case of microcliques of more than one neuron, to the target subnetworks in the LTM $\mathcal{N}_M^b$ of section 5.1 which we have rendered in terms of the associated barcode. (The colors were chosen merely for purposes of distinction.) Again, it is the responses of these target ensembles that we assume are the ones being recorded in a Tsien-like experiment. The connections are shown by the colored arrows on the left branch in the manner of axon terminals, which they may be, but we are assuming that the tuning is such that the target ensembles do not respond unless the input is appropriate. Thus if n_1 is ON, or firing, only the target ensemble corresponding to the top row responds. If the clique (n_1, n_3) is ON, for example, then the 5th ensemble from the top responds, along with its subcliques n_1 and n_3. So along with (n_1, n_3) both n_1 and n_3 respond again, driving up their response rate. And so on.

The rightmost part of the diagram separately depicts these grouped projections from the individual unitized cellular microcliques in the FCM, which are convergent at least for microcliques of more than one neuron, and mirror what could not be shown by drawings on the left hand side.

In a little more detail, consider the $2^{3-3} = 1$ appearance of the unitized triplet (n_1, n_2, n_3) (the last row of the barcode). If this microclique is ON, or firing, then it itself responds once while its subcliques also respond. Namely once each for the singletons and one each for the doublets. If the unitized doublet (n_2, n_3) is ON then each subclique responds again, so the count of responses is now:

1 for n_1, 2 for n_2, 2 for n_3, 1 for (n_1, n_2), 2 for (n_2, n_3), 1 for (n_1, n_3), 1 for (n_1, n_2, n_3).

Now suppose (n_1, n_3) is ON. Then the count becomes:

2 for n_1, 2 for n_2, 3 for n_3, 1 for (n_1, n_2), 2 for (n_2, n_3), 2 for (n_1, n_3), 1 for (n_1, n_2, n_3).

Now suppose (n_1, n_2) is ON. Then the count becomes:

3 for n_1, 3 for n_2, 3 for n_3, 2 for (n_1, n_2), 2 for (n_2, n_3), 2 for (n_1, n_3), 1 for (n_1, n_2, n_3).

Now suppose n_3 is ON: then we have

3 for n_1, 3 for n_2, 4 for n_3, 2 for (n_1, n_2), 2 for (n_2, n_3), 2 for (n_1, n_3), 1 for (n_1, n_2, n_3).

Now including n_2 and n_1 the singleton counts go up to 4 each, leaving the other counts as they were. Namely:

$2^{3-1} = 4$ responses for each of the top $\binom{3}{1} = 3$ rows of singletons;

$2^{3-2} = 2$ responses for each of the next $\binom{3}{2} = 3$ rows of doublets;

$2^{3-3} = 1$ response for the last $\binom{3}{3} = 1$ row consisting of the triplet.

Any order of the events described above will yield the same counts.

In time, the state of the FCM system will be in a superposition of basic firing patterns, namely the temporal descendants of equation (5.3.6), and so each of the events listed above will occur with a certain probability depending on the time, with the response intensity accumulating via the process outlined above. Thus, consider a map of response rates as a function of neuronal unit number in which units with similar response signatures are placed near each other on a neural unit number axis. In the above discussion we have identified the "neural units" with the ensembles targeted by the projective connections of the FCM firing patterns. In the idealization above, such a map would show blocks of different sizes corresponding to the binomial coefficients. Each block would have a corresponding amplification of the form $2^{m_1 - k}$, which, *ceteris paribus*, goes monotonically from one extreme to the other. Since binomial coefficients tend to a normal distribution, we would expect to find a large response rate in a small number of units at one end of the neural unit axis (corresponding to response accumulations which are higher powers of 2), grading into a larger expanse of medium response rates in the middle, and then grading into another smaller number of weakly responding units at the other end (lower powers of 2).

For another example, let us revert to the situation described in section 5.1 for four ($= m_1$) single cells: here our $e_i^{(1)}$ above corresponds to the a_i in that section. Note that each a_i appears in $8 = 2^{4-1}$ positions in the diagram (5.1.20). So the response of each of the top four rows would be amplified by this amount. Similarly, the next 6 rows will be amplified by a factor of $4 = 2^{4-2}$; the next 4 rows by a factor of 2; and the last row by a factor of 1. Hence a "heat map" would follow this pattern from hottest at the top to coldest at the bottom: cf. the leftmost image in Figure 5.2. Recall that this is supposed to apply to a single stimulus. More stimuli affect this structure as we shall see.

The actual scenario is of course likely to depart from this ideal one in various ways. Our idealization has assumed, in the absence of a network structure and detailed knowledge of the interaction coefficients, that each subset of neurons can be unitized to co-act and that there is an association of each of these co-acting unitized subsets of neurons to a distinct response signature. Certain subsets may fail to be unitized and co-acting and even if they are, they may not respond at all: that is, the corresponding l-coefficients may be zero. Also, of the ones that are unitized, co-acting and responsive, many may project to the same target and therefore yield the same response signature.

Another possibility is that the stimulus may stop before there has been sufficient time to recruit all the available higher-grade basic tuples as prescribed by the time advancement of which only the first two steps are shown in equation (5.3.6).

We consider these possibilities in turn. Namely:

Possibility 1. Some tuples are not unitized and/or do not respond. The binomial-sized "amplification blocks" to which they would belong would be accordingly smaller. The regular monotonic distribution of amplifications (i.e. the sequence $2^{m_1-1}, 2^{m_1-2}, \ldots$ in the ideal case, and a sequence of smaller numbers in the non-ideal case) would, in a recording of response indices as a function of unit number, be interrupted by discontinuities where these tuples are absent.

Possibility 2. Many unitized subsets of neurons may respond with the same signature. Since the corresponding units would have identical responses they would be placed close together in the kind of map we are considering. This would lead to the appearance of larger stretches of the same response index than would be expected in the ideal normally distributed case above.

Possibility 3. The *puff* stimulus may stop before the system has been able to recruit the higher grades of basic firing pattern tuples. In this case, a recording of the type we are considering would be truncated towards the low end of the amplification range (i.e. the range associated with the lower powers of 2).

Of course, in reality, any and all of these possibilities are likely to happen at the same time. Moreover, we have suppressed the differences in the probabilities inherent in the dynamics at great risk to our results.

We note that inclusion of the dynamics of a network structure among the *puff*-sensitive neurons would involve (equation (5.2.1)) at least the addition of terms of the form $tJ_{ij}a_i^\dagger a_j$ and products of them in the equations (5.3.4) to (5.3.6). Such products could, as time goes on, increase the probability that some of the sub-basic firing patterns will be removed from the ones they act on, while being replaced by others. For instance a term of the form $\frac{t^2}{2!}J_{56}J_{32}a_5^\dagger a_6 a_3^\dagger a_2$ will, in time, increase the probability that the doublet $e_2^{(1)} \wedge e_6^{(1)}$ will be removed from any firing pattern it acts on (in which the doublet is present) while it is replaced with the doublet $e_3^{(1)} \wedge e_5^{(1)}$. (Of course the higher powers of H_{Net} would induce even more such exchanges.) Thus some occurrences of those counted above are, in time, more probably knocked out, introducing striations of locally lower responsiveness in the normal distribution of amplifications described above, while others are enhanced, yielding striations of locally higher responsiveness. Moreover, the more connected the network — that is, the more non-zero exchange factors J_{ij} it has — the more variegated will be the striations appearing in a response map.

We summarize the effects we may crudely anticipate in a map of neural unit numbers against response indices (to a single stimulus) in which the neural units are arranged in proximal accordance with their similarity of response signature. We note that the amplification block sizes will be difficult to determine in the absence of knowledge about the original response index being amplified. However, some things can be ascertained. Namely:

- In the ideal case described above, for a degenerate network of b-neurons, the map of response indices would binomially distribute block sizes along the neuronal unit axis with response amplifications moving from one extreme to the other. For large sizes of network, this distribution would approach the one known as normal.
- In case of Possibility 1, there would be discontinuities in the map as a function of unit number.
- In case of Possibility 2, there would be stretches of constant response index in the map.
- In case of Possibility 3, there would be truncation towards the low amplification end of the neural unit axis.
- If the network is not degenerate, there will be instances of transposed response indices arising in the map, producing in time striations of locally different response indices and these will increase in number the more connected the network.

This crude conclusion would indubitably be vitiated by the presence of factors we have not taken into account, such as noise, etc., not to mention the simplistic interaction Hamiltonian we have postulated. However, it does seem to be generally borne out by the "heat maps" obtained in [101], Figure 2B, although this cannot be definitively asserted with the data so far collected. The relevant part of this figure is reproduced on the left of our Figure 5.2. (These FCMs depict the heat maps acquired from the basolateral amygdalas

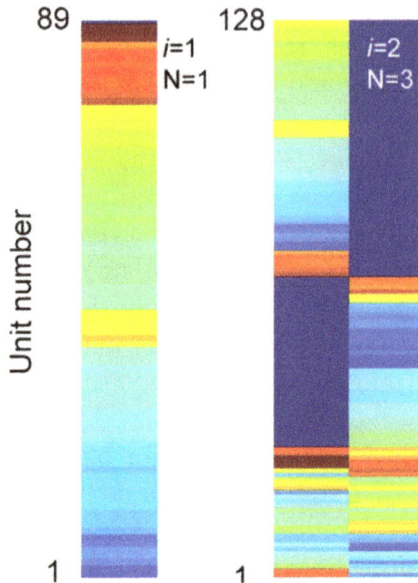

Figure 5.2 Heat maps: Single stimulus on the left; two stimuli on the right. Part of a figure as originally published in [101], licensed under CC-BY 4.0.

of mice experiencing first one taste and then two.) The maps were created from the data using an agglomerative hierarchical clustering algorithm, in which neuronal units with similar response signatures were gathered together.

We emphasize that in this section we have been treating the case of a single stimulus. The FCM in this case is the single *puff*-sensitive network or cluster we started with and it coincides with the single clique it constitutes. The basic firing patterns have a disposition that will repeat at the larger scale of more than one stimulus as we shall see in the next section. Thus, in this section a single FCM will become a single block in the FCM barcodes arising from further stimuli.

5.4   The case of more than one stimulus

Now we shall consider the case of a second fear-inducing stimulus, *quake*, say, delivered to the unfortunate mouse soon after the *puff* stimulus stops. We consider the subnetwork of $\mathcal{N}_A^b$ of m_2 *quake*-sensitive neurons and list them arbitrarily as $\{n_{m_1+1}^A, \ldots, n_{m_1+m_2}^A\}$. The corresponding states we label accordingly $\{e_1^{(2)}, \ldots, e_{m_2}^{(2)}\}$: here $e_j^{(2)} = e_{m_1+j}$ for $j = 1, \ldots, m_2$.

The new interaction Hamiltonian may now be written

$$H_I^{(2)} := \sum_{i=m_1+1}^{m_1+m_2} (l_i^{(2)}(t)a_i^\dagger + h_i^{(2)}(t)a_i). \tag{5.4.1}$$

We assume that the previous stimulus has stopped and that the system has been left in a state that has evolved as in the last section. After an interval $\Delta t_1 (\geqslant \sum \delta t_i)$ such a state will be of the form

$$\xi(\Delta t_1) = c_0^{(1)}e_0 + \sum_{i=1}^{m_1} c_i^{(1)}e_i^{(1)} + \sum_{i<j}^{m_1} c_{ij}^{(1)}e_i^{(1)} \wedge e_j^{(1)}$$

$$+ \sum_{i<j<k}^{m_1} c_{ijk}^{(1)}e_i^{(1)} \wedge e_j^{(1)} \wedge e_k^{(1)} + \cdots \tag{5.4.2}$$

which lies in the subalgebra of !A generated by the *puff*-sensitive firing patterns, and where we have suppressed the time dependence of the c coefficients. It is this state that we assume describes the state of affairs for the *puff*-sensitive subnetwork in the previous section, and accounts for the first heat map recording.

After a further time interval Δt_2 the interaction Hamiltonian $H_I^{(2)}$ is in effect and the time evolution operator is

$$T_2(\Delta t_2) := \omega(\Delta t_2)I + \sigma(\Delta t_2)H_I^{(2)}(\Delta t_2) \tag{5.4.3}$$

so that the state $\xi(\Delta t_1)$ develops into the state

$$T_2(\Delta t_2)\xi(\Delta t_1) = (\omega(\Delta t_2)I + \sigma(\Delta t_2)H_I^{(2)}(\Delta t_2))\xi(\Delta t_1) \tag{5.4.4}$$

$$= \omega(\Delta t_2)\xi(\Delta t_1) + \sigma(\Delta t_2)$$

$$\times \sum_{i=m_1+1}^{m_1+m_2} (l_i^{(2)}(\Delta t_2)a_i^\dagger + h_i^{(2)}(\Delta t_2)a_i)\xi(\Delta t_1).$$

$$\tag{5.4.5}$$

It can be seen that the initial firing pattern is repeated or re-activated and maintained by the identity term, but with the additional factor $\omega(\Delta t_2)$ affecting the overall probabilities of its summands, while the new inhibitory h-terms nullify *puff* states in the basic firing patterns of these states. (Please see the note above concerning the functions ω in the case when $\kappa < 0$.) Each non-zero excitatory l-term wedges a *quake* state to these *puff* firing patterns: thus e_0 (on the right hand side of equation (5.4.2)) becomes certain $e_i^{(2)}$ (since e_0 is the identity in the algebra); $e_j^{(1)}$ becomes certain $e_i^{(2)} \wedge e_j^{(1)}$; $e_j^{(1)} \wedge e_k^{(1)}$ becomes certain $e_i^{(2)} \wedge e_j^{(1)} \wedge e_k^{(1)}$; and so on. After the individual $e_i^{(2)}$s, these are the states that are sensitive to both puffs and quakes. (It will be convenient later to write these basic shared states in the order of the stimuli. This only involves a possible sign change and does not affect any of our arguments.)

After a further interval Δt_3 the evolution operator $T_2(\Delta t_3)$ continues to maintain the original *puff* state (now with a factor of $\omega(\Delta t_3)$) while now *quake*-only doublet firing patterns are added, and these are shared as before (via wedging) with the basic *puff* firing patterns. After sufficient time intervals, every available basic *quake* firing pattern will be paired with every available basic *puff* firing pattern. The h-terms will add further terms from which *quake* singletons have been removed, which adds to the lower rank *quake* terms but does not affect the terms already superposed.

Thus, we may now expect a map of the following "barcode" form:

$$(!A_1)_+ : \text{ certain } e_{i_1}^{(1)} \wedge \cdots \wedge e_{i_p}^{(1)} \wedge e_0 : \quad \textit{puff and no response} : \blacksquare\square$$
$$(!A_2)_+ : \text{ certain } e_0 \wedge e_{j_1}^{(2)} \wedge \cdots \wedge e_{j_r}^{(2)} : \quad \textit{no response and quake} : \square\blacksquare$$
$$(!A_1)_+ \otimes (!A_2)_+ : \text{ certain } e_{i_1}^{(1)} \wedge \cdots \wedge e_{j_r}^{(2)} : \quad \textit{puff and quake} : \blacksquare\blacksquare$$

$$(5.4.6)$$

where the black squares depict single heat maps as in the previous section, though with some changes due to such factors as different probabilities coming from the changes in the coefficients. This depicts the FCM for two stimuli, with each row corresponding to a clique. There are three cliques (N) for two stimuli ($i = 2$), illustrating the power-of-two law in this case: $N = 2^i - 1$. Note that the notation "$\wedge e_0$" and "$e_0 \wedge$" in the first two rows of (5.4.6) should be taken *cum grano salis* since in fact $\wedge e_0$, etc., just means multiplication by the unit $e_0 = 1$ in the exterior algebra. The notation merely reflects the place-holding function of the vacuum state in this context.

All the vitiating circumstances listed in the last section may still apply here.

We assume that each independent firing pattern, shared and unshared alike, have independent axonal projections to other cell assemblages whose response is recorded in such experiments. Since both the original unshared *puff* firing patterns, and the new ones involving *quake*, are superposed, they all occur with certain probabilities determined by the coefficients in the superposition. It is not asserted for instance that the unshared *puff* firing patterns and the new shared ones involving the same *puff* firing patterns are necessarily acting simultaneously.

Continuing in this way with a third stimulus in the fear class, it may be seen, after some more simple but tedious algebra as in equation (5.4.5), that the pattern continues, and produces the barcode we have depicted in (5.1.29):

$$
\begin{array}{c}
\blacksquare\square\square \\
\square\blacksquare\square \\
\square\square\blacksquare \\
\blacksquare\blacksquare\square \\
\square\blacksquare\blacksquare \\
\blacksquare\square\blacksquare \\
\blacksquare\blacksquare\blacksquare
\end{array}
\qquad (5.4.7)
$$

This FCM for three stimuli contains $2^3 - 1 = 7$ cliques. It shows an increasingly general response, reading from top to bottom. At the top are depicted the cliques that respond to the individual stimuli while at the bottom lies the clique that responds to the general case (of fear inducement, in this case). We might expect a truncation in vertical size of the more general blocks since, as time goes on, fewer of the higher-grade basic tuples are likely to be recruited, as in Possibility 3 in the last section. We note also that the barcode in the case of two stimuli is repeated as part of the barcode for the case of three stimuli.

These patterns and their vitiating circumstances repeat for any number of stimuli.

We note here that it is apparent that a repeated stimulus cannot introduce any new additions to the barcode structure since it would involve ultimately a re-application of linear combinations of products of the creation and annihilation operators for only that stimulus and therefore cannot activate any other stimulus-sensitive networks. Thus the barcode structure is maintained, though the heat map for that stimulus may change. This is again consistent with the logical view from the argument in section 5.1 (5.1.11).

The power-of-two pattern can perhaps be seen more clearly and with a little less laborious algebra as follows. It is clear that we can consider the individual fear stimulus subnetworks of our original large network $\mathcal{N}_A^b$ as separate networks. Let us consider the first two. So $\mathcal{N}_{A_1}^b$ will denote the network tuned to the *puff* stimulus, and $\mathcal{N}_{A_2}^b$ the *quake*-tuned network, with corresponding algebras of firing patterns $!A_1$ and $!A_2$. The algebra of firing patterns of the two networks taken together is then $!A_1 \otimes !A_2$. Then the time evolution operator for the first stimulus after time Δt_1 is $T_1(\Delta t_1) \otimes I_2 = (\omega(\Delta t_1)I_1 + \sigma(\Delta t_1)H_I^{(1)}(\Delta t_1)) \otimes I_2$, where I_1 and I_2 denote the identity maps on the respective algebras. Then, if the second stimulus is applied after a further interval Δt_2 we obtain for the time evolution of the whole FCM (for two stimuli):

$$(I_1 \otimes T_2(\Delta t_2))(T_1(\Delta t_1) \otimes I_2) = T_1(\Delta t_1) \otimes T_2(\Delta t_2) \qquad (5.4.8)$$

$$= (\omega(\Delta t_1)I_1 + \sigma(\Delta t_1)H_I^{(1)}(\Delta t_1)) \otimes (\omega(\Delta t_2)I_2 + \sigma(\Delta t_2)H_I^{(2)}(\Delta t_2)) \qquad (5.4.9)$$

$$= \omega(\Delta t_1)\omega(\Delta t_2)I_1 \otimes I_2 + \sigma(\Delta t_1)\omega(\Delta t_2)H_I^{(1)}(\Delta t_1) \otimes I_2$$
$$+ \omega(\Delta t_1)\sigma(\Delta t_2)I_1 \otimes H_I^{(2)}(\Delta t_2) + \sigma(\Delta t_1)\sigma(\Delta t_2)H_I^{(1)}(\Delta t_1)$$
$$\otimes H_I^{(2)}(\Delta t_2). \qquad (5.4.10)$$

The first term in the last expression reactivates the state it acts upon and so preserves the previous structure (though with a different probability), while each of the other summands corresponds to the evolution of a new clique (i.e. new row in the barcode diagram above). Each of the latter Hamiltonian operators acts on the states of the corresponding clique states $(!A_1)_+ (\cong (!A_1)_+ \otimes \mathbb{R}e_0^{(2)})$, $(!A_2)_+ (\cong \mathbb{R}e_0^{(1)} \otimes (!A_2)_+)$, and $(!A_1)_+ \otimes (!A_2)_+$, which decompose $(!A_1 \otimes !A_2)_+$ as a direct sum as discussed in section 5.1. Clearly this generalizes to any number i of stimuli, which produces the expected $2^i - 1$ Hamiltonian terms, and seems to reproduce quite closely the barcodes found in rodent brains by Tsien *et al.* [61, 101].

Thus, after the second stimulus has been registered we have, up to the unknown coefficients, the reactivation of the original *puff*-only firing pattern, a new *quake*-only firing pattern, and a new firing pattern in which basic *puff* and *quake* firing patterns are shared.

The counting argument may be applied to the *quake*-only firing pattern as it was in the last section, but this argument for the shared states is somewhat different. Consider the occurrences of $e_1^{(1)}$ among the shared groups. Each such occurrence is wedged with every available basic *quake* firing pattern and there are less than $2^{m_2} - 1$ of these in the non-ideal case of these, or that number in the ideal case. So the previous number $N_1^{(1)}$, or less than this in the non-ideal case, must be multiplied by this number.

In detail, in the ideal case, for each single shared basic firing pattern $e_i^{(1)}$:

there are 2^{m_2-1} shared basic firing patterns of the form $e_i^{(1)} \wedge e_j^{(2)}$;

there are 2^{m_2-2} shared basic firing patterns of the form $e_i^{(1)} \wedge e_j^{(2)} \wedge e_k^{(2)}$;

and so on down to $2^{m_2-m_2}$. So the total number of appearances of $e_i^{(1)}$ in the shared region is

$$\sum_{j=1}^{m_2} 2^{m_2-j} = 2^{m_2} - 1 \qquad (5.4.11)$$

and this clearly applies to every shared basic *puff* firing pattern so this is the new amplification factor which must be multiplied by the old one. (This is just a restatement of the power-of-two law as it applies to the firing patterns of the second stimulus.)

Similar arguments apply with the rôles of the shared stimuli reversed: thus the additional amplification of the shared *quake* basic firing patterns is $2^{m_1} - 1$. And so on for further stimuli: the amplifications of shared firing patterns acquire similar factors with further stimuli. Thus, the coldest bottom part of the map for a single stimulus now heats up in its two incarnations in the heat map for two stimuli where it impinges on the shared portion: note the redder lower parts of the two active segments of the first column in the rightmost image in Figure 5.2.

Thus for three stimuli the additional amplification factor, for example, for $(!A_1)_+$ in the clique $(!A_1)_+ \otimes (!A_3)_+$ is $2^{m_3} - 1$ over and above the "native" one that would be present if the first stimulus were acting alone, namely $2^{m_1} - 1$. This latter factor would still apply to the unshared reactivated region corresponding to $(!A_1)_+$ responding alone. The additional factor for $(!A_3)_+$ in this clique is $2^{m_1} - 1$; while that for $(!A_1)_+$ in the most "general" clique $(!A_1)_+ \otimes !(A_2)_+ \otimes (!A_3)_+$ is $(2^{m_2} - 1)(2^{m_3} - 1)$, and similarly for the others. (We are assuming that the base responses for different stimuli are different, so that their native amplification factors as single stimuli will yield amplified differences in the respective response patterns.) The first consequence is that the shared regions of a heat map will generally show a different pattern from one shared region to another even if they are contiguous, as seems to be the case in the Tsien *et al.* recordings. Moreover, it seems that the trend, given the many unknowns, omissions, and our idealization of the circumstances, etc., would be that the more general the clique (i.e. the more blocks it comprises), the higher the response amplification, as prior cliques are reactivated. Thus, in the case of fear stimuli, for instance, the most general response (implying a general threat of DANGER!) is the one that should stand out. The selective advantage of this scheme has been noted. It is perhaps to be remarked that it seems to have emerged from the simple axioms of our model. In this connection Figures 5.2 and 5.3 may be consulted. The latter figure depicts heat maps taken from mice in a 4 fear stimuli recording experiment: ILA = infralimbic area, RSC = retrosplenial

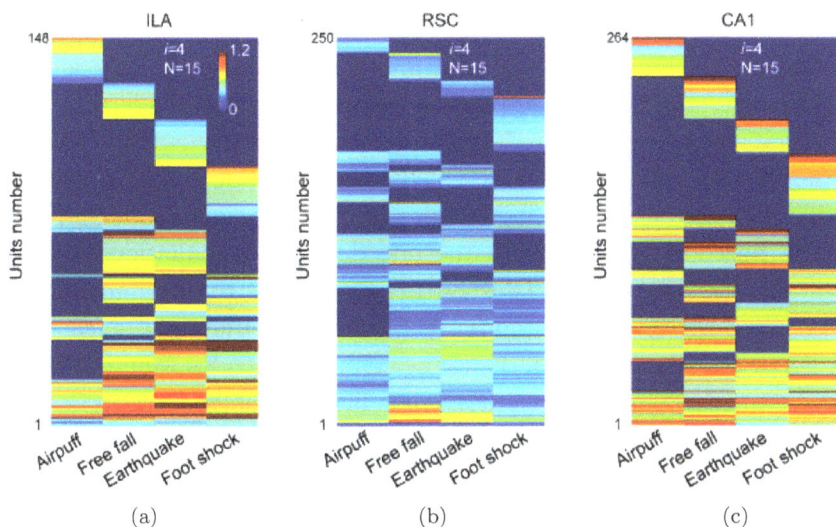

Figure 5.3 Heat maps for four fear stimuli. Figure as originally published in [101], licensed under CC-BY 4.0.

cortex, CA1 = cortical pyramidal cell cliques. In Figure 5.2 the shared first stimulus, namely the lower left of the rightmost heat map, shows a high response at the bottom, or most general part, compared with the low response in that area in the left hand heat map for this stimulus when it was the only one. In Figure 5.3 the more responsive areas in the shared regions have tended to move further down (i.e. towards greater generality) in each shared block relative to their position at the top of the single blocks, the latter being expected from the analysis in section 5.3.

We note that in Figure 5.3 the rightmost CA1 cortical pyramidal cell heat map seems to conform most closely to our idealized results: indeed, these cells are our primary models for b-neurons.

5.4.1 *Multimodal b-neurons in the FCMs*

What if our b-networks, tuned to slightly different fear modes, say, overlapped in multimodal common b-neurons, unlike in the example given? For instance let us suppose there are cells that respond to both *puff* and *quake*. Such cells would be shared among the first two networks $\mathscr{N}^b_{A_{\text{puff}}}, \mathscr{N}^b_{A_{\text{quake}}}$ and indeed would constitute their intersection. After the first fear stimulus the microscopic barcode for this single stimulus will appear as before, since the *quake* response has not yet been registered by the bimodal cells. When this second stimulus is applied, the bimodal cells respond to both stimuli so this will add a column to the first cellular or microscopic barcode and this small barcode (analogous to the one depicted in equation (5.4.6)) will expand accordingly. Clearly this generalizes to any multiplicity of modes within a multimodal cell, and any number of such cells in any other combination. Thus the cellular barcodes would develop extra lines having bigger cliques, much like the splitting of the lines of atomic spectra in the presence of an external field (the Zeeman/Stark effect in the presence respectively of magnetic/electric fields.) In both cases it is because of the activation of superpositions that had not been provoked before. Presumably these splittings inside the microscopic barcodes, which are the black squares of the big barcode, whose interiors are opaque at the big barcode resolution, do not much affect the overall structure of the big barcode. They are also likely to get absorbed in the general reactivation of the prior networks anyway.

5.4.2 *Cyclicity in a special case*

Let us return to the interaction Hamiltonian for several stimuli, namely equation (5.3.1) repeated here for ease of reference:

$$H_I = \sum_{i=1}(l_i a_i^\dagger + h_i a_i). \qquad (5.4.12)$$

Let us assume that $\kappa < 0$ is constant with $\kappa = -\varrho$, and we assume a trivial FCM network with no connections, so we can ignore the internal dynamics. Then, starting with the system being in its vacuum state $|\mathbf{0}\rangle$, after time t the system is in the state

$$T(t)|\mathbf{0}\rangle = [\cos(\varrho^{\frac{1}{2}}t)I - \varrho^{-\frac{1}{2}}\sin(\varrho^{\frac{1}{2}}t)\sum_{i=1}(l_i a_i^\dagger + h_i a_i)]|\mathbf{0}\rangle \qquad (5.4.13)$$

$$= \cos(\varrho^{\frac{1}{2}}t)|\mathbf{0}\rangle - \varrho^{-\frac{1}{2}}\sin(\varrho^{\frac{1}{2}}t)\sum_{i=1}l_i|\mathbf{1}_i\rangle. \qquad (5.4.14)$$

This state is cyclical in time with period $2\pi\varrho^{-\frac{1}{2}}$ in case the l_i-coefficients are constant or only slowly varying. The b-neurons that are firing (with states $|\mathbf{1}_i\rangle$) will fade in and out differentially, depending on the values of the l_i, while the external stimulus is active. This stimulus is external to the FCM and may linger in another part of the brain or nervous system as a residue, such as a flow of neurotransmitters, after the initial stimuli stop. Thus, different memories, or sets of firing b-neurons, may appear to be cycled through in periodic fashion. (This periodicity might be complicated by the external l_i coefficients which may themselves be changing in time.) Although much simplified, this phenomenon may be compared to the one depicted in Figures 3 and 4B of [61]. For groups of such ensembles, such cycling of singlet firing patterns would trace out patterns which appear to move on toroidal surfaces, since they will be essentially products of topological circles.

(It is to be noted that very sophisticated recent work has revealed a toroidal structure in the course of the processing of certain inputs to rat brain regions [37]. This work involves the grid cells, their grid-like matrix possibly having an influence on the toroidal structure. Thus our hypothesis of a general toroidal structure is not necessarily supported by this work.)

5.4.3 *Digression: Dopaminergic (DA) cells*

It was found by Tsien *et al.* that dopamine neurons *en masse* seem to obey ordinary Boolean logic, or at least do not produce the power-of-two barcode structure: cf. [101], particularly Figure 6C. This is consistent with our model because some or most of the dopaminergic (DA) neurons in mammals have distal axons emanating directly from a dendritic body, cf. [39], so cannot be resolved as in the bicameral case, and their logic must remain as it would be for the unicameral case, and therefore is possibly Boolean. On the other hand, it is possible to imagine the parameter window of a network of bicameral neurons moving into, or remaining permanently in, a part of the appropriate space such that the geometrical conditions described in section 1.4.2 giving rise to a Boolean logic are fulfilled.

This circumstance would depend on the nature and disposition of the receptors on the neuron membranes, known to be of particular complexity in the DA case.

Such DA neuronal assemblages are the exceptions that prove the power-of-two law.

5.5   The time course in verbal fluency

In the above, let us suppose that the time intervals between the switching on of the stimuli are equal: $\Delta t_1 = \Delta t_2 = \cdots := \Delta t$. In this case, since additional rows and columns are added at each time interval to the growing barcode, it may be regarded as a clock, measuring the time since its inception by the number of steps taken up to the time in question. Thus,

after one interval Δt we have the simplest barcode of one row consisting of one square corresponding to $(!A_1)_+$ above;

after two intervals the next entire barcode is produced (5.4.6) with 3 rows;

after three intervals the next entire barcode is produced (5.4.7) with 7 rows;

$\ldots\ldots$;

after n intervals we have the entire barcode with $2^n - 1$ rows.

Each barcode has a unique structure depending on its number of rows, or projection cliques, i.e. its length. Thus, we may take each barcode length itself as a measure of the time taken to reach it. But this length, namely the number of rows, or projection cliques, is given by the power-of-two law, where the power is the number of stimuli provoked up to that point. Consider the case of the Verbal Frequency test as discussed in [34]. In this test, the subject is required to name as many items in a given category as he or she can retrieve from memory within a given time period, such as *farm animals*. Each word recognized in a given category presumably constitutes a stimulus: this would be an example of a stimulus coming from another part of the brain. We now assume that each retrieval, processing *à la* Tsien and provoking an utterance, takes on average the same amount of time, namely our Δt above. In a little more detail, we have for instance, starting at $t = 0$:

stimulus 1 for duration Δt: word 1 (*pig*, say) retrieved from the mental "lexicon", activating the $!A_{\text{pig}}$ clique with output $(!A_{\text{pig}})_+$, producing utterance 1, "pig," and singleton barcode: $2^1 - 1 = 1$. Time taken from $t = 0$: Δt;

stimulus 2 for duration Δt: word 2 (*cow*, say) retrieved from the mental lexicon, activating the $!A_{\text{pig}} \otimes !A_{\text{cow}}$ clique with output $(!A_{\text{pig}} \otimes !A_{\text{cow}})_+$, producing utterance 2, "cow," and 3-row barcode: $2^2 - 1 = 3$.
Time taken from $t = 0$: $2\Delta t$;

and so on.

So the exponent in the power law is the number $n(t)$ of words recognized up to the time $t \propto$ number of stimulations provoked up to time t. Then we would have for some constant γ having the dimension T^{-1}:

$$\gamma t = 2^{n(t)} - 1 \qquad (5.5.1)$$

from which we obtain

$$n(t) = (\ln 2)^{-1} \ln(1 + \gamma t). \qquad (5.5.2)$$

This would only apply to those runs in which the time intervals may be evenly spaced, i.e. not necessarily all of them. This closely resembles the result quoted in [34] for 80% of the VF measurements. Since

$$n'(t) = (\ln 2)^{-1} \left(\frac{\gamma}{1 + \gamma t} \right) \qquad (5.5.3)$$

the slope of the decline curve at $t = 0$ is $(\ln 2)^{-1}\gamma = 1.442\ldots\gamma$. Presumably γ could be determined if there is enough data to determine the slope of the curve near zero.

We note that if the factor γ is not constant but a function $\gamma(t)$ of t, assumed at least differentiable near $t = 0$, equation (5.5.3) becomes

$$n'(t) = (\ln 2)^{-1} \left(\frac{\gamma'(t)t + \gamma(t)}{1 + \gamma(t)t} \right) \tag{5.5.4}$$

near $t = 0$ so that in this case the slope of the decline curve at $t = 0$ is $(\ln 2)^{-1}\gamma(0)$.

This logarithmic decline with time in the number of retrievals is evidence for Tsien's Law in certain human brain regions associated with language processing.

5.6 Forgetting

The process of forgetting memories is as important as the process of acquiring them, and of course there is an oceanic literature in the subject. (See for example [25, 54] for reviews.) There are roughly speaking two general modes of forgetting:

1. Failure of the retrieval process, which may leave the actual memory engrams intact though inaccessible;
2. Destruction of the engrams themselves, either by the death or inactivation of the neurons involved, or the loss of coherence of the engrammatic structure.

(In our model there is a bit of overlap between these two modes, since our version of retrieval itself overlaps somewhat with our version of acquisition, as noted in section 4.2.3, both being represented by interpreted sequents.)

We shall briefly address two examples of the first of these and one of the second, in the context of our model. Both of the examples in the first group contain important observations regarding the implementations of the connectives $\otimes$ and $\oplus$ in the network context. We will return to them in the next chapter.

5.6.1 *Failure of retrieval I:* $\otimes$ *may not be implementable*

The tensor product of the state spaces of N networks comprises superpositions of the states of the basic unitized, or co-acting, N-tuples of their nodes (or in our case neurons or parts of neurons), and we have assumed that these describe the result of possible "substrate" connections implemented among actual neurons (and generally inhibiting their firing) by such signaling modes as flows of neurotransmitters, hormones, interneural networks, etc. When any or all of these effects are operating, the tensor product of the state spaces represents the state space of a presumed actual network. However, as in all things biological, these effects may be absent among the networks under consideration, as we have noted. For instance, there may be a deficit of the inhibitory neurotransmitter GABA, or a problem with the interneural substructure, etc. If this happens, then the tensor product may not correspond to an actual network. We say that the state space, or formula it is interpreting, is not *implementable*. So we have the situation, perhaps unusual in pure logic, that the connective $\otimes$ may not always be implementable. In this case, a proof in our calculus will, at the point at which $\otimes$ is invoked, depart from applicability and should be suspended. (An added complication arises since our calculus is timed, and this failure of $\otimes$ may be intermittent.)

Such a circumstance would prove catastrophic to most of our purported proof fragments and their applicability, and would lead to failures of pattern completion, and a lot else. Since, for b-networks, the fundamental identity equation (B.1.8) mixes $\oplus$ with $\otimes$, we are required to try to sort these matters out. This sort of issue arises in the next section. A more comprehensive appraisal will be attempted in the next chapter. (We have postulated [84] that the failure of $\otimes$ seems to give rise to schizophrenia-like anomalies. We will take this up again in Chapter 6.)

5.6.2 *Failure of retrieval II: Neurogenesis*

Suppose we have a retrieval operation

$$!A \vdash_{t_0} !D \tag{5.6.1}$$

as described in our earlier discussion following equation (4.2.5). Thus we assume the sequent would normally be τ_c^A-persistent at t_0 so that the second LTP theorem applies to maintain the retrieval operation until a later time t_F. Now assume that a new b-neuron n_a appears on the scene at some time after t_0 and connects to $\mathcal{N}_D^b$ at some later time $t_1 \in [t_0, t_F]$ to yield a sequent $!a \vdash_{t_1} !D$. (Neurogenesis seems to happen in the hippocampi of several species where it seems to facilitate forgetting, among other effects [25].)

Then we have

$$
\frac{\dfrac{!A \vdash_{t_1} !D \qquad !a \vdash_{t_1} !D}{!A, !a \vdash_{t_1} !D \otimes !D} R\otimes \qquad !D \otimes !D \vdash_{t_1} !D \; \text{Cut}}{\dfrac{!A, !a \vdash_{t_1} !D}{!A \otimes !a \vdash_{t_1} !D} L\otimes} \qquad (5.6.2)
$$

At this point we come to the crux of one of the issues surrounding the idea of a Gestalt, which is intimately related to the problem of the promiscuity of the connective $\oplus$. Namely, if $\mathcal{N}_A^b$ and n_a (the latter being considered as the minimal b-network) are to be considered as lying in the same Gestalt, then the cluster of output nodes of $\mathcal{N}_A^b$ and that of n_a should be sufficiently similar in their behavior that each cluster may be considered to be a subcluster of a *single* system. In which case we can *implement* this single system or network, by just taking the two together and regarding it as a single network whose state space is $A \oplus a$. Then we have $!A \otimes !a \equiv !(A \oplus a)$ and the formula to the left of the turnstile in the concluding segment of the last sequent proof fragment above is itself the banged form of another sequent, namely $A \oplus a$. Then the LTP theorems may apply and the sequent may continue its existence, perhaps proceeding with the retrieval in some form.

On the other hand, n_a may be so different from the neurons in $\mathcal{N}_A^b$ that the two networks cannot be considered to be parts of the same network. For instance, a neuron may have ion channels, etc., on its surface that respond very differently from those on the surfaces of the neurons in the other network. Then n_a and the neurons in $\mathcal{N}_A^b$ will behave very differently when exposed to the same flux or bath of certain neurotransmitters, etc. In this case neither component network can be considered part of a single network containing the other, and the cluster whose state space is $A \oplus a$ *cannot be implemented*. That is to say, we deny the use of the space $A \oplus a$ as the state space of the cluster. Therefore $!A \otimes !a$ *cannot be implemented as a b-network*, i.e. realized as a banged type, and the LTP theorems, whose implementation requires this, may no longer apply. In this case the retrieval operation may fail.

(The upshot for our struggle towards a definition of the Gestalt notion is that if two networks share a Gestalt and are disjoint then their $\oplus$ may be implemented as a network. We will return to this in Chapter 6.)

We may note that the kind of similarity or dissimilarity we have posited here must be definable, if it is definable at all, only up to a certain degree. Thus the neurons in the set of assemblages tuned to *fear* responses as above must be sufficiently dissimilar to be able to distinguish among the various fear stimulus types, but sufficiently similar to allow the implementation of $\oplus$ among the types. (Note that in fact the individually tuned fear assemblages above (*puff*, *quake*, etc.) are not only disjoint (by assumption) but also must be in the same Gestalt, namely the fear Gestalt.)

In the next chapter we will attempt to address these problems, and will find an approximation to this sort of two-fold distinction within the mathematics of the model.

5.6.3 *Loss of engrammatic cohesion in short term memory loss*

Memories may presumably also be lost or de-commissioned through processes of chronic homeostasis. A re-ordering process is presumably counteracted by some process that restores the previous order when the re-ordering agency is removed. If the Tsien process of structuring incoming memories in a hierarchical form is symbolized by a sequence of expanding barcodes, then the opposite process must symbolically take these barcodes down. This can be interpreted in (at least) two ways. Firstly, the FCM b-neurons may themselves be subject to loss by death or more likely some other form of inactivation: this would lead in our setup presumably to the loss of the short term memories acquired through the Tsien process. The ensembles of neurons forming the targets of the projective connections, symbolized by the barcode in Figure 5.1, may themselves remain intact, though they would become inactive. Secondly, these latter neuron ensembles may independently suffer de-commissioning or loss, leaving the FCM b-neurons stranded with no targets for their projective connections. This would presumably be associated with longer term memory loss. We shall try to treat the first of these — loss of short term memory via loss of FCM b-neurons — in some detail in our model. (For a review of short term memory mechanisms cf. [54].)

There is another way that these short term memories may be lost and that is for some agency to inactivate the actual projective connections themselves, while leaving the neuron ensembles in both the FCMs and their projective targets intact. Neurogenesis seems to be an example of this possibility. It seems that there are too many unknowns regarding this possibility to allow a treatment within the confines of our model but it should be borne in mind.

We shall address the first mentioned process — namely the loss of STM via the loss of FCM b-neurons — shortly, but first we must consider how to determine memory size. Recall from the discussion in the last paragraph of section 4.2.3 that memory resides in the disposition of certain synaptic and extrasynaptic connections, namely the ones leading into a working memory "module" such as the network $\mathcal{N}_D^b$ or a long term memory module such as $\mathcal{N}_M^b$, say. In the case at hand we will take these to be the projective connections from the b-neurons in FCMs represented in Figure 5.1 by the colored arrows. Thus, the total number of such arrows or connections would be a measure of *memory content*. We will regard the targets for these projective connections, represented by the rows of the barcode, as being part of the network strands that comprise the engrams, possibly residing in working memory or long term memory.

We shall, as earlier, first treat the case of a single memory established by a single incoming stimulus *à la* Tsien. Using the notation as before, each b-neuron n_i has 2^{m_1-1} projective connections to the micorocliques involved in maintaining the memory. Here m_1 is the number of b-neurons in the single FCM and the connections for each b-neuron are the $2^{3-1} = 4$ arrows per b-neuron shown on the left in Figure 5.1 for $m_1 = 3$. So the total number of such connections in this case is

$$C_1 := m_1 2^{m_1-1}, \qquad (5.6.3)$$

which of course is also the number of occupied, i.e. black, squares in the barcode. Some remarks are in order here. Firstly, note that we are only counting the assumed projective connections to the FCM targets, not the many other possible connections, and nor are we counting anywhere the number of neurons involved in the projective targets or their possible interconnections. The idea is that we are trying to count only those parameters that may be involved in the short term memory engram itself. Secondly, we note that the loss of one b-neuron reduces by half the number of these connections from each remaining b-neuron. This is because each target of a projective connection from a unitized k-tuple of b-neurons (such as n_i, (n_i, n_j), (n_i, n_j, n_k), etc.) requires exactly the k inputs to operate: if one is missing this target ensemble goes "offline" and the other connections to it are silenced. Thus the engrammatic structure is vitiated.

Now suppose that the b-neurons in this FCM go out of commission, i.e. effectively lose their connections, either by the operation of some sort of deactivating mechanism such as a flux of neurotransmitter (such as dopamine expressed by "forgetting cells" [25] or interneuron involvement), or the interference effects of new incoming stimuli, or cell death (which is much less likely), at a constant rate of r b-neurons per unit time. Then the number of connections remaining after time t is

$$C_1(t) = (m_1 - rt)\, 2^{(m_1-rt)-1} \qquad (5.6.4)$$
$$= 2^{m_1-1}(m_1 - rt)\exp(-(\ln 2)rt) \qquad (5.6.5)$$
$$= 2^{m_1-1}m_1(1 - (r/m_1)t)\exp(-(\ln 2)rt) \qquad (5.6.6)$$
$$= C_1(0)(1 - (r/m_1)t)\exp(-(\ln 2)rt). \qquad (5.6.7)$$

So the fraction of memory remaining after time t is

$$F_1(t) := C_1(t)/C_1(0) = (1 - (r/m_1)t)\exp(-(\ln 2)rt). \qquad (5.6.8)$$

Note that $F_1(t) = 0$ when $t = m_1/r$ and decreases up to that point. (A proof of this last assertion is left as an elementary exercise.) In this it is unlike exponentially declining curves which never reach zero. However, as r/m_1 decreases, the time until $F_1(t)$ vanishes completely lengthens, while at the same time the curve becomes more like an exponential. For a fixed r this happens as m_1 increases. So the more FCM b-neurons involved in maintaining the memory, the more like an exponential this relative "forgetting" curve becomes.

Note again that we are not counting the possible number of neurons that are firing during the process of recall, but only certain connections projecting from the FCM neurons to their targets. In this microscopic or minimal case, the minimum total number of neurons firing, and presumably detectable in an *in vivo* experiment, would occur if each occupied square, representing an ensemble of neurons making up a microclique in this case, were to comprise a single neuron. Then the minimal total number of neurons involved would be:

number of neurons in the FCM + number of black squares in the barcode $= m_1 + m_1 2^{m_1 - 1}$.

That is:

$$T_1 := m_1(1 + 2^{m_1 - 1}). \tag{5.6.9}$$

CAVEAT!
This is assuming that the projective target ensembles (in this case single neurons) do not overlap.

These formulas are quite easy to generalize to N memories of incoming experiences established in this way, thanks to the resolutional self-similarity of the structure. Thus, Figure 5.1 will look the same for the case in which the circles representing single b-neurons on the left are replaced by whole assemblages, each now tuned to a *different* input or discrete experience in the Gestalt, while the arrows would now represent the entire previous collection of connections, and each black square would represent an entire "microscopic" barcode of the previous type. Thus with N such established tuned networks or clusters, labeled as $\mathcal{N}_i$, where this contains m_i b-neurons, $\mathcal{N}_i$ has 2^{N-1} — the total number of b-neuron connections to the large barcode from the single "block" $\mathcal{N}_i$ among N of them. So

$\mathcal{N}_1$ has $2^{N-1}(m_1 2^{m_1-1})$ b-neuron connections to the large barcode;
$\mathcal{N}_2$ has $2^{N-1}(m_2 2^{m_2-1})$ b-neuron connections to the large barcode;
...

So, for the whole ensemble, or FCM, the total number of b-neuron connections of the type we are considering, which maintain the set of N memories, is

$$C_N := 2^{N-1} \sum_{i=1}^{N} m_i 2^{m_i-1} \qquad (5.6.10)$$

$$= 2^{N-2} \sum_{i=1}^{N} m_i 2^{m_i}. \qquad (5.6.11)$$

Let us now suppose that the b-neurons in the whole FCM go out of commission as above, at a rate of r b-neurons per unit time. We shall simplify matters by assuming as before that r is constant and that the de-commissioning process is uniformly distributed over the whole ensemble (at least for the time the process operates). Then, the total number of b-neurons lost from the whole ensemble by time t is rt and so the number of b-neurons lost on average by each $\mathcal{N}_i$ is rt/N. Therefore, the total number of connections remaining by time t is

$$C_N(t) := 2^{N-2} \sum_{i=1}^{N} (m_i - rt/N) 2^{m_i-rt/N} \qquad (5.6.12)$$

$$= 2^{N-2} \left(\sum_{i=1}^{N} (m_i - rt/N) 2^{m_i} \right) 2^{-rt/N} \qquad (5.6.13)$$

$$= 2^{N-2} \left(\sum_{i=1}^{N} m_i 2^{m_i} - \left[\frac{r \sum_{i=1}^{N} 2^{m_i}}{N} \right] t \right) \exp\left(-\left[\frac{r \ln 2}{N} \right] t \right) \qquad (5.6.14)$$

$$= 2^{N-2} \sum m_i 2^{m_i} \left(1 - \left[\frac{r \sum 2^{m_i}}{N \sum m_i 2^{m_i}} \right] t \right) \exp\left(- \left[\frac{r \ln 2}{N} \right] t \right)$$

$$\tag{5.6.15}$$

$$= C_N(0) \left(1 - \left[\frac{r \sum 2^{m_i}}{N \sum m_i 2^{m_i}} \right] t \right) \exp\left(- \left[\frac{r \ln 2}{N} \right] t \right). \tag{5.6.16}$$

So the fraction of memory remaining by time t in this case is

$$F_N(t) = \left(1 - \left[\frac{r \sum 2^{m_i}}{N \sum m_i 2^{m_i}} \right] t \right) \exp\left(- \left[\frac{r \ln 2}{N} \right] t \right) \tag{5.6.17}$$

and we have $F_N(t) = 0$ when

$$t = \frac{N \sum_i m_i 2^{m_i}}{r \sum_i 2^{m_i}} \tag{5.6.18}$$

and again we have $F_N(t)$ decreasing until this time. Also again, we have approximation to an exponential decrease as the time until depletion grows longer.

The formula for the minimal number of neurons involved including those in the $\mathcal{N}_i$ s and their projective targets, or cliques, is likewise (somewhat abusing the language):

number of neurons in the FCM + number of black squares in the big barcode × total minimal number of neurons in these black squares, that is

$$T_N = \sum_i^N m_i + \sum_i^N \text{minimal number of neurons in all projective}$$

$$\text{targets of } \mathcal{N}_i \tag{5.6.19}$$

$$= \sum_i^N m_i + \sum_i^N \left(N 2^{N-1} \times \text{minimal number of projective}$$

$$\text{(single neuron) targets of } \mathcal{N}_i) \tag{5.6.20}$$

because there are $N2^{N-1}$ targets/black squares
in the big barcode

$$= \sum_i^N m_i + \sum_i^N N2^{N-1}m_i2^{m_i-1} \qquad (5.6.21)$$

$$= \sum_i^N m_i + N2^{N-2}\sum_i^N m_i2^{m_i}. \qquad (5.6.22)$$

CAVEAT!
As above. The total minimum increases so fast with the m_i that either there are generally very few FCM b-neurons involved, or there must in reality be sharing or overlap/multimodality among the target ensembles (or cliques), or likely a combination of these two strategies, assuming our arguments are correct. Ignorance of the numbers involved compels us to ignore this latter possibility for the small examples we discuss below.

These formulas can be much simplified if we assume that the $\mathcal{N}_i$ are all the same size, say $m_i = m$. Then we obtain

$$F_N(t) = \left(1 - \left[\frac{r}{Nm}\right]t\right)\exp\left(-\left[\frac{r\ln 2}{N}\right]t\right); \qquad (5.6.23)$$

$$T_N = Nm(1 + N2^{N+m-2}). \qquad (5.6.24)$$

In this case the lifetime of the memory is Nm/r.

Many of these constants seem to be unknown as of this writing but we may produce a possible simple toy example with the following choices, which may not be too outlandish. Firstly, let $N = 1$. It would be difficult in practice to parcel out a particular array of memoranda into separate items, so this seems a reasonable choice, corresponding to a single discrete memory, of an event like a click, a flash of light or a verbal syllable. We take $m := m_i = 10$ since it is the projective connections that we are counting towards a measure of memory content and even a few neurons will have very many of these, namely for the choice made above there are $m2^{m-1} = 5120$, and we invoke the principle of the parsimony of nature. In this context, our interpretation of Tsien's FCM-tuned components (our $\mathcal{N}_i$) resembles the idea of ensembles of narrowly tuned *concept cells*, also known as "grandmother" or "Jennifer Aniston" cells in the extreme — but probably mythical — case of such an ensemble comprising a single neuron: cf. [75]. The problem with fewer neurons in this capacity would be one of redundancy: if a few go out, would not the whole memory be lost? This might be true if it were neurons that count but our argument implies that it is rather the connections that count. And the loss of a few of b-neurons in an ensemble such as an FCM would seem to generally entail a relatively negligible loss of projective connections emanating from the ensemble.

Note that the small number of FCM neurons counted here will only be a very small fraction of the entire complex of neurons that fire during this process, since they are projectively connected to the target ensembles each of which may contain very large numbers of neurons which are not counted in this scheme but will be firing. In fact, the minimal total number of active neurons involved, according to equation (5.6.9), is in this example $T_1 = 5130$ and in reality the total number is likely to be very many multiples larger. (To reiterate, this does not take into account overlaps). So, the number of FCM b-neurons involved in this example is only about 0.2 per cent of this minimum total active neuron count. (Of course, merely adding one b-neuron to the FCM more than doubles the minimal total number of neurons involved, to 11275.) This will be borne out in the following example. Namely, with these choices, and a guess of r at one neuron every 3 seconds, or $r = 0.33$ neurons per second, equation (5.6.24), which reduces to the first one we obtained, equation (5.6.8), becomes

$$F_1(t) = (1 - 0.033t)\exp(-[0.33\ln 2]t) \qquad (5.6.25)$$

with t measured in seconds. Cf. Figure 5.4.

This would indeed be a short term memory: after 30.3 seconds it is all gone, with half of it having gone after about 2.611 seconds! However, this brief lifetime of a short term memory seems at least consistent with known results (cf. [71]). Note that such a rapid decline may be vitiated by repeatedly retrieving the memory and thus setting

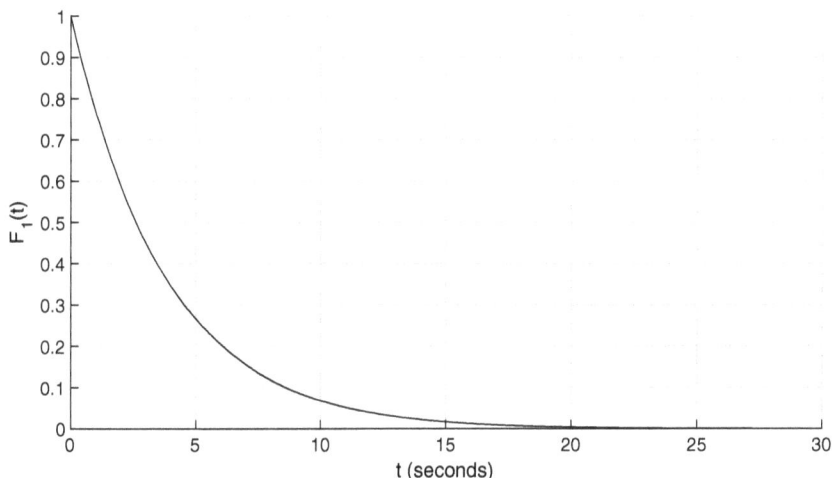

Figure 5.4 Fractional short term memory forgetting curve, $F_1(t)$, the fraction remaining after time t for one memorandum, a 10 b-neuron FCM, and a de-commissioning rate of 0.33 FCM b-neurons per second.

the stage for the conditions of the LTP theorems to hold and the initiation of long term memory consolidation to proceed, at least as far as our model is concerned. This phenomenon seems to be observed and is known as "rehearsal." Conscious or very likely unconscious, it is presumably the process by which certain short term memories become long term. Note that our model seems to lean towards the idea of *sparse neural coding*, since, as noted, it is the connections that we are counting towards a measure of memory content and even a few neurons can have very many of these.

Here the lifetime of a short term memory is proportional to the number of b-neurons in the FCM. So with half that number, the lifetime would halve. But the relative or fractional change can be very small indeed. In the above example, halving the number of b-neurons, m, changes the time until half the memory is lost from 2.611 seconds to 2.344 seconds, approximately. So the redundancy built into the number of connections seems to be borne out.

This argument does not depend on the order of de-commissioning and so any order is possible. Thus the loss of half a single memory, as above, does not specify which pieces of the memory go first.

It could be an overall fade-out with half going in the first three or so seconds and the rest more slowly, finally going completely after 30 seconds. (But please see the following section.)

As to the fraught issue of the subjective experience of short term memory loss, presumably it is a function of this sort of proportional loss of the original "amount," since we do not immediately experience the actual size of the networks nor their degree of connectivity, but only their effects at many removes.

Memory formation (acquisition and loss) in larval zebrafish has very recently been found to be associated with reciprocal changes in synapse numbers, rather than strength of connections, in a manner very similar to our postulates concerning memory size allocation and forgetting via connection "de-comissioning" as above [27].

5.6.4 *Longer term memory effects*

Although the arguments above do not depend on the order of the projective connections fading out, our earlier arguments would tend to favor fading in the order: first in, first out. That is, the projective connections to the more specialized cliques will fade out before the projective connections to more general ones, since response intensity increases with generality, as we argued in section 5.4. In detail, this response intensity presumably entails the frequency of firing along the projective connections from FCM b-neurons to cliques: the amplifications we referred to in section 5.4 resulted from the more frequent firing of subensembles of FCM b-neurons. These amplification/frequencies go up as the receiving cliques become more general. Thus, via the LTP process, the connections involved become stronger, and therefore the more general long term memories, which now reside in the more general clique synapse patterns, presumably last longer despite the loss of the originating FCM neuron ensembles. The upshot is that the recall of general features would be more rapid than the recall of special ones. Thus one might more readily retrieve the general look of an artist's paintings, style of writing or composing, or scenes from a movie actor's movies, before the specific name of the artist itself is retrieved, if it can be retrieved at all.

5.7 Conclusions

In this chapter we have argued that the basic tenets of Tsien's Theory of Connectivity follow immediately from our model, despite the many unknowns. Moreover, salient conclusions from our logical machinations consist with the corresponding conclusions from the physics-like face of our model, namely:

- The activation by the current stimulus reactivates the prior networks (though with a different interpreted Hilbert space map). The logical expression of this is to be found in the discussion in section 5.1 following (5.1.11) while the physics-like expression is to be found in equation (5.4.5) and its temporal descendants;
- The repeat of any stimulus does not activate any new network. The logical expression of this is to be found in the proof fragment (5.1.11) while the physics-like explication is to be found in a paragraph following soon after the barcode (5.4.7).

A reason for this consistency might be that the physics-like dynamics is driven by an external Hamiltonian. This externality is "structural" and therefore also falls under the purview of the logic.

Tsien makes the following testable predictions on the basis of his Theory of Connectivity [101]:

1. Cognitive universality: the power-of-two logic should apply across a wide range of stimulus types (fear, taste, etc.);
2. This logic should be prevalent across many circuits, regardless of their variation;
3. Modulatory neurons, such as the DA ones, use a different logic;
4. The specific to general organization should be developmentally prefigured, rather than being formed after learning;
5. This logic is implemented in the cortex vertically via the differential assignment of specific to general cliques in distinct laminar layers. This vertical implementation of FCMs has the advantage of being replicated via horizontal surface expansion, rather than cortical depth. This presumably explains the selective advantage

of the folded laminated structure that has evolved for cortices confined to rigid boxes;

6. Species conservancy: this organization should be conserved across different animal species.

Since the quite recent formulation of this theory (as of this writing) there seems to have been some experimental support for predictions

1, 3, 4, 5, 6 above: mainly from the careful experiments of Tsien *et al.*, the last being necessarily always partial, and in these cases the experiments were confined to rodents;

2, 6 above: 2 may overlap with 1 so evidence for it may provide evidence for 2. In the case of 6 there has been additional evidence for the phenomenon in the brains of macaques, as referenced above, and also evidence for it in humans, via language processing tests, though this is rather indirect.

As far as our model is concerned it predicts that the power-of-two law and its operation will follow ineluctably if the neurons are bicameral; if the networks have non-Boolean parameter windows; if there are enough substrate connections (chemically, physically or electrically implemented) to effect $\otimes$-product states of the networks; and if the other axioms hold. These are rather mild restrictions.

The predictions

1, 2, 4, 6 above are automatically fulfilled under these circumstances;

3 above would be fulfilled at least for DA neurons if they often have the non-standard gross morphology discussed above;

5 above is partially fulfilled by the very nature of sequents, which reveals an implicitly layered structure for network connections.

The Tsien phenomenon also leads to a rudimentary theory of forgetting for short term memories. The salient points that emerge from our model and may possibly be subject to experiment include the following.

- Neurons or ensembles of neurons involved in maintaining the traces of short term memory — these are the ones represented by the barcode in our picture — are often *multimodal*: that is to say, single ensembles of these neurons, or single neurons, respond to multiple modalities. This would be a consequence of the overlapping or sharing of the projective targets, or cliques, of the FCMs. As we have noted, such multimodal assemblages or single neurons would have high efficiency and implement sparse coding. The tuned FCM neurons are likely to constitute only a small percentage of the total number of neurons required to maintain the memory;

- Short term memories decay at rates which are close to exponential at the outset but become less so at the ends of their lives, when they disappear completely. (This loss of memory is more likely to be due to the death or de-commissioning of synapses rather than that of neurons and this notion has driven our calculations in section 5.6.3: cf. [27]). These lifetimes can be extended by "rehearsals" in the presence of which the decay rates will more strongly resemble exponentials, at least until the memories vanish completely. These rehearsals may or may not be voluntary or conscious;

- Independently of the Tsien phenomenon, neurogenesis may disrupt memory retrieval by interfering with the connections between parts of the engrams involved.

Our derivation of the power-of-two law has its origin in — and is an expression of — the logic of subspaces of a vector space that we found to model the logic of hidden variables, rather than the Boolean logic of subsets of a set. This is because the exterior algebra is a *graded* structure, this grading not being immediately apparent from the UMP version of its definition (cf. section B.1.3). The grading is via the dimensions of its base space's subspaces. So it might be said that the entire subspace structure of the base vector or Hilbert space is classified by the exterior algebra just as it is in quantum logic itself (cf. the Plücker Embedding, section B.1.4). This aspect of the exterior algebra seems to have been recognized by its discoverer H. Grassmann before 1845, more than half a century before even the embryonic quantum theory of Planck arose (circa 1900) and almost a century before the discovery/invention of quantum logic itself by von Neumann and G. Birkhoff in 1936. It was also known at least to A. N. Whitehead who gave an account of it in his *A Treatise on Universal Algebra* (1898).

We may note finally that the Tsien process transforms sets of linear sequential stimuli into hierarchical structures. Some consequences of this phenomenon will be investigated in the next chapter.

Chapter 6

A General Syntax of Retrieval

Combinators with rules of use generate certain patterns. These are generally deemed *well-formed* and the study of well-formedness and its absence in the context of linguistics, formal logic, and other disciplines, has been hotly pursued for decades if not centuries. Such studies generally come under the rubric *syntax*.

It is apparent that our combinators $\otimes$, $\wedge$ and $\oplus$ must also generate a syntax of some kind. A first question, for any purported syntax, is: what are the syntactic *units* or *atoms*? In the case of natural spoken languages, this turned out to be a non-trivial issue. It is similarly a problem for the protosyntax we shall find ourselves confronted with by our model. In fact this is just the first of a number of striking correspondences we shall find between our model's syntax and the syntactic patterns revealed by linguists executing the Chomskian Minimalist Program. These correspondences prompt the following.

CONJECTURE
There is a general syntax of memory retrieval which underlies, or gives rise to, the syntax of (human) language as a consequence or effect.

We claim no originality for this conjecture: it seems to be implicit at least in biolinguistic investigations from the very beginnings of the subject. All we hope to do here is contribute what evidence there may be for it that arises from our model. Although we shall hypothesize a general scheme, we are borne back ceaselessly to comparisons with the linguistic model of syntax. The specific bio- or neurolinguistic enterprise has indeed vastly and exponentially expanded, both experimentally and theoretically, in the last two or three decades, largely due to the increasing sophistication of *in vivo* imaging techniques, and the fillip these advances have given to theory. For outstanding references to some of the topics touched on here we strongly recommend [2, 33, 91].

The truth of this conjecture would imply, among other things, that there is a basic innate syntactical structure common to all languages, and possibly other cognitive processes. The syntactic differences between languages, and the other processes should they exist, are attributable to superficial "local" variations. (This is not to mention the complex issues involving the interactions of syntax with semantics, which we must necessarily leave aside in this discussion.)

In this chapter we shall therefore make a tentative start on uncovering a general syntax for memory retrieval.

CAVEAT!

In this chapter we push the boundaries of our model perhaps further than is warranted by the model itself, in the hope that the risk is worth the venture. This being said, the results are tentative and provisional at best.

The layout of the chapter is as follows. In section 6.1 we posit a minimal circuit to effect general recursive syntax processing. In section 6.3 we again approach the issue of classification of networks according to postulated Gestalts and in section 6.4 we propose our general syntax rules, work some examples and offer a discussion. In section 6.5 we compare our general syntax with an outline of the Chomskian Minimalist Program's basic tenets. We discuss the question of why, among the primates at least, is *homo sapiens* the only one to have an innate recursive language facility. A final subsection considers the consequence to language production of the vitiation of one of our combinators, namely the tensor $\otimes$. We argue, as in [84], that this produces effects mimicking those found in schizophrenia patients.

6.1 A minimal syntax circuit

We begin at what for us will be the outermost level and work inwards to specify possible syntactic atoms and their combinations.

As noted, our main extant model of syntax is of course afforded by linguistic syntax, but it should be kept in mind that we are aiming for a more general kind of syntax. Consequently we are forced to invent our own terminology, though we shall point out the linguistic analogues as they arise.

A very general minimal circuit to effect *recursive* syntax processing may be depicted as in Figure 6.1.

Recursion in this context means that amalgams previously obtained via syntactic processing may be reused as syntactic units.

This diagram is schematic, topological, and therefore does not reflect actual distances.

We note also that it is meant to reflect only a *productive* process or processes, not processes of *comprehension*, which are beyond the remit of our model. In the case of language this would mean that we are concentrating on the production of syntax, not its comprehension, although in reality these two processes are probably shared or combined. (For this reason most neuroanatomical models of language circuits tend to do both comprehension and production in tandem: we are restricted here to attempting only production, in the sense of producing syntactic structures.)

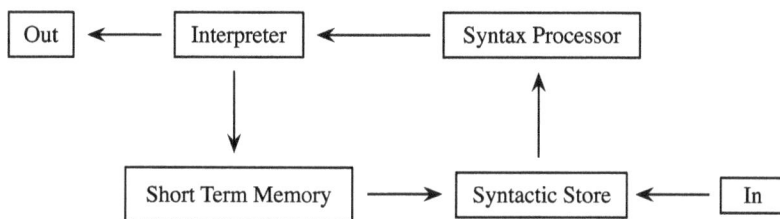

Figure 6.1   Minimal circuit for recursive syntax processing.

With these provisos (and a few more to come) the boxes in Figure 6.1 are described below.

In

This represents some input to the Syntactic Store either from the outside, via the sensorium (such as the auditory pathway in the linguistic case), or from some source within the brain;

Syntactic Store

This is partly the analogue of the *lexicon* in linguistic syntax. Namely, the collection of syntactic units and other categorial and contextual information about them, that are lodged in some from of long term "archival" memory and available to the Syntax Processor. However, we are compelled to assume a more complex rôle for it here since it must be able to filter and/or pre-process the input in some sense. For instance, in the linguistic case, this area must also be capable of some semantic involvement, such as comprehension of input, which could differentiate between verbal and non-verbal stimuli, and process the correct input into recognizable lexical form. In order to effect recursion it must have at least one other input internal to the circuit, via which it receives certain already processed syntactic units, as shown at the bottom of the diagram. Namely input from a **Short Term Memory** which may be either very closely associated with it or integral to it: cf. next paragraph;

Syntax Processor

The network or set of networks assumed to perform the syntactic combinations our rules will specify. Its input is drawn from the Syntactic Store (vertical up arrow in the right) via some protocol that remains unknown to our system, and may be variable. To effect recursion the Syntax Processor must also have access to the contents of some sort of **Short Term Memory**, or STM (lower left of the diagram), which may be part of the Syntactic Store, or very closely associated with it, or similarly, it — the STM — might be very closely associated with or part of the **Interpreter**. Since we cannot know how long any particular memorandum lasts, we shall assume for simplicity that this STM acts in the manner of a last-in-first-out *stack*. Moreover, we shall allow transfers of items in this STM to the presumed archival LTM part of the Syntactic Store. This arrangement seems the minimal needed to effect recursion, since not only must recent amalgams be available to the Syntax Processor after interpretation, but so also must older ones, which will, as conjectured, become part of the Syntactic Store (such as gender forms, plurals, idioms, etc., in the linguistic case);

Interpreter

The network or set of networks assumed to effect the task or tasks that the syntax processing is preparation for: integration, analysis, parsing and/or linearizing (or sequencing) of the syntactic product transmitted in hierarchical form from the Syntax Processor, among other things (with concomitant speech production in the linguistic case for instance). It cannot be part of our model which does not stray into such areas. The Interpreter should have at least two outputs: one or more to the unknown regions just alluded to (speech, action, etc.) — the leftmost horizontal arrow to **Out** — and one or more to the STM mentioned above — the left vertical down arrow. Here it is assumed that as part of its function the Interpreter can flag which amalgams should be kept for reuse or permanent ensconcement in the Syntactic Store, so this pathway is required.

The left L-shaped part of the diagram — the left down arrow and the bottom right pointing arrow — would be crucial for recursion, the *punctus crucis* being the STM and its connections. And we note again that this STM may be associated closely with either of its neighbors in the diagram, without changing the operation of the circuit. Of course, the rightmost up arrow would be crucial for any kind of syntax at all. This observation will have consequences as we shall see (section 6.5.2).

With the possible exception of the rightmost horizontal (left pointing) arrow from In to the Syntactic Store, transmission along the arrows would have to be very reliable, and have other properties, in order to maintain the integrity of the syntactic combinations as they are conducted through the circuit. We should remark here that these transmissions, with the possible exception of the effect of the L-shaped left part, effect *non-destructive copying* of syntactic units. As we shall argue presently, our syntactic units will be tensors of firing patterns and we have already considered what sort of pathways would reliably transmit these: namely, if there is any physical distance among the modules, they should be white matter tracts (section 4.1.4). (Note that this is not cloning of quantum-like states, in the usual physics-like understanding of the word, since they are states in different Hilbert spaces, and, moreover our Hamiltonians are not orthogonal.) So we hypothesize that in case any physical distance exists between the actual modules depicted in the diagram, the connecting arrows correspond to such tracts in actual brains, since they would be selectively advantageous. By the same token, should the boxes in the diagram correspond to areas in actual brains, they need not be localized as shown, though such localizations would be selectively advantageous for speed and efficiency.

Although the details of the Interpreter network or complex of networks, should it exist, are beyond the remit of our model, the recursive step involving the leftmost L-shape may be hypothesized to follow the Hebbian pattern of our LTP theorems at least in part: frequent use of a particular circuit may strengthen it and eventually make permanent a frequently used pathway. In our picture this could be the mechanism effecting the move of syntactic elements from the STM to the long term Syntactic Store along the bottom right pointing arrow.

In this chapter our concern will be mainly with the possible inner workings of the Syntax Processor. A large burden of functionality has been left to other modules: it is no coincidence that our diagram resembles the bed of Procrustes. Appropriately, what we shall produce here is a torso of a syntax rather than its complete body. To supply the ambient functions is work for another time and place.

We will add pertinent model-based elements to effect and complement this simple structure as we proceed, but now we turn to an attempt to build a general syntax on the basis of the model.

6.2 Multimodality, again

As we have noted (section 2.3.2) *modality* in neuro-speak refers to the nature of sensory input, and since there is necessarily an involvement of the sensorium in memory processing — the taste of the madeleine, the well-known evocativeness of odors, etc. — we are forced to engage with the possibility of modalities and multimodalities in a consideration of memory acquisition and processing.

Let us review some of the issues involved. We suppose that unicameral nodes are allowed to have state spaces with a multiplicity of states. For b-neurons, this applies to their output nodes. Then

- the state spaces of b-networks remain Fermi-Dirac spaces/exterior algebras, but a multimodal output node state, formerly written as for instance e_i^A, is now replaced by a set of subindexed ones, and each state of the node is a superposition of these, as illustrated in section 2.3.4 for the bimodal case;
- there are now additional efficacies or exchange terms, creation/annihilation operators, and a more complex indexing scheme. Otherwise the bicameral Hamiltonians, both internal and external, remain in the same form as before. Firing patterns evolve into firing patterns, as before, though these have some internal superpositional structure reflected in the subindexing;
- the discussion of Tsien's theory in Chapter 5 remains unchanged, except for possible "fine structure" splittings within the smallest cellular barcodes (section 5.4.1).

The benefits are, as noted, the multiplicity of uses promoting sparse coding. In addition an analogue of time division multiplexing of axonal connections may be implemented, and the $\oplus$ connective may be applied to intersecting networks whose common nodes are all sufficiently multimodal. We noted also that this may not be as contingent a condition as it appears since networks may have common nodes *because* these nodes are multimodal.

We should also note that such multimodal neurons *in vivo* might be expected to have a multiplicity of dendritic arbors to handle the many possible inputs.

Despite the attractiveness of the idea of multimodality for neurons, as of this writing the only experimental evidence for their existence lies in the sensoria of a few species: some retinal cells and macaque auditory cells (in the *inferior collicus*). We should note that the idea of a more general modality, or "broad tuning," of neurons outside the sensorium has been mooted for years [18].

6.3   Gestalts II

We shall make another pass at pinning down the postulated notion of Gestalt, which we shall take to mean a family $\mathscr{G}$ say, of networks which are supposed to be unified in some sense by a common or uniform set of attributes or features, and for which we stipulate the rules given below. First we will find it convenient to introduce some definitions.

Definition 6.3.1. Given two sets X and Y, X and Y are in an inclusion relation, or *inclusional*, if $X \subseteq Y$ or $Y \subseteq X$.

Definition 6.3.2. A collection of n networks *satisfies the $\oplus$-condition* if the intersection of their sets of nodes is either empty or *fully n-modal*, i.e. consists entirely of n-modal nodes.

(For the import of this last definition, we refer to section 2.2.4, and note again that networks are likely to share nodes *in virtue of the fact* that the nodes are multimodal, though this may not always be the case.)

We shall continue to use the term *implemented* in reference to Hilbert or state spaces to mean that the space in question may be realized, in a specified fashion, as the state space of a network, where we shall allow multimodal nodes into our networks unless explicitly excluding them.

CAVEAT!
For any $\mathscr{N}_A$ the family $\mathscr{S}_A$ of all subnetworks of $\mathscr{N}_A$ will turn out to be a Gestalt, and is the smallest, or generated one, containing the network: if $\mathscr{N}_A \in \mathscr{G}$ then $\mathscr{S}_A \subset \mathscr{G}$. It is improperly self-defining, or

trivial, and we must exclude such Gestalts from our population of them. So in what follows the word *Gestalt* will refer to a *proper* Gestalt, i.e. a Gestalt that is not the trivial one generated by a network. We will generally denote a generic proper Gestalt by $\mathscr{G}$.

In formulating the rules posited below, our interest is in the behavior of the operators $\otimes$ and $\oplus$ upon spaces of states and we have therefore refrained from a separate discussion of the case in which networks may intersect in ordinary unimodal nodes. A collection of such networks taken together determines a single network each of whose components are subnetworks: the set of nodes of this network is just the union of the nodes of the component ones, with the connections as they were. Its state space is the Hilbert space generated by the union of the basic nodal states of the components. It is in general neither the internal nor the external direct sum of the component spaces.

In the following, $\mathscr{G}$ denotes a proper Gestalt.

Definition 6.3.3. For a network $\mathscr{N}_A$ let $\mathfrak{G}(\mathscr{N}_A) := \{\mathscr{G} : \mathscr{N}_A \in \mathscr{G}\}$. This set is not empty by G1 below.

Rules for Gestalts

G1. For every non-empty $\mathscr{N}_A$ there exists a (proper!) Gestalt $\mathscr{G}$ such that $\mathscr{N}_A \in \mathscr{G}$;

G2. $\mathscr{N}_A \in \mathscr{G}$ iff $\mathscr{N}_A^b \in \mathscr{G}$;

G3. $\mathscr{N}_A \in \mathscr{G}$ iff *any* non-empty subnetwork of $\mathscr{N}_A$ lies in $\mathscr{G}$;

G4. $A \otimes B$ may be formed or is *selectable* if $\mathfrak{G}(\mathscr{N}_A)$ and $\mathfrak{G}(\mathscr{N}_B)$ are inclusional, the ordering being determined by the direction of the inclusion: i.e. $A \otimes B$ if $\mathfrak{G}(\mathscr{N}_A) \subset \mathfrak{G}(\mathscr{N}_B)$. (This order is purely conventional: it could as well go the other way.) The (un-symbolized) compound system, of which the $\otimes$ product represents the states, shall be assigned the ambient or larger of the above inclusional sets as its set of ambient Gestalts. If this inclusion is an equality, the order is ambiguous though the assignment of Gestalts is not;

G5. For a finite set of networks $\{\mathcal{N}_{A_i}\}$, $\bigoplus_i A_i$ may be implemented (by the network we have denoted $\mathcal{N}_{\bigoplus_i A_i}$ in section 2.2.4) if the $\mathcal{N}_{A_i}$ satisfy the $\oplus$-condition and there exists a Gestalt $\mathscr{G}$ such that each $\mathcal{N}_{A_i} \in \mathscr{G}$. (This sharing of a Gestalt is our criterion for the component networks to be considered as part of the "same" network.) In this case, $\mathcal{N}_{\bigoplus_i A_i} \in \mathscr{G}$. Thus $\mathfrak{G}(\mathcal{N}_{\bigoplus_j A_j}) = \mathfrak{G}(\mathcal{N}_{A_i})$ for all i.

In G4 we recall that the $\otimes$ product represents the space of states of the compound system describing superpositions of possible substrate connections among their nodes. The possible ambient Gestalts of the two underlying networks are unchanged, so if for instance we have $\mathfrak{G}(\mathcal{N}_A) \subseteq \mathfrak{G}(\mathcal{N}_B)$ then if $\mathcal{N}_A \in \mathscr{G}$ so also is $\mathcal{N}_B \in \mathscr{G}$. Since the $\otimes$ product represents unitized co-acting pairs of nodes, each such state must partake of both sets of Gestalts so we attribute the larger set to $A \otimes B$, if there is one, or the single set if they are equal.

A remark on notation: we have denoted the bicameral version of a network $\mathcal{N}_A$ by $\mathcal{N}_A^b$. Thus we denote also by n_i^A the trivial subnetwork of $\mathcal{N}_A$ whose only node is n_i^A, with no connections to any other node, and by $(n_i^A)^b$ the bicameral version of this network, namely the trivial b-network whose only node is the b-neuron of which the node n_i^A is the output node (which may be multimodal), with no external connections.

CAVEAT!

In this discussion and those following, we shall make the simplifying assumption that the multimodality of a node, should it obtain, does not arbitrate the Gestalt it belongs to as a trivial one-node network in its own right. That is to say, the possible multiple modalities of a node do not cause it to lie in different Gestalts: the different modalities do not take the node out of any Gestalt it may lie in. All the modalities "belong" to the same Gestalt. This assumption removes a possibly troublesome ambiguity in the assignment to, and of, Gestalts. It also seems to be borne out in the few known *in vivo* examples, such as the monkey auditory cells in [18], which, though tuned to different frequencies, presumably operate entirely within the *sound* or *hearing* Gestalt.

Before discussing their motivation, we note some immediate consequences of these rules.

Proposition 6.3.1.

1. If $\mathfrak{G}(\mathscr{N}_A)$ and $\mathfrak{G}(\mathscr{N}_B)$ are inclusional, then there exists a Gestalt $\mathscr{G}$ such that $\mathscr{N}_A \in \mathscr{G}$ and $\mathscr{N}_B \in \mathscr{G}$.
2. Suppose $\{\mathscr{N}_{A_i}\}_{i=1}^N$ is a family of subnetworks of a network $\mathscr{N}_C$. Then $\mathfrak{G}(\mathscr{N}_{A_i}) = \mathfrak{G}(\mathscr{N}_{A_j}) = \mathfrak{G}(\mathscr{N}_C)$ for all i, j and similarly for their associated b-networks. Thus each pair of state spaces is selectable by G4.
3. Suppose $\{\mathscr{N}_{A_i}\}_{i=1}^N$ is a family of subnetworks of a network $\mathscr{N}_C$ and are mutually disjoint. Then for any subset of indices $\bigotimes_i !A_i$ is selectable, $\bigoplus_i A_i$ is implemented, so $\bigotimes_i !A_i = !(\bigoplus_i A_i)$, and $\mathscr{N}_{\bigoplus_i A_i}$ is a subnetwork of $\mathscr{N}_C$.
4. If $\mathscr{N}_A$ and $\mathscr{N}_B$ are not disjoint, then $\mathfrak{G}(\mathscr{N}_A) = \mathfrak{G}(\mathscr{N}_B)$. Thus G4 holds and $A \otimes B$ is selectable. (This is the case regardless of the modalities of the nodes in the intersection.)

Proof. Statement 1 follows immediately from G1.

Statement 2 is immediate from G2 and G3.

For statement 3, it follows from 2 that $\mathfrak{G}(\mathscr{N}^b_{A_{i_1}}) = \mathfrak{G}(\mathscr{N}^b_{A_{i_2}})$ so $!A_{i_1} \otimes !A_{i_2}$ is selectable by G4 and since these are disjoint networks it follows from 1 that G5 holds for $\mathscr{N}_{A_{i_1}}$ and $\mathscr{N}_{A_{i_2}}$. Thus $A_{i_1} \oplus A_{i_2}$ is implemented by $\mathscr{N}_{A_{i_1} \oplus A_{i_2}}$ which is also a subnetwork of $\mathscr{N}_C$ (and $!A_{i_1} \otimes !A_{i_2} = !(A_{i_1} \oplus A_{i_2})$). So $\mathfrak{G}(\mathscr{N}^b_{A_{i_1} \oplus A_{i_2}}) = \mathfrak{G}(\mathscr{N}^b_{A_{i_3}})$ by 2, so by G4 $!(A_{i_1} \oplus A_{i_2}) \otimes !A_{i_3}$ is selectable. Moreover, $\mathscr{N}_{A_{i_1} \oplus A_{i_2}}$ and $\mathscr{N}_{A_{i_3}}$ are disjoint so again G5 holds for them so $(A_{i_1} \oplus A_{i_2}) \oplus A_{i_3}$ is implemented and

$$!(A_{i_1} \oplus A_{i_2}) \otimes !A_{i_3} = !((A_{i_1} \oplus A_{i_2}) \oplus A_{i_3}) \quad \text{or} \tag{6.3.1}$$

$$!A_{i_1} \otimes !A_{i_2} \otimes !A_{i_3} = !(A_{i_1} \oplus A_{i_2} \oplus A_{i_3}) \tag{6.3.2}$$

and so on for all $\mathscr{N}_{A_{i_j}}$.

For statement 4, suppose there exists a node n, say, in both $\mathscr{N}_A$ and $\mathscr{N}_B$. Then if $\mathscr{N}_A \in \mathscr{G}$, $n \in \mathscr{G}$ by G3, where here we consider the node by itself to constitute a trivial subnetwork, so $\mathscr{N}_B \in \mathscr{G}$ also by G3. So $\mathfrak{G}(\mathscr{N}_A) \subseteq \mathfrak{G}(\mathscr{N}_B)$ and similarly vice versa, on interchanging A and B. □

Note that 4 says that if two networks intersect, then the purely Gestalt structure cannot tell them apart, and also that, from G4, they are selectable — i.e. have enough "common features" to be put together via tensoring. This supports the intuition that if they have some nodes in common then they may be more widely connectable via a substrate connection symbolized by $\otimes$.

Although obvious, the following statements are worth encapsulating in a couple of lemmas for later reference.

Lemma 6.3.1. *If $\mathscr{N}_A$ and $\mathscr{N}_B$ are* not *disjoint then:*

1. $\mathfrak{G}(\mathscr{N}_A^b) = \mathfrak{G}(\mathscr{N}_B^b)$ *so G4 holds and $!A \otimes !B$ is selectable (but $A \oplus B$ may not be implemented).*
2. *If the intersection of $\mathscr{N}_A$ and $\mathscr{N}_B$ is fully bimodal then in addition G5 holds so $A \oplus B$ is implemented (by $\mathscr{N}_{A \oplus B}$) and $!A \otimes !B = !(A \oplus B)$. Thus $!A \otimes !B$ is a banged type.*
3. *If the intersection of $\mathscr{N}_A$ and $\mathscr{N}_B$ is not fully bimodal then $A \oplus B$ is not implemented and $!A \otimes !B$ may not be a banged type.*

If $\mathscr{N}_A$ and $\mathscr{N}_B$ are disjoint *then:*

4. *Neither the selectability of $!A \otimes !B$ nor the implementability of $A \oplus B$ can be asserted without further conditions being met. Therefore nor can the bangedness of $!A \otimes !B$ be asserted.*
5. *G4 holds iff G5 holds: then $A \oplus B$ is implemented and $!A \otimes !B = !(A \oplus B)$.*

Proof. All of these assertions are special cases of assertions proved above, except the converse in 5. So, if $\mathscr{N}_A$ and $\mathscr{N}_B$ are disjoint and G5 holds, then each is a subnetwork of the network $\mathscr{N}_{A \oplus B}$ so G4 holds by Proposition 6.3.1, part 2. □

Of course the fundamental identity $!A{\otimes}!B\;=!(A\oplus B)$ is always true for vector spaces, but in our applications to networks, if $A\oplus B$ is not implemented (tacitly as the state space of a specific actual network) then the logical or mathematical machinations involving this identity no longer necessarily apply to our networks and this should be noted. A couple of important consequences of the failure to implement $A\oplus B$ in this way is that the logical type $!A{\otimes}!B$ is no longer necessarily the ! of another type (i.e. banged), and therefore is no longer usable in such applications as the LTP theorems, etc., and that the $\wedge$ combinator may no longer be available among elements of $!A$ and elements of $!B$. Such combinations exist mathematically of course, but may no longer reliably represent firing patterns. This means that in the application of the logic, etc., and in other contexts, $\otimes$ products of firing patterns in $!A$ with firing patterns in $!B$ cannot be reliably reduced to $\wedge$ products. So some care is required in these applications.

We summarize our conclusions in a further lemma.

Lemma 6.3.2. *For any networks $\mathcal{N}_A$ and $\mathcal{N}_B$ it is certain that $!A{\otimes}!B$ is selectable and $A\oplus B$ is implemented, so that the former is the banged version of the latter, if any single one of the following conditions holds:*

1. $\mathcal{N}_A$ and $\mathcal{N}_B$ are not *disjoint and their intersection is fully bimodal.*
2. $\mathcal{N}_A$ and $\mathcal{N}_B$ are *disjoint and either G4 or G5 holds. (If one holds the other will hold.)*
3. *G4 and the $\oplus$-condition hold.*

Thus, certain tensor products of firing patterns may themselves be realized as firing patterns (of b-networks), via the fundamental identity (or logical equivalence). That is to say, certain states in the form of tensors of firing pattern states may themselves be realized as firing patterns of other b-networks (the external direct sums), effected by replacing the $\otimes$s by $\wedge$s. Others, presumably the majority, may not be so realized. At the risk of further increasing our burden of nomenclature, it will be convenient to introduce a term for tensor products of firing patterns that are not thus realizable as firing patterns.

Definition 6.3.4. A tensor product of firing patterns that is *not* realizable as a firing pattern itself via the fundamental identity shall be called a *compound* tensor.

CAVEAT!
An important exception to this will be the inclusion of the vacuum state, or state of no firing, which is the unit of the exterior algebra of firing patterns of all b-networks, to the pantheon of compound tensors. We do this for reasons very similar to the reasons that modern number theorists do not admit the number 1 as a prime: it has an infinite number of factors, never mind that they are all 1 itself. In our case, it represents the state of no firing so is therefore not itself a state of firing. If this seems to be a solipsism, note also that for any firing pattern ξ, $1 \wedge \xi = 1.\xi = \xi$ in the exterior algebra, so nothing changes for $\wedge$ combinations of 1 with firing patterns, while for any tensor Ξ, $1 \otimes \Xi$ is a very different state from Ξ. This unit, which we shall continue to write as 1 in the syntax rules to follow, and representing the vacuum state, will be used as a place holder, to represent potential inputs, or the "holes" left when the components of a tensor fade from memory (cf. Example 10 below). It may be thought to be associated with any Gestalt.

To motivate the choice of these rules for Gestalts, let us note first that we are assuming the membership or otherwise of a network in a Gestalt is likely to be determined by intrinsic features common to its *nodes*, and in the b-network case, to its *output nodes*, since these are the entities which presumably determine the network's distinctive qualities. These considerations motivate the rules G2 – G3, G1 being taken as axiomatic.

G3 is a strong limitation on the nodes of a network. It implies that the membership of a network in a Gestalt depends entirely on the intrinsic nature of the nodes and conversely, as illustrated by the Proposition 6.3.1, statement 2, if we take $\mathcal{N}_{A_i} = n_i^A$ there for any network $\mathcal{N}_A$.

To motivate G4 we return to the Proustian taste of a madeleine. In that scenario we posited a "childhood" Gestalt $\mathscr{G}_{\text{childhood}}$ which corralled various networks whose features were associated in the narrator's mind with his childhood. For instance memories of a person: Aunt Léonie; a beverage: her tea; a plant species: the waterlilies; etc., all evoked, or *selected*, by the taste of the madeleine, a confection. The associated networks must also belong to their own Gestalts, presumably $\mathscr{G}_{\text{people}}$, $\mathscr{G}_{\text{beverages}}$, $\mathscr{G}_{\text{plants}}$, etc., as well as to $\mathscr{G}_{\text{childhood}}$. Presumably, once established in childhood, the association of, say, the taste of a madeleine with Aunt Léonie, must be ever present so that each time a madeleine comes to mind, as perhaps the result of a taste of one, there must be associated inextricably with it the memory of Aunt Léonie, putting the network $\mathscr{N}_{\text{Aunt Léonie}}$ bearing her determining or essential features in any Gestalt that the madeleine's network is in. That is $\mathfrak{G}(\mathscr{N}_{\text{madeleine}}) \subseteq \mathfrak{G}(\mathscr{N}_{\text{Aunt Léonie}})$. This is if, as in the novel, the taste of the madeleine *selects* the memory of Aunt Léonie, and the others. This, we assume, is the mode of selection which limits the choices to be made in the proof fragment implementing pattern completion in section 4.2.1, and hence limits the choices of which state spaces may be selected for tensoring together. (The order counts in a tensor because changing it produces a different state, so that it would not be the case that a reminder of Aunt Léonie would be the selective agency, in which case we would not have $\mathfrak{G}(\mathscr{N}_{\text{Aunt Léonie}}) \subseteq \mathfrak{G}(\mathscr{N}_{\text{madeleine}})$, so that the inclusions that count for unambiguous tensoring should ideally be strict. Unfortunately they are often an equality and therefore the choice of order is ambiguous. Later we shall disambiguate this, but only by fiat.) And similarly in general, at least for each pair. The general case of more $\otimes$ products may require more constraints to be met.

The rule G5 is self-explanatory in view of the discussion in section 2.2.4.

The discussion of Tsien's Theory of Connectivity in Chapter 5 also demonstrates some of these rules. For instance, we have assumed there that each component is a subnetwork of a single network and are mutually disjoint, thus Proposition 6.3.1, part 3 applies and the fundamental identity, etc., may be used (though the Tsien Power-of-two Law does not depend on it).

Our very first use of the fundamental identity in section 2.3.1 is also now retroactively justified by Proposition 6.3.1, part 3, since there the disjoint subnetworks of a given unicameral network are the single nodes of the network itself. (Here, $(n_i^A)^b$, as in $\mathfrak{G}((n_i^A)^b) = \mathfrak{G}(n_i^A)$, is just the associated b-neuron itself, considered as a trivial subnetwork of $\mathcal{N}_A$.)

Returning to the context of pattern completion as in section 4.2.1, we must now impose restrictions required by the Gestalt rules. The generalization is that the A_i networks there are required to have their Gestalts $\mathfrak{G}(\mathcal{N}_{A_i})$ lying in a *chain* relative to the partial order $\subseteq$. That is to say, for each pair i, j we have that $\mathfrak{G}(\mathcal{N}_{A_i})$ and $\mathfrak{G}(\mathcal{N}_{A_j})$ are inclusional. Then any tensor product of the spaces $!A_i$ already so selected and tensored together will inherit the largest set of Gestalts which will be that associated with the rightmost member of the product. The next one selected will have a $\mathfrak{G}$-set either larger than this current larger one or smaller than it. If it is smaller it may be selected and tensored on the left of the current product; if it is larger it may be tensored on the right, and its $\mathfrak{G}$-set now replaces the previous one. The resulting tensor product, which inherits the largest $\mathfrak{G}$-set, may be rearranged in any order up to isomorphism of vector spaces. However, differently ordered tensors actually represent different states, despite this ambient isomorphism of spaces, and these different states may be alternatives that should be superposed. This will have consequences later: cf. section 6.4.1.

Our definition of this notion of Gestalt is necessarily rather vague in practice, since it must depend on unknown factors which probably vary from individual to individual. We shall adopt the view that their presence nevertheless rises to consciousness in the sense that we are aware of such Gestalts as those associated with childhood, personhood, and others such as those associated with the sensorium, fear, etc. Moreover, such Gestalts may vary in time, with variable and plastic boundaries. Their passage to consciousness may be evinced by a process akin to the heightened responsiveness of the most general part of the barcode in the Tsien paradigm of memory acquisition.

Thus in the example of the fear Gestalt in Chapter 5, the most responsive networks are those that are the most general, advertising the activation of members of the fear Gestalt. The boundaries of such a Gestalt are not fixed since, for example, new fearsome experiences may be encountered in the future.

Note that for a network $\mathcal{N}_A$ the generating Gestalt $\mathcal{S}_A$ corresponds to the microscopic or cellular level barcode in the Tsien paradigm. Our postulated proper Gestalts may thus be regarded as the next step in generality beyond the most general covered by the full clique of the cellular barcode.

It is possible to envisage collections of Gestalts with finer distinctions among them, but we shall venture no further along this limb. Although we may be conscious in a vague way of Gestalts, we do not seem to be conscious of the networks *per se*. We do, however, seem to be conscious of the firing of subsets of networks, as many decades of experiment and observation have revealed. So we must go deeper in seeking the syntactical atoms.

6.4   Cognitive syntactic atoms and molecules

As noted earlier, the coins of our realm are *firing patterns*, at particular times, of b-networks. These are (possible) states of the network which are superpositions of *basic* firing patterns, namely the states of its (possible) subsets of co-firing b-neurons at the time in question. The coefficients in the superposition relate to the probabilities that the associated co-firing subset will be "observed" when the network is in that superpositional state or firing pattern. How is this "observation" made and who or what makes it? There are as many answers to this question as there are episystems of the network. One of them may be an NMR device recording an fMRI of the network *in vivo*. Might another be the "consciousness" of the owner of the network? Some firing patterns are deemed to be of *cognitive significance*: namely, those that are eigenstates of the dynamics, presumably because of their temporal stability. And presumably, if they are thereby raised to conscious awareness, they may do so while still in a superpositional form, since eigenstates often have degeneracies. Such a firing pattern would appear as a multiplexed mixture of states of different basic co-firing subsets of the network, and indeed of many networks, depending on the timing and nature of the coefficients/probabilities in the superposition, and of the observation, such as an fMRI. Or such a state may be the basic firing pattern of a single subset, i.e. not a superposition of more than one basic firing pattern. Since we must ascribe cognitive significance to superpositions (since they can appear as eigenstates), the minimal syntactic atom should be the general superpositional firing pattern. Of course there is a mathematically more primitive syntactic atom, namely the basic firing pattern, but there seems to be no way of distinguishing cognitively between a general superpositional firing pattern and a basic one. Consequently we must adopt the general superpositional firing patterns as our *cognitive* syntactic atoms, if only provisionally.

At this point we are driven onto the horns of a dilemma produced by our overarching doctrine of hiding the variables. For, ordinary theories of syntax are in a sense classical, in that the atoms or units are not generally *states*, and when they are, they are not generally quantum-like. The issue is how to treat the non-classical combinator of superposition, which in our formulation reduces to the internal summation of states $(+)$ or the external direct sum $(\oplus)$ of states, when this is appropriate. We shall adopt our logical interpretation of superposition as a form of disjunction and apply it to situations that are ambiguous in the sense that the cases involved in the superposition are all *possible* with certain probabilities reflected in the coefficients. We remain generally ignorant of these coefficients since they are among the variables we are opting to hide: which is to say, they are beyond the reach of our Syntax Processor. (In the case of linguistic syntax at least these hidden variables are likely to arise from unknowable external sources, such as grammatical parsing, speech processing, etc., rather than the inner cellular ones of our earlier approach.)

Thus, we assume there is available a population of firing patterns retrievable by memory processes, which constitute our cognitive syntactic atoms. These cannot be "cognitively" reduced further, although they are themselves superpositions of basic units.

We are left with $\otimes$ and $\wedge$ with which to syntactically combine our cognitive atoms, positing superposition when it is appropriate.

(The problem of ascertaining the dividing line between the atomic unit and the next higher compound unit, or molecule, in a syntactic hierarchy is a problem also in the linguistic context, perhaps for the same reason.)

We note that a firing pattern by itself contains no information about the connectivity structure of any ambient network. Thus, in endowing such a state with cognitive significance, we are tacitly assuming that it is only the actual firing of the neurons involved that is registered cognitively and not the details of the connectivity that may have given rise to the state. Thus the neurons fired by the thought of Aunt Léonie may also fire at the taste of her tea cake, though the particular connections (wired or wireless) may effect such a firing differently. (This is inherent in the implementation of the sequent calculus **GN**, in which the maps or wiring diagrams represented by the sequents may change with each iteration.)

Finally, let us note that the apparently necessary adoption of firing patterns as syntactic units is in keeping with our doctrine of hiding variables. For, the number of neurons involved in a typical firing pattern is very large indeed, and merely listing all the basic firing patterns appearing in an expansion such as that found on the right hand side of equation (6.4.2) below might take up a whole lifetime or more.

6.4.1 *Syntax for firing patterns and tensors*

We shall continue to use lowercase Greeks to denote firing patterns other than the vacuum 1 (which is not a firing pattern, and will count as a "compound" tensor for our purposes here) but shall generally drop the reference to time: its presence should be kept in mind, however. As noted we will be interested in the firing patterns that are ON, which is to say not entirely OFF: that is, those of the form

$$\xi_A = \sum_{p \geqslant 0} \sum_{i_1 < \cdots < i_p} \alpha_{i_1 \ldots i_p} e_{i_1}^A \wedge \cdots \wedge e_{i_p}^A \tag{6.4.1}$$

$$= \alpha_{i_0} 1 + \sum_{i_1} \alpha_{i_1} e_{i_1}^A + \sum_{i_1 < i_2} \alpha_{i_1 i_2} e_{i_1}^A \wedge e_{i_2}^A + \cdots \tag{6.4.2}$$

in which not all the $\alpha_{i_1 \ldots i_p}$, for $p > 0$, are zero, and which, by assumption, belongs to the state space of at least one b-network, $\mathcal{N}_A$ say. As noted, such a firing pattern may belong to other networks. That is to say, the same nodes n_i^A whose basic states appear in ξ_A may also belong to a b-network $\mathcal{N}_B$, in which they will be labeled n_i^B, with a corresponding expression

$$\xi_B = \sum_{p \geqslant 0} \sum_{i_1 < \cdots < i_p} \alpha_{i_1 \ldots i_p} e_{i_1}^B \wedge \cdots \wedge e_{i_p}^B. \qquad (6.4.3)$$

Clearly, both ξ_A and ξ_B are merely labels for the same firing pattern, so we shall identify them and just write ξ. This will be understood when we say that ξ "belongs" to the two or more networks.

Thus, such a shared firing pattern comes with two levels of categorization: namely, the family of networks whose state spaces it may belong to, and the families of Gestalts these networks may belong to. For the latter we note that for a given ξ all the networks that it belongs to have the nodes appearing in ξ in common, therefore they all have a non-empty intersection and therefore by Proposition 6.3.1, part 4, they share Gestalts. It will prove convenient to invent names and notations for these categories, but first we note the following. For a firing pattern ξ, denote by $[\xi]$ the Hilbert space spanned by the basic nodal states appearing in ξ. Let $\mathcal{N}_{[\xi]}$ denote the trivial network whose nodes are the nodes whose basic states are the ones appearing in ξ. Then the state space of $\mathcal{N}_{[\xi]}^b$ is $![\xi]$ and $\xi \in ![\xi]$. Thus $[\xi]$ would be the smallest state space containing ξ. This definition is trivially self-defining in the same sense that the Gestalt of subnetworks of a network is, and we similarly disallow such self-generated state spaces from our population of them. The state spaces that are not of this form we accordingly call *proper*.

Our productions, or syntactic molecules, will be (well-formed) tensor products of atoms, namely firing patterns. These atoms are also tensors, but it will be convenient to distinguish the atoms from the molecules and we shall continue to do so. Thus the term *tensor* will generally refer to a tensor product of more than one atom, though it will be convenient to include atoms among tensors: we shall be careful to advertise this when it is done.

Definition 6.4.1. For an atom ξ:

1. $\mathscr{E}(\xi) := \{A \; proper : \xi \in !A\}$. We shall call this set the set of *essential features* or *e-features* of ξ. It is not empty by assumption. Here A is assumed to denote the state space of the output nodes of a b-network;
2. For any $\mathscr{N}_A^b$ that ξ belongs to, let $\mathfrak{g}(\xi) := \mathfrak{G}(\mathscr{N}_A^b)$. (As noted, this is the same set for all such $\mathscr{N}_A^b$.) We shall call this set the set of *global features* or *g-features* of ξ.
3. Note that it follows from G4 that for atoms ξ and η, if their tensor product is well-formed, $\mathfrak{g}(\xi \otimes \eta) = \mathfrak{g}(\eta)$. Thus for compound tensors Ξ and Θ, if their tensor product is well-formed, we recursively define $\mathfrak{g}(\Xi \otimes \Theta) := \mathfrak{g}(\Theta)$. The e-features of a compound tensor are not defined.

The conditions for composing firing patterns now follow with the proviso that we shall superpose ambiguous productions as appropriate.

Lemma 6.4.1. *For firing patterns ξ and η:*

1. *If the g-features of ξ and the g-features of η are inclusional then the tensor product of ξ and η may be formed in the order specified by the inclusion if the inclusion is strict. If the inclusion is an equality, then the order is ambiguous;*
2. *If the e-features of ξ and the e-features of η are inclusional then $\xi \wedge \eta$ may be formed. In this case the order is immaterial, and the inclusion (of e-features) need not be strict. The e-features of $\xi \wedge \eta$ are those of the* smaller *inclusion, namely those spaces containing both ξ and η;*
3. *If the last condition holds, then $\mathfrak{g}(\xi) = \mathfrak{g}(\eta)$ so the order of the tensor product of ξ with η becomes ambiguous, though it may be*

discarded in favor of the exterior product. Since the exterior product disambiguates, it should take precedence over the tensor product.

Proof.

1. Suppose $\mathfrak{g}(\xi) \subset \mathfrak{g}(\eta)$. Then by hypothesis there exists A such that $\xi \in !A$, so this inclusion implies that if $\mathscr{G} \in \mathfrak{G}(\mathscr{N}_A^b)$ $(= \mathfrak{g}(\xi))$ then $\mathscr{G} \in \mathfrak{g}(\eta) = \mathfrak{G}(\mathscr{N}_B^b)$ for some B with $\eta \in !B$. Thus $\xi \otimes \eta \in !A \otimes !B$ may be formed since the latter space is selectable by G4. A similar argument applies if $\mathfrak{g}(\eta) \subset \mathfrak{g}(\xi)$.

2. Note that from Lemma 6.3.2 the attribute common to firing patterns ξ, η having a well-formed exterior product $\xi \wedge \eta$ is that there is an ambient (proper, implementable) state space that they both belong to. (Mathematically, any two vector spaces are always subspaces of a canonical space, namely the external direct sum of the two spaces, but as we have seen this construct may not always be implementable, and therefore not always available to us.) Moreover, this must apply to *all* such ambient spaces since otherwise we would have the toxic ambiguity that $\xi \wedge \eta$ is sometimes well-formed and sometimes not. Thus we should have either $\mathscr{E}(\xi) \subseteq \mathscr{E}(\eta)$ or $\mathscr{E}(\eta) \subseteq \mathscr{E}(\xi)$ to guarantee an unambiguous well-formedness of $\xi \wedge \eta$. The order in this product is immaterial since $\xi \wedge \eta$ and $\eta \wedge \xi = -\xi \wedge \eta$ define the same state or firing pattern.

3. This is self-explanatory. □

This lemma prescribes how our atoms may be combined. It is a misfortune of nomenclature that our atoms may sometimes combine (via $\wedge$) to form other atoms (i.e. firing patterns). (There is a similar problem in linguistic syntax, as we have noted. Morphemes, which are supposed to be "atoms" of words, may be difficult to distinguish from whole words, as words may be difficult to distinguish from clauses or other so-called constituents.)

Another point to note is that in the instances of strict inclusionality of g-features, the larger set determines the corresponding g-features of the combination. This consists well with the idea that

Gestalts are generalizations of Tsien's FCMs. In the latter, as we saw in the last chapter, it is the most general cliques that are the most responsive.

We are almost in a position to posit tentative syntactic rules for tensors. However, there are some provisos that apply in our case, that are generally absent in the extant cases of syntax, for example, linguistic syntax. To wit:

- There will be many possible ambiguous productions to be super-posed, and such superposition may not always be possible. We assume that in reality there are criteria outside the scope of our model that would cut down the number of possibilities that may exist. In the linguistic case, these would include for instance rules of semantics.
- One of these ambiguities shall be laid to (uneasy) rest as part of our syntax rules. Namely, the ambiguity of the order of tensor products in case the sets of g-features are the same. This generally results in local mirror images of the structures involved and we shall assume a particular bias in our system, namely the one espoused in G4, which we shall refer to as the *left hand rule* or LHR. We resort to this mainly to avoid a treatment of a huge proliferation of alternative productions which have the same structural properties of interest.
- In the cases of other ambiguities we shall regard them as leading to possible branches of production and treat them separately until some sort of resolution is achieved at the end of each branch, at which point which they will be superposed if different.

CAVEAT!
It is important to note that there are effectively two types of ambiguity in connection with tensor products. There is firstly the ambiguity of the order of the components within a tensor product if $\mathfrak{g}(\xi) = \mathfrak{g}(\eta)$: this is ambiguous vis-à-vis G4, and it is this ambiguity we are suppressing by fiat by adopting the LHR. (This applies also to compound tensors.) Doing otherwise, and maintaining the ambiguous cases, would produce families of binary trees which are

laterally symmetric with respect to sister interchange. Although in our formulation these trees represent different states, they are not different in a certain hierarchical structural sense, which reflects the vertical tree structure, and which is the attribute of interest in syntactical considerations. To wit, given a tree we shall assume that the set of trees with sisters variously interchanged will appear with the same probability as the initial tree: the differences between these trees are not relevant to the syntactic process. It is only the change in the hierarchical or vertical structure that is of syntactic relevance, and all of these cases will be covered if only a one handed rule is adopted. We have arbitrarily chosen the LHR. (In the linguistic context, there are languages which seem to adhere to the LHR, others that adhere to a similar right hand rule, and some that use both.) The other ambiguity occurs in the presence of the LHR and is what may be called a *hierarchical, structural* or *derivational* ambiguity which would be reflected in multiple strict inclusions of the same of g-features of a tensor, as in Examples 6, 9 and 10 below. The ambiguity here lies in the *order of derivation*, clearly of prime syntactic importance. It gives rise to different structural possibilities which change the trees vertically. If the ambiguity of order arises in this derivational/hierarchical manner then we treat each branch separately, superposing what conclusions emerge. The (provisional) intuition here is that the different *structurally derived* states dominate the merely mirrored ones in their contributions to a final superposition. This is consistent with the general run of linguistic syntax which similarly generally concerns itself with such "vertical," or hierarchical, structural possibilities, rather than mirrored versions, although it must be admitted that the rules here are not hard and fast [91].

We shall introduce (in rule S3) a perhaps not too artificial addition to this treatment of the second form of ambiguity for $\otimes$ products, namely we shall give precedence to actions which reduce ambiguity. Thus in cases in which we have an ambiguous choice between inclusionality of sets of g-features that are equalities and inclusionalities that are not, we give precedence to the latter (Examples 3 and 10 below).

Some further notation is needed, namely one that displays the list of features of the atom or tensor involved. For atoms ξ we shall attach unordered lists in square brackets, first of the e-features and then of the g-features, as in

$$\xi[A_1^\xi, \ldots, A_n^\xi; \mathscr{G}_1^\xi, \ldots, \mathscr{G}_m^\xi], \tag{6.4.4}$$

where $\mathscr{E}(\xi) = \{A_i^\xi\}$ and $\mathfrak{g}(\xi) = \{\mathscr{G}_j^\xi\}$, and for compound tensors Ξ as in

$$\Xi[\mathscr{G}_1^\Xi, \ldots, \mathscr{G}_k^\Xi], \tag{6.4.5}$$

where $\mathfrak{g}(\Xi) = \{\mathscr{G}_i^\Xi\}$.

(Note that in equation (6.4.4) *every* A_i^ξ is in *every* $\mathscr{G}_j^\xi$.)

Though indexed, these feature lists are meant to be unordered, merely listing the relevant respective set elements. We shall sometimes later omit these feature lists, in part or in whole, if they become clear from the context. Some care must be exercised in the atomic case since the two sets of lists are not independent of each other, as we shall see shortly.

A point to note here is that the features we have attributed to these tensors in the square brackets have to be attached engrammatically to them in some way, thus heavily loading the memory capacities involved. This is where multimodality may be selectively advantageous in its attributes of efficiency.

We shall posit the rules for a full recursive syntax first, since this is presumed to reflect the normal run of human processing, and discuss the possible modes of the failure of recursion later (sections 6.5.2 and 6.5.3).

Recursive syntax rules for tensors and atoms

For the convenience of the reader we repeat the syntax circuit diagram Figure 6.1 below:

```
┌──────┐      ┌─────────────┐        ┌──────────────────┐
│ Out  │ ◄─── │ Interpreter │ ◄───── │ Syntax Processor │
└──────┘      └─────────────┘        └──────────────────┘
                     │                         ▲
                     ▼                         │
          ┌───────────────────┐      ┌──────────────────┐      ┌──────┐
          │ Short Term Memory │ ───► │ Syntactic Store  │ ◄─── │  In  │
          └───────────────────┘      └──────────────────┘      └──────┘
```

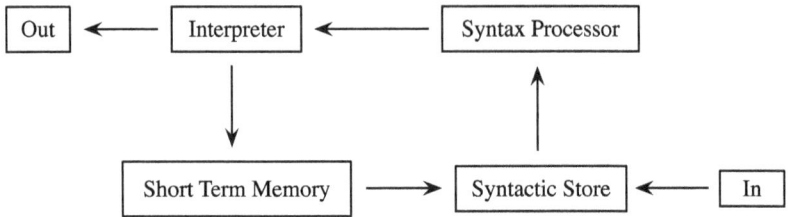

The recursive step is built into assumptions concerning the input syntactic units, which allows for some already processed output to be recycled: these are compound tensors, "stored" in the engrammatic sense, in the Syntactic Store. Thus the input at a particular moment is a finite sequence of compound tensors and/or atoms in which the order does not count, copied into the Syntax Processor via the rightmost up arrow in the diagram according to some protocol which may select recent impermanent additions from the associated STM (bottom arrow). Since repeats are allowed, such sequences are not quite sets: they are usually called *multisets*. These multisets, and those to follow, will be of variable length.

S1. Retrieve (into Syntax Processor) the current multiset from Syntactic Store;

S2. Repeatedly execute all possible $\wedge$ products between pairs of any atoms in the current multiset, each time separately removing the used atoms involved from the current multiset while adding the new atom, unless zero results: in this case do nothing (zero is not an atom). For definiteness, let us insert the new entry at the end of the old multiset. Multiple alternative new multisets may form. This is an ambiguity so they are to be treated separately. (For atoms ξ and η: if the e-features of ξ and η are inclusional then the exterior product $\xi \wedge \eta$ may be formed, the product order being immaterial, and this product takes precedence over $\otimes$. In this case $\mathfrak{g}(\xi) = \mathfrak{g}(\eta) = \mathfrak{g}(\xi \wedge \eta)$. The e-features of $\xi \wedge \eta$ are those of the smaller inclusion.);

S3. For each resulting multiset, repeatedly execute all possible tensor products of pairs of entries when these entries are at different positions in the current multiset. Remove the used tensors from that multiset while adding the new tensor, again at the end, for definiteness. Ambiguities are to be treated as before to produce separate streams of multisets. (For tensors and/or atoms Ξ and Θ: if the g-features of Ξ and Θ are inclusional then the $\otimes$ product of Ξ and Θ may be formed, for simplicity, in the order dictated by the direction of the inclusion, and the g-features assigned to it are those of the larger as in G4. In case of derivational ambiguity, in which there are strict inclusionalities, these get preference over equalities. If the choices are all equalities all streams must be followed.);

S4. If a stream of such multisets fails to terminate in a single tensor or atom, that stream is suspended pending further input, or eventually discarded. The tensors and/or atoms remaining at the ends of the successfully terminating streams are copied to the Interpreter, where they are superposed. The resulting tensor is processed by the Interpreter and may provoke some action via the Out interface (such as vocalization, or speech in the linguistic case) and may or may not be deemed by the Interpreter to be worthy of being copied by the Interpreter to the STM. Either or both may happen. The whole process may or may not terminate at this point. If it has not terminated, go to S5.

S5. All streams may fail to terminate in a single tensor and the final multiset or sets of multisets may remain suspended indefinitely or discarded as above. We are not assuming that this necessarily entails termination. The Syntax Processor may still retrieve a new input multiset, which may or may not include recent productions returned from the Interpreter, overwriting the previous one. Therefore, at this step, go to S1.

As noted, we are not assuming the entire process stops at the failures mentioned above: in nature it may go on, sometimes leading to what might appear in humans and other animals to be maladies. These will be discussed later. At the moment we must leave the process in abeyance if these failures occur.

Even if these failures do not occur, the process may not terminate, since the input multiset may not decrease with each iteration (as in Example 4 below). This is an aspect of this kind of recursion. For instance, in linguistics, sentences in a recursive language may theoretically go on forever (and often seem to). This is a feature, not a bug. In these cases we can only offer some sort of extraneous stopping rule, such as a time-out. Or, as in standard minimalist linguistic syntax, a countdown. On the other hand our process does not have an obvious termination criterion, and this is possibly a drawback, or possibly not. In nature there seems to be nothing obviously holding back runaway mental catastrophes.

In S2, we amalgamate atoms via $\wedge$ first because this product takes precedence over $\otimes$ when it can be formed. Then we remove the amalgamated atoms from the current multiset because they have been "used up" or syntactically processed.

Similar processing occurs at S3 with $\otimes$. It is important to note that, at S3, the internal structure of compound tensors become inviolable: the $\otimes$ combinator cannot "look inside" them. Also, note that although S3 does not allow in-place self-tensor products, which would lead to infinite streams of multisets, finite repetitions are of course not denied to the input multisets. Any finite repetition of an atom in a multiset will survive the ravages of S2 (giving zero) to be replaced at S3 by a repeated tensor product.

The step S4 we assume is achieved by the Interpreter or its Out interface for uses beyond the boundaries of our model: integration, parsing, action, speech, etc. We shall also assume that superposition, when there is ambiguity, is performed in/by the Interpreter and is always possible. For instance, under the reasonable assumption that it is an integral network, the faithful copies it holds of the syntactic productions will all be states of the same network and therefore superposable. We have argued that superposition — $+$ and/or $\oplus$ — should be denied our Syntax Processor: it must be done elsewhere. Once superposition of these ambiguous syntactic units are formed, they will collapse upon retrieval, which event we assume to occur at least as a result of transmitting the superposed state through the arrows/tracts out of the Interpreter. Thus, a recycled version of the superposed ambiguous states may land up in the Store as one of the alternative component productions with a certain probability.

And thence, unambiguously to the Syntax Processor, if retrieved. The upshot of this assumption will be explored later: cf. Examples 6, 9 and 10. (Since we are assuming that that the collapse of such superpositions occur along these tracts and not the rightmost up arrow/tract, damage to the former tracts would presumably have detectable effects: cf. section 6.5.1).

The boundaries of our syntax rules are determined by the boundaries of our model circuit diagram above: the input at one boundary, the output at the other. The last act, of possible superposition, is perched on the boundary itself.

These rules seem to be the simplest available given our setup: first amalgamate the atoms, then combine the results in every way allowed by the naturally arising combinatory rules for $\otimes$, then superpose the alternate states.

6.4.2 *Examples*

We shall consider first a series of 10 template-like examples, later concretizing some of them, and one concrete example, namely Example 11.

EXAMPLE 1
S1. Input multiset:

$$\xi[A; \mathscr{G}_1], \ \eta[B; \mathscr{G}_1, \mathscr{G}_2]. \tag{6.4.6}$$

S2. No change since the e-features are not inclusional;

S3. Since $\mathfrak{g}(\xi) = \{\mathscr{G}_1\} \subset \{\mathscr{G}_1, \mathscr{G}_2\} = \mathfrak{g}(\eta)$ we have the tensor

$$(\xi \otimes \eta)[\mathscr{G}_1, \mathscr{G}_2] \tag{6.4.7}$$

as the successful conclusion of the only stream. (We emphasize in this first example that we are applying the LHR by fiat, and that the other order of tensoring is being omitted not only as a labor-saving maneuver but also because it is not hierarchically or structurally different from the one selected. This comment will be tacitly assumed in further examples of our application of the LHR.)

S4. Final tensor copied to the Interpreter. Superposition not needed.

S5. Termination assumed.

EXAMPLE 2
S1. Input multiset:

$$\xi[A; \mathscr{G}_1], \ \eta[B; \mathscr{G}_1, \mathscr{G}_2], \ \zeta[A; \mathscr{G}_1]. \tag{6.4.8}$$

S2. Since $\mathscr{E}(\xi) = \mathscr{E}(\zeta)$ we can form $(\xi \wedge \zeta)[A; \mathscr{G}_1]$ to obtain the next multiset

$$\eta[B; \mathscr{G}_1, \mathscr{G}_2], \ (\xi \wedge \zeta)[A; \mathscr{G}_1]. \tag{6.4.9}$$

S3. Since $\mathfrak{g}(\xi \wedge \zeta) = \{\mathscr{G}_1\} \subset \mathfrak{g}(\eta) = \{\mathscr{G}_1, \mathscr{G}_2\}$ we can form, as the successful conclusion of the only stream, the tensor

$$((\xi \wedge \zeta) \otimes \eta)[\mathscr{G}_1, \mathscr{G}_2]. \tag{6.4.10}$$

Comparing this to the result of Example 1 we notice that when an atom with exactly the same features (namely ζ) as one already present (namely ξ) is added, indicating a type of plurality, the result changes by the insertion of the new similarly featured atom into the middle of the tensor concluding Example 1.
S4 and S5 as above.

EXAMPLE 3
It is worth considering what happens when the last atom does not merely have the same features as the first one but is actually identical to it. Namely:

S1. Input multiset:

$$\xi[A; \mathscr{G}_1], \ \eta[B; \mathscr{G}_1, \mathscr{G}_2], \ \xi[A; \mathscr{G}_1]. \tag{6.4.11}$$

S2. No change since $\xi \wedge \xi = 0$.
At the next step we have an ambiguity because we can have $\mathfrak{g}(\xi) \subset \mathfrak{g}(\eta)$ *and* $\mathfrak{g}(\xi) = \mathfrak{g}(\xi)$. In this case the second option is an equality of g-features so we give precedence to the first.

S3. $(\xi \otimes \eta)[\mathscr{G}_1, \mathscr{G}_2]$ can be formed so the next multiset for this stream is

$$\xi[A; \mathscr{G}_1], \quad (\xi \otimes \eta)[\mathscr{G}_1, \mathscr{G}_2]. \tag{6.4.12}$$

Now $\mathfrak{g}(\xi) \subset \mathfrak{g}(\xi \otimes \eta)$ so we can form

$$(\xi \otimes (\xi \otimes \eta)[\mathscr{G}_1, \mathscr{G}_2] \tag{6.4.13}$$

which successfully concludes this cycle. S4 is not needed and we again assume termination so S5 is also not needed.

(Note, as an exercise, that if we had gone ahead and processed a second stream with $\mathfrak{g}(\xi) = \mathfrak{g}(\xi)$ instead of $\mathfrak{g}(\xi) \subset \mathfrak{g}(\eta)$ we would have arrived at the same conclusion and superposition would have been fruitless in this case.)

EXAMPLE 4

The mere repetition of a syntactic unit may be easily implemented without using the recursive part of the circuit: that is to say, the lower left L-shaped portion. For instance, if any part of it is *damaged*. The unit may be stored permanently in the Syntactic Store (as in a *lexicon*), non-destructively retrieved by the Syntax Processor, copied to the Interpreter and then to Out without being copied back to the Store, and this may be repeated. If, as in a special case, the processing of Ξ by Out entails a vocalization, such as *Woof!*, then the result may be repeated without any recursion being necessary. If the recursive part is damaged, or absent, the unit's repetition may be the *only* effect possible. (Psychological readers may replace *Woof!* with *Tan!* in this example: cf. section 6.5.1).

On the other hand, a cumulative sort of repetition, such as a count of some kind, seems to require recursion. The following example illustrates this.

Here we shall assume that the initial input Ξ is lodged, as above, in long term (syntactic) storage, and remains there after initial retrieval.

S1. Input multiset:

$$\Xi. \tag{6.4.14}$$

S2. No change.

S3. No change.

S4. Ξ is copied to the Interpreter and possibly processed to activate Out. (In the case of language, this may lead to speech production and an utterance: perhaps *Kitty!*). At the same time a copy is transmitted to the STM part of the Syntactic Store.

S5. Non-termination. Go to S1.
New Input is retrieved from the Syntactic Store where the initial tensor has remained after the initial non-destructive copying to the Syntax Processor.

S1. So the new input multiset is:

$$\Xi, \Xi \tag{6.4.15}$$

(the first Ξ from the STM, the second from the long term part of the Syntactic Store.)

S2. No change.

S3. $\Xi \otimes \Xi$ may be formed and the next multiset is just

$$\Xi \otimes \Xi \tag{6.4.16}$$

itself.

S4. $\Xi \otimes \Xi$ is transmitted by the Interpreter to the STM part of the Store, which replaces the previous copy of Ξ on the top of the stack, where it joins the long term copy of Ξ, and is possibly processed to activate Out. (*Kitty! Kitty!*)

S5. Non-termination. Go to S1.

S1. New Input multiset:

$$\Xi \otimes \Xi, \Xi \tag{6.4.17}$$

S2. No change.

S3. $\Xi \otimes \Xi \otimes \Xi$ may be formed and the next multiset is just

$$\Xi \otimes \Xi \otimes \Xi \tag{6.4.18}$$

itself.

S4. $\Xi \otimes \Xi \otimes \Xi$ is transmitted by the Interpreter to the STM part of the Store, which replaces the previous copy of $\Xi \otimes \Xi$ on the top of the stack, where it joins the long term copy of Ξ, and is possibly processed to activate Out. (*Kitty! Kitty! Kitty!*)

S5. Non-termination . . .

And so on. If, in the case of language, the Out output produces speech, this may produce repeated utterances, as shown, and continues until some event interrupts the process. For instance, the long term Ξ may die out. Various other failures might occur and we shall consider these separately later.

EXAMPLE 5
S1. Input multiset:

$$\xi[A; \mathscr{G}_1, \mathscr{G}_2], \quad \eta[A, B; \mathscr{G}_1, \mathscr{G}_2]. \tag{6.4.19}$$

Here we note that because $\mathscr{N}_A^b$ and $\mathscr{N}_B^b$ both contain the nodes appearing in η we must have $\mathfrak{g}(\eta) = \mathfrak{g}(\mathscr{N}_A) = \mathfrak{g}(\mathscr{N}_B) = \mathfrak{g}(\xi)$, from Proposition 6.3.1, part 4.

S2. $\mathscr{E}(\xi) = \{A\} \subset \{A, B\} = \mathscr{E}(\eta)$ so we can form the atom

$$(\xi \wedge \eta)[A; \mathscr{G}_1, \mathscr{G}_2] \tag{6.4.20}$$

as our successful conclusion to the only stream.

S3. No change.

S4 and termination assumed.

EXAMPLE 6
S1. Input multiset:

$$\xi[A; \mathscr{G}_1, \mathscr{G}_2], \quad \eta[A, B; \mathscr{G}_1, \mathscr{G}_2], \quad \zeta[B; \mathscr{G}_1, \mathscr{G}_2]. \tag{6.4.21}$$

Here again ζ must share the g-features of the first two atoms because its e-feature is B.

In this example we have $\mathscr{E}(\xi) \subset \mathscr{E}(\eta)$ *and* $\mathscr{E}(\zeta) \subset \mathscr{E}(\eta)$ so we have a bifurcation at S2 into two streams that must be treated separately.

Stream 1

S2. $(\xi \wedge \eta)[A; \mathcal{G}_1, \mathcal{G}_2]$ can be formed producing the next multiset for this stream:

$$\zeta[B; \mathcal{G}_1, \mathcal{G}_2], \quad (\xi \wedge \eta)[A; \mathcal{G}_1, \mathcal{G}_2]. \qquad (6.4.22)$$

S3. Since all the g-features are shared the tensor

$$(\zeta \otimes (\xi \wedge \eta))[\mathcal{G}_1, \mathcal{G}_2] \qquad (6.4.23)$$

can now be formed successfully concluding Stream 1.

Stream 2

S2. Now $(\zeta \wedge \eta)[B; \mathcal{G}_1, \mathcal{G}_2]$ can be formed producing the next multiset for this stream:

$$\xi[A; \mathcal{G}_1, \mathcal{G}_2], \quad (\zeta \wedge \eta)[B; \mathcal{G}_1, \mathcal{G}_2]. \qquad (6.4.24)$$

S3. This yields the successful conclusion of Stream 2, namely the tensor

$$(\xi \otimes (\zeta \wedge \eta))[\mathcal{G}_1, \mathcal{G}_2]. \qquad (6.4.25)$$

S4. The Interpreter superposes $\zeta \otimes (\xi \wedge \eta)$ and $\xi \otimes (\zeta \wedge \eta)$ and may copy a collapsed result to the Syntactic Store and/or pass it on to the Out interface.

Unfortunately we do not know the coefficients/probabilities involved, so all we can say is that these two states may manifest with certain probabilities which may change in time. It might appear that one state manifests at some times and the other at other times.

It will prove illuminating to express these two states in terms of binary parse trees, dropping the common g-feature list, as follows:

$$\zeta \otimes (\xi \wedge \eta) \qquad (6.4.26)$$

and

$$\xi \otimes (\zeta \wedge \eta) \qquad (6.4.27)$$

(Note that in this case there is indeed a structural, vertical, difference between the two final states/trees.)

Superposition of the root states means, very roughly speaking, that sometimes one tree manifests and sometimes the other one manifests, with varying probabilities which are beyond the remit of our model. These could range from one tree obtaining and the other one not, to the alternation between one and the other at frequencies necessarily not given by our model.

We note that the lower left leaf of the first tree, has, in the second tree, changed places with the upper left leaf position (and *vice versa*).

Such (collapsed) manifestations in our model of the syntax circuit, as noted, may land up in the Store if recycled. There they appear as one of the alternative trees, in the manner described. Thus the Syntax Processor will, upon retrieval, retrieve one tree or the other with the probabilities determined by the coefficients of the superposition. So sometimes one tree may be processed and sometimes the other. This ultimately will give rise to what, in the linguistic case, is called *movement*. This will be discussed further: cf. Example 10.

S5. We again assume termination so this is not needed.

EXAMPLE 7
S1. Input multiset:

$$\Xi_1[\mathscr{G}_1], \ \eta_1[A, B; \mathscr{G}_1, \mathscr{G}_2], \ \zeta_1[B; \mathscr{G}_1, \mathscr{G}_2]. \qquad (6.4.28)$$

This example is similar to the last one, except that the initial multiset contains a compound tensor, possibly recently recycled to the Store, and the entries bear subscripts, to be explained in the next example.

S2. We can form $(\eta_1 \wedge \zeta_1)[B; \mathscr{G}_1, \mathscr{G}_2]$ to obtain the next multiset

$$\Xi_1[\mathscr{G}_1], \quad (\eta_1 \wedge \zeta_1)[B; \mathscr{G}_1, \mathscr{G}_2]. \tag{6.4.29}$$

S3. Since $\mathfrak{g}(\Xi_1) \subset \mathfrak{g}(\zeta_1 \wedge \eta_1)$ this yields the successfully concluding tensor

$$(\Xi_1 \otimes (\eta_1 \wedge \zeta_1))[\mathscr{G}_1, \mathscr{G}_2]. \tag{6.4.30}$$

S4 is not needed and we assume termination.

Note that in the last two examples, the atom η forms a sort of link between the other units, and in the last case it acts like a linguistic hyphen which moreover causes the concluding tensor to inherit ζ_1's second g-feature.

EXAMPLE 8
S1. Input multiset:

$$\Xi_2[\mathscr{G}_1, \mathscr{G}_2], \quad \eta_2[A, B; \mathscr{G}_1, \mathscr{G}_2, \mathscr{G}_3], \quad \zeta_2[B; \mathscr{G}_1, \mathscr{G}_2, \mathscr{G}_3]. \tag{6.4.31}$$

Note here that the input compound tensor's g-features match the last example's output g-features.

S2. We can form $(\eta_2 \wedge \zeta_2)[B; \mathscr{G}_1, \mathscr{G}_2, \mathscr{G}_3]$ to obtain the next multiset

$$\Xi_2[\mathscr{G}_1, \mathscr{G}_2], \quad (\eta_2 \wedge \zeta_2)[B; \mathscr{G}_1, \mathscr{G}_2, \mathscr{G}_3]. \tag{6.4.32}$$

S3. Since $\mathfrak{g}(\Xi_2) \subset \mathfrak{g}(\eta_2 \wedge \zeta_2)$ this yields the successfully concluding tensor

$$(\Xi_2 \otimes (\eta_2 \wedge \zeta_2))[\mathscr{G}_1, \mathscr{G}_2, \mathscr{G}_3]. \tag{6.4.33}$$

S4 is not needed, and we may assume termination or not.

Here the compound input tensor at S1 could have been any available with matching g-features, such as the possibly just formed output of the last example, which would have been ripe for the plucking from the top of the STM stack, and would have yielded, at the last step above, the output

$$((\Xi_1 \otimes (\eta_1 \wedge \zeta_1)) \otimes (\eta_2 \wedge \zeta_2))[\mathcal{G}_1, \mathcal{G}_2, \mathcal{G}_3]. \tag{6.4.34}$$

We note here that the parentheses, though not needed mathematically, are left in since they reflect the hierarchical/structural aspects of the production, which are of syntactic interest. This structure is also revealed by the corresponding tree or trees. This sort of recursion goes by the name *embedding* in the linguistic context and so we shall call it likewise in the general case.

Thus the tree corresponding to equation (6.4.33) is (omitting the feature lists):

$$\tag{6.4.35}$$

while the tree corresponding to equation (6.4.34) is:

$$\tag{6.4.36}$$

EXAMPLE 9
This example demonstrates another interchange.

S1. Input multiset:

$$\xi[A; \mathcal{G}_1, \mathcal{G}_2], \ \Theta[\mathcal{G}_1], \ \eta[B; \mathcal{G}_1, \mathcal{G}_2], \ \zeta[A, B; \mathcal{G}_1, \mathcal{G}_2]. \tag{6.4.37}$$

Here $\mathscr{E}(\xi) \subset \mathscr{E}(\zeta)$ and $\mathscr{E}(\eta) \subset \mathscr{E}(\zeta)$ so we have an ambiguity and must execute both streams separately.

Stream 1
S2. $(\xi \wedge \zeta)[A; \mathscr{G}_1, \mathscr{G}_2]$ can be formed producing the next multiset for this stream:

$$\Theta[\mathscr{G}_1], \ \eta[B; \mathscr{G}_1, \mathscr{G}_2], \ (\xi \wedge \zeta)[A; \mathscr{G}_1, \mathscr{G}_2]. \tag{6.4.38}$$

Here we have $\mathfrak{g}(\Theta) \subset \mathfrak{g}(\xi \wedge \zeta)$, $\mathfrak{g}(\Theta) \subset \mathfrak{g}(\eta)$ and $\mathfrak{g}(\eta) = \mathfrak{g}(\xi \wedge \zeta)$. Removing the equality ambiguity still leaves us with the ambiguity of two strict inclusions, so we are faced with another branching into two substreams. We shall pursue the first of these and leave the second as an exercise. Then we have the new multiset

$$\eta[B; \mathscr{G}_1, \mathscr{G}_2], \ (\Theta \otimes (\xi \wedge \zeta))[\mathscr{G}_1, \mathscr{G}_2] \tag{6.4.39}$$

and finally

$$(\eta \otimes (\Theta \otimes (\xi \wedge \zeta)))[\mathscr{G}_1, \mathscr{G}_2] \tag{6.4.40}$$

which concludes this substream.

Stream 2
S2. $(\eta \wedge \zeta)[B; \mathscr{G}_1, \mathscr{G}_2]$ can be formed producing the next multiset for this stream:

$$\xi[A; \mathscr{G}_1, \mathscr{G}_2], \ \Theta[\mathscr{G}_1], \ (\eta \wedge \zeta)[B; \mathscr{G}_1, \mathscr{G}_2]. \tag{6.4.41}$$

Proceeding as before we conclude the corresponding substream with the tensor

$$(\xi \otimes (\Theta \otimes (\eta \wedge \zeta)))[\mathscr{G}_1, \mathscr{G}_2] \tag{6.4.42}$$

leaving the corresponding other substream as a corresponding other exercise.

S4. The Interpreter superposes the concluding tensors and may copy a collapsed result to the Syntactic Store and/or pass it on to the Out interface. The possibly alternating trees for the two worked expamples are depicted below in abbreviated form.

(6.4.43)

and

(6.4.44)

The alternation resulting from this superposition interchanges the first left leaf with the third.

Termination may or may not occur.

EXAMPLE 10

Here is another form of interchange, this time applied only to compound tensors.

S1. Input multiset:

$$\Xi[\mathscr{G}_1], \ \Theta[\mathscr{G}_1,\mathscr{G}_2], \ \Psi[\mathscr{G}_1,\mathscr{G}_2]. \qquad (6.4.45)$$

Here we have $\mathfrak{g}(\Xi) \subset \mathfrak{g}(\Theta)$, $\mathfrak{g}(\Xi) \subset \mathfrak{g}(\Psi)$ and $\mathfrak{g}(\Theta) = \mathfrak{g}(\Psi)$ but the first two, being strict, take precedence. Their ambiguity compels us to bifurcate into two streams.

Stream 1

S2. We can form $(\Xi \otimes \Theta)[\mathscr{G}_1,\mathscr{G}_2]$ to obtain the next multiset

$$\Psi[\mathscr{G}_1,\mathscr{G}_2], \ (\Xi \otimes \Theta)[\mathscr{G}_1,\mathscr{G}_2] \qquad (6.4.46)$$

and then conclude this stream with

$$(\Psi \otimes (\Xi \otimes \Theta))[\mathscr{G}_1, \mathscr{G}_2]. \qquad (6.4.47)$$

Stream 2

This stream similarly concludes with

$$(\Theta \otimes (\Xi \otimes \Psi))[\mathscr{G}_1, \mathscr{G}_2]. \qquad (6.4.48)$$

The remaining steps are as before. In this case the two alternating trees are:

(6.4.49)

and

(6.4.50)

As before, any tensors with the correct g-features could be used here. Let us suppose that the Ψ entry presented to the Syntax Processor is a place holder, or "hole," of the right g-feature constitution, which will be denoted by $1[\mathscr{G}_1, \mathscr{G}_2]$. That is, we take $\Psi = 1$, which we have allowed as a compound tensor, in the input multiset. This could arise because the stack is empty or because a set of firing patterns has faded out. Then the trees above may be expressed as:

(6.4.51)

and

$$(6.4.52)$$

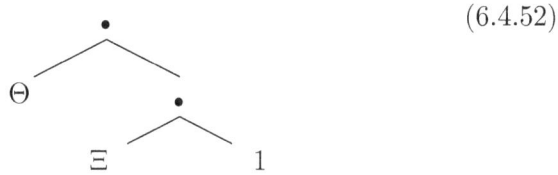

Since 1 represents the vacuum, or syntactic place holder, the alternation induced by superposition implements a *move* of the Θ leaf as shown. This is essentially a *movement* in the sense of the linguistic minimalist syntax model: cf. the next section.

EXAMPLE 11

This example *non-linguistically* concretizes Example 7. Here we go back to Proust's narrator and take $\mathscr{G}_1 = \mathscr{G}_{\text{childhood}}$ and $\mathscr{G}_2 = \mathscr{G}_{\text{tea things}}$, where the last Gestalt refers to those things associated in the memory of the narrator with tea, its preparation, etc. Suppose $\Xi_1[\mathscr{G}_1]$ denotes the childhood memory of Aunt Léonie's hand, surely a memory more complex than an atomic one and associated entirely with the childhood Gestalt. Let $\eta_1[A, B; \mathscr{G}_1, \mathscr{G}_2]$ represent the memory of her teaspoon, a simpler atomic tensor or firing pattern, which has two e-features, one pertaining to childhood tea *utensils* (A, say) and the other to childhood tea *eatables* (B, say). Likewise an atom $\zeta_1[B; \mathscr{G}_1, \mathscr{G}_2]$ represents a crumb of madeleine. Both of these atoms have both Gestalts as their g-features. As in Example 7, the algorithm produces the tensor

$$(\Xi_1 \otimes (\eta_1 \wedge \zeta_1))[\mathscr{G}_1, \mathscr{G}_2]. \qquad (6.4.53)$$

Since this is supposed to be a non-linguistic syntactic unit we must not express it linguistically but can do it in the form of a *rebus* as in Figure 6.2.

We note that the order of the two atoms ∧-ed together inside the inner parentheses does not matter, and the other order, of the $\otimes$ combination, was chosen by default, and reflects the prejudices of an anglophone writer, reading and writing left to right. This is supposed to be a memory perhaps retrieved by a voluntary action

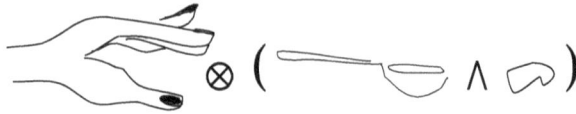

Figure 6.2   The rebus from Example 11.

of some kind — involving the In box — not necessarily provoked as in the *locus classicus* of the flood of involuntary memories provoked by the original taste of the madeleine. The ∧ combinator is more tightly binding than the ⊗ operator over which it takes precedence, so we conclude that the spoon is more strongly associated in memory with the crumb than this ∧-ed combination is with Aunt Léonie's hand. Of course, this may be obvious from the way the example has been cooked up, but the crucial point is that it has been the "machine of a computing technique," to paraphrase Wittgenstein, that has produced a syntactic object out of a distributed hierarchical arrangement of inputs.

6.4.3   *General properties of the syntax*

Despite the semantic and other limitations of our syntax model, there are some general properties that can be observed. Namely

1. The syntax is recursive: syntactic products already formed may be returned for reprocessing;
2. There is a combinator, ⊗, which acts to combine a pair of either kind of syntactic unit (compound tensor or atom). This combinator's operations are determined by the distribution of g-features of the component syntactic units;
3. There is a combinator, ∧, which applies only to atoms, and produces other atoms. Its operations are determined by the distribution of e-features of the component atoms;
4. There are symmetries associated with both of these combinators. Namely, both act in a binary manner to produce binary trees according to the distribution of e- and g- features. The order of the factors of a ∧-ed set of atoms does not count, whereas the order of a ⊗-ed set of tensors may or may nor count;

5. The different outcomes of derivational ambiguities are superposed and this superposition may manifest upon return to the syntax processor, or elsewhere, as alternation among the different choices of alternate outcomes (with unknown frequencies or sets of frequencies). Since the place holding unit 1 is allowed as a non-atomic compound tensor, this may reduce to the appearance of movement within a resulting syntactic unit;

6. Such movements will generally break whatever symmetry results from the derivational constraints imposed by the original distribution of features;

7. If only atoms and their $\wedge$-ed products (which are also atoms) are involved, then no derivations are ambiguous, since the order in such a combination is immaterial, and so no alternation or movement is possible. That is to say, there is no movement inside an atom;

8. There can be no embedding of compound tensors inside atoms (since compound tensors cannot be $\wedge$-ed with atoms);

9. No clear distinction can be made between those syntactic objects deemed to be *cognitively* atomic and those deemed to be cognitively compound. However this division is accomplished, the syntactic rules must be the same for both classes.

Most if not all of the difficulties of linguistic syntax reside in the complexities of the analogue of the Move operation, to which we are denied access by the limitations of our simple model. In our model these would amount to access to, or possession of, an intimate knowledge of the coefficients that go into our surmised superpositions, which are accomplished by the Interpreter and modules beyond it. In particular, in the linguistic case, there are semantic, lexical and grammatical constraints that drive the relative probabilities of the possible tree structures obtained by Movement.

If Procrustes had had an interest and had been more observant, the study of human anatomy might have advanced thousands of years earlier than it actually did. Like him, this writer will leave the harder parts to others.

In the next section, we will briefly draw comparisons with the one exemplar whose syntax has been most thoroughly investigated, namely language, and concretize some of our examples.

6.5 Language

We reiterate that we address ourselves here only to syntactical issues, and moreover only to the rules pertaining to syntactical productions. In the linguistic case this means that we cannot address problems of semantics, comprehension, or audition, all of which lie beyond the remit of our model. This is limiting since, in reality, all of these processes are intermingled to a greater or lesser extent.

The field of linguistic syntax is currently dominated by the ongoing Minimalist Program of Noam Chomsky. Of the vast literature on it we have singled out two introductory references, namely [2] and [91], and we will draw mainly on the latter for examples.

It is customary to distinguish the syntactic atoms from compound forms. Words are thought to be made up of (cognitively?) atomic units called *morphemes*. Empirically derived syntax rules for the construction of words are then laid out: this study goes by the name *morphology*.

Words congregate into bigger units called, variously: constituents, phrases, clauses and sentences. Units and subunits are assigned certain categories of different levels, often called *syntactic features*. Examples include parts of speech such as Nouns, Verbs, Determiners Adjectives, and types of phrases such as Verb Phrases, Determiner Phrases, Tense Phrases, etc. These categorizations enter into the syntax rules.

In the early 1990s Chomsky proposed a radically simplified model of (recursive) syntax, at least at the clausal level, which called for only two basic combinatorial operations out of which all languages emerge, the variations being essentially superficial to the basic structure. This structure, he conjectured, must originate in some neurobiological property of humans brains, which our nearest primate relatives lack, since it has become clear that they do not have innate recursive abilities. They can be taught a limited form of recursion, but they never develop true language, whereas humans usually master recursive syntax by the time they are two to three years old.

In simplified form the Minimalist Program specifies the following, at the clausal level.

1. The syntax is recursive: syntactic products already formed may be returned for reprocessing;
2. The syntactic atoms are *morphemes*, assumed to be the output from the morphological level (to be discussed later). These are assigned categorizations or features of certain kinds;
3. Morphemes are combined to form bigger molecular structures. The act of combining them is called Merge. This operation produces binary parse trees, and recursiveness manifests as the ability to combine trees already grown to form bigger trees. There are a few rules concerning modes of selection that constrain the resulting tree topology, which involve the various feature assignments and which, with the exception to follow, we omit for the purposes of this simplified discussion. Our exception is the principle known as Locality of Selection (LOS): this pertains to the assignment of features to the Merged entity and how such assignments relate to the features of those entities being Merged;
4. There is another operation called Move, which can take one part of an established tree and move it to another location in the tree. Insofar as LOS is a symmetry, Move breaks it. This operation depends on a variety of constraints which vary across languages and contributes greatly to the differences among them;
5. It is conjectured that there is little or no Movement within morphemes.

Words and their own atoms (morphemes) are supposed to be drawn from a *lexicon* which must also contain information concerning the features and contexts of these units through which the rules of selection, etc., are computed. The lexicon must also have certain analytical attributes promoting comprehension since it must be able to discriminate among inputs, among its other possible functions.

Over the course of many decades, brilliant work by linguists has established this pattern, or something very like it, as the basic structure underlying all languages. (This writer knows of only one language that seems to be an outlier, though of course there may be others. This is the indigenous Brazilian language known as Perahã whose recursiveness is debated. This language only seems to have about ten morphemes, and has no words for colors or numbers.)

It seems only to have been noticed later that the morphological rules (of word construction) were essentially identical to the more global rules for syntax at the clausal level. In the words of the authors of [91] (Introduction to Chapter 12, p. 332):

> "We will conclude that there is a single computational engine, driving both syntactic and morphological composition."

Clearly the general properties of this unified "morphosyntax" bear a striking resemblance to the general properties of our syntax of retrieval outlined above. Here we assume that morphemes, words, etc., correspond to firing patterns and their tensor products.

Our primary conjecture is that this is not a mere resemblance but that the general syntax underlies linguistic syntax, this being a "macroscopic" manifestation of it.

The main difference is that our general syntax has two inter-related combinators $\otimes$ and $\wedge$. We hypothesize that the more intimate $\wedge$ combinator cannot generally be differentiated in linguistic studies from its closely associated $\otimes$ companion: only its interaction with $\otimes$ rises to the level of awareness sufficiently to drive the mechanism of Merge that is detected in language. Concomitant with this is the hypothesis that the categories posited by linguists may correspond to a mix of our essential and global features which likewise rise only in interaction to the level of cognitive awareness.

Whatever the status of these hypotheses it is possible to find concrete linguistic instantiations of some of the template-like examples given above. We note first that linguistic interpretations have already been given to Example 4.

Applications of EXAMPLES 1, 2 and 3

These examples deal with *plurals*. Some languages are rich in the
variety of strategies they use to pluralize nouns, and these strategies
vary widely among languages. Some languages, such as the Bantu
languages of Southern Africa, use prefixes: in Lesotho there is one
impala, and many *izimpala*. Standard Arabic is rich in different kinds
of affixes to denote plurals. For instance, one doctor is *tabeeb*, more
than one is *atibaa'*, a dog is *kalbun*, more are *kilaabun*, one girl is *fatat*,
more girls are *fatayat*. Compare this last to Example 1 (dropping the
feature lists):

$$\xi,\ \eta\ \longrightarrow\ \xi \otimes \eta$$
$$fat,\ at\ \longrightarrow\ fatat$$

Example 2 pluralizes this by inserting the pluralizing affix in the
middle of the tensor:

$$\xi,\ \eta, \qquad\qquad \zeta \qquad\qquad \longrightarrow\ (\xi \wedge \zeta) \otimes \eta$$
$$fat,\ at,\ (\text{another } fat\text{-ish atom})\ ay\ \longrightarrow\ fat\text{-}ay\text{-}at$$

The *kilbun* pluralized to *kilaabun* is similar. (English has something
similar: mouse, mice; man, men; goose, geese.)

   Some languages take the pluralizing affix literally as the noun, or
a part of it, itself repeated. This is called *reduplication*. It is absent
in English but many languages have versions of it. One mentioned
in [91] is the Philippine language Agta in which one leg is *takki* and
many legs are *taktakki*. This example seems to perfectly instantiate
our Example 3, with *tak* ↔ ξ and *ki* ↔ η.

Application of EXAMPLES 7 and 8

There are many examples of affixes changing the grammati-
cal/syntactic category of the words they are affixed to. An example
from English is exhaustively treated in [91]: namely derivations from
the noun *nation*. Here, for example, we have this noun acquiring
the suffix - *al* to be become the adjective *nation-al* and then this

adjective acquiring the suffix *-ize* to yield the verb *nation-al-ize*. This may be compared to our Examples 7 and 8 with $\Xi_1 \leftrightarrow$ *nation*, $\eta_1 \wedge \zeta_1 \leftrightarrow$ *-al*, $\eta_2 \wedge \zeta_2 \leftrightarrow$ *-ize*. Thus $(\Xi_1 \otimes (\eta_1 \wedge \zeta_1)) \otimes (\eta_2 \wedge \zeta_2) \leftrightarrow$ *nationalize*. Here we are including the bridging atoms η_i as component morphemes, namely the hyphens, though they are unvoiced. Note the progression of g-features in our example as suffixes are added. This presumably corresponds to the change in linguistic category per suffix of the derived words.

Applications of EXAMPLES 6, 9 and 10

The Move operation in linguistic syntax is generally extremely complex. Example 10, with $\Psi = 1$, realizes possibly the simplest case, which is most easily described by an example from French, given in [2]:

Georges mange.

Georges eats.

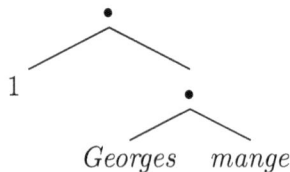

Pascale fait manger Georges.

Pascale makes Georges eat.

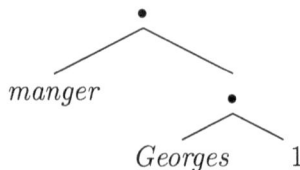

Here the verb moves in front of the subject. In this example, we surmise that the advent of *Pascale fait* has the effect of changing the coefficients (or weights) in the superposition so as to certify the choice of this alternate. This would appear to a linguist as one of the rules of French syntax.

Another simple case is illustrated by Example 6, namely what is called *subject-auxiliary inversion*, which arises in yes/no questions.

Thus

$$\text{George} \otimes (\text{will} \wedge \text{run})$$

$$\text{Moves to}$$

$$\text{Will} \otimes (\text{George} \wedge \text{run})?$$

Example 9, though apparently simple, seems to have a rather complex linguistic interpretation involving what are called *islands*: syntactic units that get stranded and cannot move from their position. We will not follow this example further but consider these two English sentences, both grammatically correct:

> I discussed which plant belongs with which pot.
> I discussed which pot belongs with which plant.

This ends our very brief discussion of the immense subject of linguistic syntax and the few examples we may have been able to address with our basic general syntactic torso.

Are similar circuits involved in other pattern-related cognitive areas? Music and the arts generally come to mind. Likewise mathematics. This writer does not know the answer. However, there seems to be a hint of such a system arising in the study of the neuroanatomy of arithmetic or algebraic operations: cf. [58, 93].

6.5.1 *Neuroanatomy is linguistic destiny*

In recent decades there has been a convergence of the subject of linguistic syntax and neuroanatomical investigations of language in the brain, the latter advancing largely as a result of the development of new imaging and other non-invasive methods of inspection. This convergence — virtually a marriage — of these two disciplines is

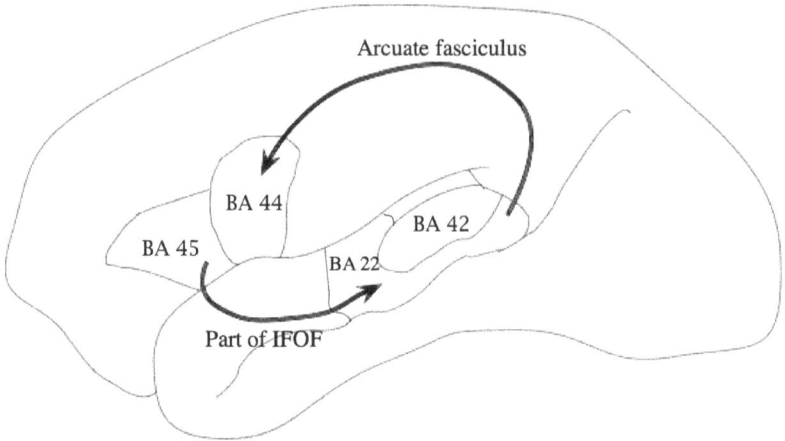

Figure 6.3 Regions of the brain relevant to language processing.

surely one of the most significant scientific advances of the present era. We briefly address this in relation to our models, starting with some diagrams.

Figure 6.3, based on many sources, depicts in simplified and approximate form the regions in the left hemisphere of a human brain which are most relevant to language processing. The front/anterior is to the left; BA (Brodmann Area) 45 ↔ *pars triangularis*; BA 44 ↔ *pars opercularis*; BA 22/42 ↔ posterior middle temporal gyrus or pMTG; IFOF = inferior frontal occipital fascicle. The IFOF tract has a branch, not shown, continuing to the right towards the posterior occipital region. The arrowed curves represent white matter tracts and may go in both directions so may have arrows at their other ends: the arrows shown indicate the directions of interest to us. Also, it should be noted that there are other, probably subordinate, nearby tracts, such as the *uncinate fasciculus* (a ventral tract not shown) and the *superior longitudinal fasciculus* (a dorsal tract not shown): the latter tract may be involved but probably is not. Tractology is still a much debated topic.

Figure 6.4 depicts the areas in the human brain corresponding to the modules in the original circuit schematic, to which it may be compared. For ease of reference we again repeat this original syntax

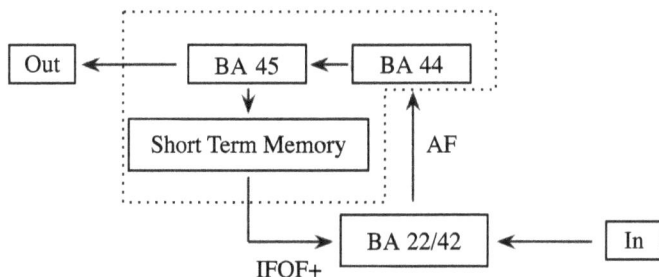

Figure 6.4   Areas in the human brain corresponding to modules in a circuit.

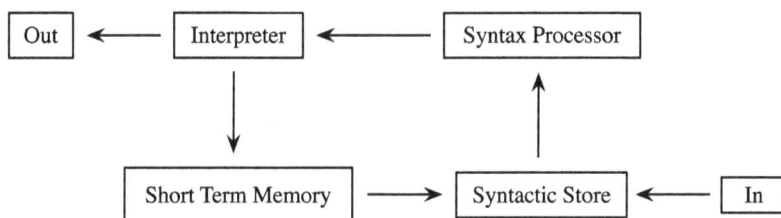

Figure 6.5   Repeat of the original syntax circuit.

circuit following Figure 6.4 as Figure 6.5. AF = *arcuate fasciculus*. IFOF+ = IFOF + other tracts, not shown, that may be involved. The boxes bounded by the dotted lines seem to act in very close association or as a unit.

Before discussing these diagrams it should be emphasized that the apparent precision of the identification of the boxes with brain areas should be taken with handfuls of salt.

The areas denoted BA 44/45 coincide roughly with what is known as Broca's Area while the areas denoted BA 22/42 coincide roughly with what is known as Wernicke's Area. Again roughly speaking, Wernicke's Area is heavily involved in language comprehension while Broca's is heavily involved in language production, though both are involved in both processes.

A comprehensive model of language processing has recently been proposed by W. Matchin and G. Hickok [66]. This model covers both comprehension and production, as it properly should, since the two processes intermingle. As noted, we can only properly address the production component. In this, their model is not dissimilar to ours as instantiated by Figure 6.4. The most salient points of contact are:

1. BA 22/42 (the posterior middle temporal gyrus or pMTG) seems to have a hierarchical lexical-syntactical function similarly to our surmised Syntactic Store;
2. Strong evidence is evinced for the involvement of BA 44/45 in speech production, particularly BA 44, correlating with our surmised Syntax Processor;
3. The Matchin-Hickok model specifies that in production the computational task is to convert non-sequential conceptual information into hierarchical structures, performed in the pMTG, and then translate these into sequences of morphemes, which is accomplished via the *pars triangularis*. This describes in almost exact detail the path around our circuit from the Syntactic Store (pMTG) to the Interpreter (BA 45);
4. "... these frontal resources may be harnessed to facilitate predictions in some situations by preactivating lexical-syntactic representations stored in temporal cortex in a top-down fashion."

This last point of contact seems to imply a stack-like working memory, specialized to linguistic memoranda, somewhere in the frontal region of interest, which corresponds to our three boxes bounded by dotted lines in Figure 6.4. More detail is supplied in [65]. In this latter reference it is made specific by hypothesizing that it is the *pars triangularis*, BA 45, that underlies language-specific working memory system, comprised of retrieval operations specialized for syntactic representations. This may be effected by tractal connections to other locations, which would exist at a finer resolution than our crude box diagrams can muster. Moreover, there seems to be the implication also that there may be some multimodal activity involved, described in terms of domains of generality or specificity.

The upshot is that there seems to be general agreement between our model, as instantiated in Figure 6.4, and the production component of the Matchin-Hickok model, with the following proviso.

The brain areas corresponding to the three boxes in the diagram bounded by dotted lines work in very close association with each other, both via connections and also via physical proximity. The stack-like Short Term Memory box may correspond to a brain area which is part of BA 45 or proximal to it, and specialized to syntactic representations. Consequently, we cannot be certain that the order of the small arrows shown within the dotted boundary reflects the actual order of precedence in brains, and can only draw conclusions which do not rely upon this internal order.

Broad conclusions we can make first involve certain types of aphasia. Consider first damage inside the area bounded by the dotted lines. If this involves damage to the STM stack then nothing is returned to the Syntactic Store for reprocessing, recursion is curtailed, and victims of this damage may find themselves being able to utter only the words or morphemes left in the non-current part of the Store, as in our Example 4 above. This seems to have been the fate of the unfortunate patient of Broca's, Louis Victor Laborgne, known as M. Tan since *tan* was the only utterance he was capable of. Post mortem, he was found by Broca to have a lesion in the general area now known for the latter, though in fact the lesion may have been elsewhere, albeit nearby. Damage to other parts of the bordered area will also of course result in speech production problems. Since we cannot address semantic or other issues dealt with by the bordered boxes, we cannot speak to these effects. Damage to the IFOF and other possibly associated tracts not shown would presumably also cut off connection to the stack and curtail recursion, and would be indistinguishable from the extreme form of Broca's aphasia exemplified by M. Tan.

The *arcuate fasciculus* seems to be fundamental to all language processes. Damage to it would result in the failure of the Syntax Processor (among the bordered boxes) to receive the output from the Syntactic Store in the proper or expected order. This could happen if some or all of its axonal fibers are disrupted so that the mappings interpreting the white matter sequents in section 4.1.4 are no longer one-to-one: this is one possible example of a white matter tract deficit relative to our model. Then the victim would not be able to correctly repeat sequences heard, though the comprehension of them (via the Syntactic Store) would not be impaired: the victim would be aware of the errors. This seems to reproduce so-called, but well named, *conduction* aphasia.

This brings us to our main application, namely to the problem of why only us.

6.5.2 *Why Only Us*

This is the title of a work by Robert Berwick and Noam Chomsky [12] and the absence of a question mark is significant. This work gives an answer to the question implied, the Us being humans and the others being non-humans, specifically non-human primates, our closest relatives. The answer is also given in [33] and it is basically this: non-human primates do not have innate recursive language because their homologue of the AF does not reach their homologue of BA 44.

This is an attractive and compelling argument, though it conflicts with our model. For, it is apparent that non-human primates do have a primitive sort of non-recursive syntax and can be taught recursion up to a point, though it seems to take a very large training effort of, for instance, macaques, while it seems almost innate in young humans [32]. Taken together, these findings — the falling short of the *arcuate* and the proto-syntactic abilities of macaques — seem inconsistent with our model. For then, in our model, if the *arcuate* homologue falls short of the BA 44 homologue, there can be no syntactic processing at all, regardless of the amount of training, since no retrieval into the Syntax Processor is possible. The second of these findings does not conflict with our model if the macaque *arcuate* does in fact reach the BA 44/45 homologue. And there is evidence that in fact, *pace* the others, it does indeed reach that far [15]. If so, the question becomes: why do macaques then only acquire recursive syntactical abilities, albeit partial ones, only after very heavy training?

An answer, suggested by our model, is this: it is the *other* end, the *BA 22/42 (pMTG) homologue* end, of the *arcuate* homologue that does not go far enough in macaques and presumably other non-human primates. That is to say, it does not go far enough into the area represented by our Syntactic Store box to be able to retrieve the input from the stack-like Short Term Memory. There is compelling anatomical evidence that in fact this tract does not penetrate as far in that direction in macaques as it does in humans while also being less voluminous [78]; for more on the comparisons of ventral and dorsal AF connections among humans and chimpanzees, see [79]. Thus it is only tenuously attached, if at all, to the surmised stack input.

This answers the question since, firstly, retrievals by the Syntax Processor are now implemented, so some syntactic processing is possible, though it will not be recursive since the returning input from the STM stack is not available. However, secondly, upon rigorous training, the connections of this lower end of the tract could be strengthened via Hebbian learning to make at least some contact with the stack input. So some recursion could be manifested, after intense training, by macaques.

The stack-like STM, should it exist, may have evolved for other purposes, such as keeping short term track of a sequence of events leading up to a possibly lethal encounter with a predator, which may soon happen again. Such an arrangement would bestow a selective advantage upon an individual. Then it could have been recruited, via a slightly mutated proximal synaptic connection, by the proto-lexical/syntactic pMTG homologue, which was presumably populated with innate call templates prototypical of the species and wired into the vocalizing system. Long term morpheme-like sequences — *calls* — might now be ensconced as part of a proto-lexicon. Even a tenuous such connection would ramify the selective advantage of such a system, since warning calls could now be delivered to the group. These connections, even if tenuous, would also enable a primitive kind of combinatorial syntax which is not fully recursive, owing to the ventral tract deficit. In fact such things are found in the vocalization of monkeys — vervets, Dianas, Campbell's — and also chimps, bonobos, and baboons. (The literature on the calls of these species is vast. For good examples cf. [70, 74].) Since the damage involved in M. Tan's version of Broca's aphasia also involves a deficit in the stack connection, it is not surprising that it resembles an animal's cry.

6.5.3 The failure of ⊗ and the language of schizophrenia

In earlier work ([85]; see also [83]) it was suggested that the failure to implement the ⊗ combinator in humans leads to schizophrenia-like deficits, such as context blindness and exaggerated von Domarus errors, but also led to better outcomes for some logical operations that are prone to context distraction in neurotypicals. The upshot is that schizophrenia patients have to rely more on a **OL**-like logic than on a **GN**-like logic. In this subsection we shall briefly consider one linguistic deficit that a failure of ⊗ may effect.

Clearly, the complete failure everywhere of our ⊗ would be a complete catastrophe for the user or owner of networks such as the ones we have been considering, so we shall assume that such failures are intermittent both in spatial location among a set of networks, and possibly in time too. This is consistent with our understanding of the mechanism behind the operation of the substrate connections implementing ⊗: namely, it acts via a flow of neurotransmitter(s) and/or the operation of interneurons (which orchestrate coordination among ensembles of neurons). In the latter case, as noted in Chapter 3, it may be an undersupply of GABA, rendering the interneurons inoperative, or an oversupply of glutamate, which would overexcite neurons, preventing the formation of ⊗, since self-connection via this operator may terminate the neurons involved. (Among other possible neurotransmitters, neuropeptides, hormones, etc., dopamine is most likely also to be involved, as it is in virtually everything. As we mentioned earlier, the treatment of dopaminergic cells themselves lies outside our model's purview since they are generally not morphologically bicameral.)

The general effect of vitiation of ⊗ would be to curtail the formation of contexts, which is a central feature of schizophrenia, and to promote memory retrieval deficits, as in the now vitiated pattern retrieval algorithm of Chapter 4. In those areas (or times) that this happens, the b-networks involved could still support the operation of the ∧ operator since it is inherent in the behavior of an ensemble of fermionic entities such as our b-neurons, regardless of any external substrate or even interaction. So, as noted, the logic involved would in this case necessarily be closer to the ones that are **OL**-like than to the ones that are **GN**-like. The upshot is that schizophrenia patients may be prone, by force of circumstance, to use a logic which is more quantum-like, and less context dependent, than that used by neurotypicals. This seems to explain certain anomalies in their logical processes [83].

The two levels of categorization found in our syntax rules are a vestige of the presence of these two logics. We shall attempt to gauge the effect of a deficit in the operation of the combinator $\otimes$ on the linguistic application of our general syntax rules. This would first vitiate the application of rule S3. Let us suppose that $\otimes$ is completely denied. In this most extreme case only atoms could be formed from other atoms, and recall that there is no Movement within an atom. No clausal hierarchies are possible, though some morphology might survive. That is to say, word formation for words not requiring the input of $\otimes$ combinations could be uttered, but there can be no clausal or grammatical complexity whatsover. Our example of the use of this combinator in word formation was in the operation of affixing morphemes. So we would predict that in this extreme case the utterable words, at least, lack affixes whatever else they may lack (cf. [21] §3.3 in this respect). At the next step we assume that the $\otimes$ combinator is less vitiated and may sometimes apply. In this case we may have some clausal complexity but it will still be limited since most of the tensors involved will be atomic and so most of the combinators will be $\wedge$s. Moreover, if the tensor embedded via recursion is not itself an atom, the embedding will fail, since combining a compound tensor can only be done with $\otimes$: no compound tensors, which must include clauses, can be embedded inside atoms. So in this case there will be less complexity or depth of clausal embedding. Moreover, since there will be fewer alternative syntactic choices, there will be less Movement. So what trees may emerge will be fragmented and shallow, manifesting a lack of linguistic cohesion and complexity (as measured by depth of trees produced), accompanied by a concomitant poverty of word production. Further, degrees of vitiation presumably produce a spectrum of defective speech from most defective to least.

Under these assumptions concerning the $\otimes$ operator, our predictions for the production of speech by schizophrenic patients then take the form of the following problems.

First there are the obvious ones having to do with vitiation of context-forming abilities, which would include:

1. Concretism: the inability to appreciate formulaic statements, or idioms, instead taking them literally, owing to the loss of context;
2. Derailment: the loss of conceptual focus in time, as the context of the previous topics is lost;
3. Other similar such contextual problems, such as neologism: the creation of non-standard words from morphemes since the standard context is lost, and tangentiality, which is similar to derailment;
4. Problems of retrieval from short and long term memory, which will affect verbal fluency.

These sets of problems, although present among schizophrenic patients, are not necessarily characteristic of that malady. The next one apparently is, thanks to the brilliant work of DeLisi [26]: see also [21]. It is the one we have arrived at above, on the basis of our general syntax, namely:

5. Lack of sentence complexity, lack of clausal cohesion and embedding depth, and impoverishment of vocabulary.

6.6 Conclusions

In this chapter we have tried to find evidence, based on our network model, in support of the conjecture that there is a general syntax underlying memory retrieval that gives rise, for instance, to the syntax of (human) languages, as encapsulated in the so-called Minimalist Program. We have also considered such questions as why humans are the only primates to have innate recursive syntax, and how vitiation of the tensor combinator might underlie the basic language deficit that has been found to be characteristic of schizophrenia.

A determination of whether or not we have found such evidence is left as an exercise for the interested reader.

Part III

Appendices

Appendix to Chapter 1

A.1   Some logical and mathematical results

A.1.1   *Modal identities*

It will be convenient to first record some *modal identities*. We repeat officially the definitions of *proximity space* and *orthogonality* space given earlier.

Definition A.1.1. A *proximity space* is a pair $\langle W, \approx \rangle$ where W is a set and $\approx \subseteq W \times W$ is a relation on it, called a *proximity*, which is *reflexive* and *symmetric*. That is to say, for $x, y \in W$, $x \approx x$ and $x \approx y$ implies $y \approx x$.

Definition A.1.2. An *orthogonality space* is a pair $\langle W, \perp \rangle$ where W is a set and $\perp \subseteq W \times W$ is a relation on it, called an *orthogonality*, which is *irreflexive* and *symmetric*. That is to say, for $x, y \in W$, $x \not\perp x$ and $x \perp y$ implies $y \perp x$.

For $x \in W$, $Y \subseteq W$ we write $x \perp Y$ iff $x \perp y \; \forall y \in Y$ and define $Y^{\perp} := \{x \in W : x \perp Y\}$. We shall generally write, for $w \in W$,

$$\{w\}^{\perp} \quad \text{as} \quad w^{\perp}, \tag{A.1.1}$$

$$\{w\}^{\perp\perp} \quad \text{as} \quad w^{\perp\perp}, \tag{A.1.2}$$

etc.

Note that each proximity space $\langle W, \approx \rangle$ determines an orthogonality space $\langle W, \perp \rangle$ where $x \perp y$ iff $x \not\approx y$, and conversely each orthogonality space $\langle W, \perp \rangle$ determines a proximity space $\langle W, \approx \rangle$

where $x \approx y$ iff $x \not\perp y$. Note also that $\emptyset^\perp = W$ and $W^\perp = \emptyset$. (For the first of these, suppose $\exists x \in W$ such that $x \notin \emptyset^\perp$. Then there would have to be an element in $\emptyset$ that is not orthogonal to x. But there are no elements in $\emptyset$ so this is impossible. The result follows.)

Let $\langle W, \approx \rangle$ denote a proximity space, with $\langle W, \perp \rangle$ denoting the associated orthogonality space. We have defined in section 1.2.1 for each subset $E \subseteq W$:

$$\Diamond E := \{w \in W : \exists v \in E \text{ such that } w \approx v\} \qquad (\text{A.1.3})$$

and noted that $E \subseteq \Diamond E$. Dually, we define $\Box E$ as

$$\Box E := (\Diamond E^c)^c. \qquad (\text{A.1.4})$$

It is convenient also to define

$$S_v := \Diamond v = \{w \in W : w \approx v\}, \qquad (\text{A.1.5})$$

the "sphere" around v in the terminology of D. Lewis, so that

$$\Diamond E = \bigcup_{v \in E} S_v. \qquad (\text{A.1.6})$$

Then we have:

Proposition A.1.1. *For $\langle W, \approx \rangle$ as above and $E, F \subseteq W$:*

M1. $\Diamond(E \cup F) = \Diamond E \cup \Diamond F$
M2. *For a family $\mathscr{F}$ of subsets of W, $\Diamond(\bigcap_{F \in \mathscr{F}} F) \subseteq \bigcap_{F \in \mathscr{F}}(\Diamond F)$*
M3. $\Diamond E = (E^\perp)^c := E^{\perp c}$
M4. $\Box E = E^{c\perp} = \{w \in W : S_w \subseteq E\}$
M5. $E \subseteq E^{\perp\perp}$
M6. $E \subseteq F$ *implies* $F^\perp \subseteq E^\perp$
M7. $E^{\perp\perp\perp} = E^\perp$
M8. $\Box\Diamond\Box E = \Box E$
M9. $\Diamond\Box\Diamond E = \Diamond E$
M10. $(E \cup F)^\perp = E^\perp \cap F^\perp$
M11. $E^\perp \cup F^\perp \subseteq (E \cap F)^\perp$

Proof. The proofs of these are mainly immediate or elementary. For instance, for M3, we note that $w \notin \Diamond E$ implies $w \not\approx x$ for all $x \in E$

so that $w \perp E$ showing that $(\Diamond E)^c \subseteq E^\perp$ from which it follows that $(E^\perp)^c \subseteq \Diamond E$. The reverse inclusion is obvious. The proof of M4 follows since

$$
\begin{aligned}
\Box E :&= (\Diamond E^c)^c \\
&= ((E^c)^{\perp c})^c \\
&= \{w \colon w \perp E^c\} \\
&= \{w \colon v \notin E \text{ implies } v \not\approx w\} \\
&= \{w \colon v \approx w \text{ implies } v \in E\} \\
&= \{w \colon S_w \subseteq E\}.
\end{aligned} \tag{A.1.7}
$$

From M6 applied to M5 we obtain $(E^{\perp\perp})^\perp \subseteq E^\perp$. But from M5 we also have $E^\perp \subseteq (E^\perp)^{\perp\perp}$, and M7 follows.

M8, M9 and M10 follow immediately from M3, etc., and M11 follows from M3 and M2. □

A.1.2 *Modal propositions*

For an orthogonality/proximity space as above Dalla Chiara *et al.* [24] define a *proposition* to be a subset $X \subseteq W$ satisfying:

if $x \in W$ is such that $\forall y \approx x, \exists z \in X$ such that $y \approx z$, then $x \in X$. (A.1.8)

Equivalently, X is a proposition iff

$$
S_x \subseteq \Diamond X \text{ implies } x \in X. \tag{A.1.9}
$$

(Note that since $\Diamond X = \bigcup_{x \in X} S_x$ the reverse implication always holds.)

Then we have:

Proposition A.1.2.

1. X *is a proposition iff* $\Box\Diamond X = X^{\perp\perp} = X$.
2. X *is a proposition iff* $x \notin X$ *implies* $\exists y \approx x$ *with* $y \perp X$.
3. *For any* $Y \subseteq W$, $Y^\perp$ *is a proposition.*
4. *If* $\mathscr{C}$ *is a family of propositions, then* $\bigcap \mathscr{C}$ *is a proposition.*
5. *If* Y *is a proposition, then* $X \subseteq Y$ *iff* $\Diamond X \subseteq \Diamond Y$.

Proof.

1. From M4 the condition labelled (A.1.9) implies $\Box \Diamond X \subseteq X$, and since always $\Box \Diamond X = X^{\perp\perp}$ from M3 and M4, the rest follows from M5.
2. This follows immediately from (A.1.9) which is equivalent to $x \notin X$ implying $S_x \not\subseteq \Diamond X = X^{\perp c}$.
3. Immediate from M7.
4. $x \notin \bigcap_{C \in \mathscr{C}} C$ implies $\exists C_0 \in \mathscr{C}$ such that $x \notin C_0$. Thus $S_x \not\subseteq \Diamond C_0$, since C_0 is a proposition: cf. proof of 2 above. So $S_x \not\subseteq \bigcap_{C \in \mathscr{C}} (\Diamond C)$ from which it follows that $S_x \not\subseteq \Diamond(\bigcap_{C \in \mathscr{C}} C)$ in view of M2.
5. Implication in one direction is immediate; the other follows from M3, M5 and M6. □

Other authors call subsets F of an orthogonality space with the property that $F = F^{\perp\perp}$ — in other words those sets we have followed Dalla Chiara *et al.* in calling propositions — *regular* sets.

It is useful to note that it follows from M5 and M6 that for a subset $E \subseteq W$, $E^{\perp\perp}$ is the *smallest proposition* containing E.

A.1.3 *Ortholattices*

Definition A.1.3. An *ortholattice* is a bounded lattice $\langle L, \sqcup, \sqcap, 0_L, 1_L, (\ )' \rangle$ where the unary operation $(\ )'$, called *orthocomplementation*, satisfies:

$$complementarity : \forall a \in L, \ a \sqcap a' = 0_L, \ a \sqcup a' = 1_L \qquad (A.1.10)$$

$$unitarity : a'' = a \qquad (A.1.11)$$

$$antitonicity : a \sqsubseteq b \ \textit{iff} \ b' \sqsubseteq a'. \qquad (A.1.12)$$

An ortholattice is said to be *complete* if arbitrary subsets of it have meets and joins.

It is easy to show that De Morgan's laws hold in any ortholattice, namely

$$a \sqcup b = (a' \sqcap b')' \qquad (A.1.13)$$

and its dual form. A complete ortholattice satisfies the complete versions of the De Morgan laws. Examples include all Boolean algebras, which are precisely the distributive ones, and the lattices of closed subspaces of Hilbert spaces. A canonical class of ortholattices is obtained as follows.

Definition A.1.4. The set of propositions in the orthogonality space $\langle W, \perp \rangle$ is denoted $R(\langle W, \perp \rangle)$.

Lemma A.1.1. $R(\langle W, \perp \rangle)$ *is a complete ortholattice under the partial order given by set inclusion, the meet being set intersection and the orthocomplement being* $\perp$.

Thus the join is given by

$$E \sqcup F = (E^\perp \cap F^\perp)^\perp = (E \cup F)^{\perp\perp}. \qquad \text{(A.1.14)}$$

This result follows immediately from the earlier identities, etc.

This class of examples is canonical in light of the following beautiful Stonean Theorem of Goldblatt [40].

Theorem A.1.1. *Any ortholattice* L *is isomorphic to an orthosublattice of* $R(\langle W_L, \perp_L \rangle)$ *for some orthogonality space* $\langle W_L, \perp_L \rangle$. *If* L *is complete then this isomorphism is complete and onto.*

This theorem also shows that each ortholattice may be embedded into a complete one.

Since this theorem is the cornerstone of all the models we discuss here, it is worth describing the construction. For a given ortholattice L, W_L is the class of *proper filters*. A proper filter is an upward closed subset of L, closed also under finite meets, which does not contain 0_L. For $x, y \in W_L$ the orthogonality is given by

$$x \perp_L y \text{ iff } \exists a \in L \text{ such that } a' \in x \text{ and } a \in y. \qquad \text{(A.1.15)}$$

At this point we recall the following notations. For sets X and Y, X^Y denotes the set of all maps from Y into X and $\mathbf{2}$ denotes the two-element set, often with its structure as the simplest Boolean algebra implied. Thus $\mathbf{2}^X$ is in one-to-one correspondence with the *power set* of X, namely the set of subsets of X, and will be used as such.

The map

$$\phi_L \colon L \to \mathbf{2}^{W_L}, \tag{A.1.16}$$

given by

$$\phi_L(a) := \{x \in W_L \colon a \in x\} \tag{A.1.17}$$

embeds L into $R(\langle W_L, \perp_L \rangle)$. Goldblatt characterizes the image of L under ϕ_L: it coincides with the family of those propositions in W_L that are *clopen* in the topology on W_L that has as a subbase the sets $\phi_L(a)$, $a \in L$, and their complements.

(A proper filter is like a "world" that validates each of its members. That is to say, if $a \in x$ is interpreted as "a is true in world x" then everything that can be inferred from a, that is, those elements above it in the lattice, should also be true in the world x, which is the case if x is a filter, and $\sqsubseteq$ is construed as a form of implication. Under this reading, $\phi_L(a)$ being a set of worlds could be interpreted as an actual proposition of a certain type, and Goldblatt's Stonean Theorem does indeed yield a class of Kripkean models that characterize orthologic, as described in section 1.2.2.)

Another class of ortholattices has already been met in section 1.2.1, namely the lattices of "parts", denoted $\mathrm{Part}(\langle W, \perp \rangle)$, of orthogonality spaces $\langle W, \perp \rangle$. This class was introduced and named by J. L. Bell ([10, 11]). Here is the official definition.

Definition A.1.5. $\mathrm{Part}(\langle W, \perp \rangle)$ denotes the family of subsets of W of the form $\Diamond E$ for subsets E of W. With ordinary set union as join, and orthocomplement given by

$$(\Diamond E)' := \Diamond (\Diamond E)^c, \tag{A.1.18}$$

$\mathrm{Part}(\langle W, \perp \rangle)$ is a complete ortholattice.

It follows that the meet is given, for any subsets E, K of W by:

$$\Diamond E \sqcap \Diamond K = ((\Diamond E)' \cup (\Diamond K)')' \tag{A.1.19}$$

$$= ((\Diamond(\Diamond E)^c) \cup (\Diamond(\Diamond K)^c))' \tag{A.1.20}$$

$$= ((\Diamond E^\perp) \cup (\Diamond K^\perp))' \qquad \text{from M3} \qquad (A.1.21)$$
$$= (\Diamond (E^\perp \cup K^\perp))' \qquad \text{from M1} \qquad (A.1.22)$$
$$= \Diamond (\Diamond (E^\perp \cup K^\perp))^c \qquad\qquad (A.1.23)$$
$$= \Diamond ((E^\perp \cup K^\perp)^\perp) \qquad \text{from M3} \qquad (A.1.24)$$
$$= \Diamond (E^{\perp\perp} \cap K^{\perp\perp}) \qquad \text{from M10.} \qquad (A.1.25)$$

It is not hard to show that this meet is indeed the largest subset of $\Diamond E \cap \Diamond K$ of the form $\Diamond (\ )$: that is to say, if $\Diamond L \subseteq \Diamond E \cap \Diamond K$ then $\Diamond L \subseteq \Diamond (E^{\perp\perp} \cap K^{\perp\perp}) \subseteq \Diamond E \cap \Diamond K$.

In what follows, we shall generally drop the reference to the orthogonality relation in the notations for the lattices of propositions and parts, assuming it is unambiguous or understood, and just write $R(W)$ and Part W.

The following theorem, due to J. P. Rawling [76], reveals the pertinent fact concerning the lattice of parts.

Theorem 1.2.5 (recalled)
Given an orthogonality space $\langle W, \perp \rangle$:

$$\Diamond: R(W) \to \text{Part}\, W \qquad\qquad (A.1.26)$$

is an isomorphism of complete ortholattices. Its inverse is

$$\Box: \text{Part}\, W \to R(W). \qquad\qquad (A.1.27)$$

Proof. For any $E \subseteq W$:

$$\Diamond (E^\perp) = E^{\perp\perp c} \qquad\qquad (A.1.28)$$
$$= E^{\perp cc\perp c} \qquad\qquad (A.1.29)$$
$$= (\Diamond E)^{c\perp c} \qquad\qquad (A.1.30)$$
$$= \Diamond (\Diamond E)^c \qquad\qquad (A.1.31)$$
$$= (\Diamond E)' \quad \text{from equation } (A.1.18) \qquad (A.1.32)$$

so orthocomplementation is preserved.

For any family E_i of elements of $R(W)$:

$$\Diamond\left(\bigsqcup_i E_i\right) = \Diamond\left(\left(\bigcap_i E_i^\perp\right)^\perp\right) \qquad \text{De Morgan in } R(W) \qquad (\text{A.1.33})$$

$$= \left(\bigcap_i E_i^\perp\right)^{\perp\perp c} \qquad\qquad\qquad\qquad (\text{A.1.34})$$

$$= \left(\bigcap_i E_i^\perp\right)^c \qquad\qquad \text{from Proposition A.1.2,}$$
$$\qquad\qquad\qquad\qquad\qquad \text{statements (3) and (4)} \quad (\text{A.1.35})$$

$$= \bigcup_i E_i^{\perp c} \qquad\qquad\qquad\qquad\qquad (\text{A.1.36})$$

$$= \bigcup_i (\Diamond E_i) \qquad\qquad\qquad\qquad\qquad (\text{A.1.37})$$

which is the join in Part W. Thus, when restricted to $R(W)$, $\Diamond$ is a homomorphism of complete otholattices. Moreover, for $E \in R(W)$, $\Box\Diamond E = E^{\perp\perp} = E$ so on $R(W)$:

$$\Box\Diamond = 1_{R(W)} \qquad\qquad\qquad (\text{A.1.38})$$

(1_X denoting the identity map on the set X) and for $A = \Diamond F \in$ Part W,

$$\Diamond\Box A = \Diamond\Box\Diamond F \qquad\qquad\qquad (\text{A.1.39})$$
$$= \Diamond F \qquad \text{by M9} \qquad\qquad (\text{A.1.40})$$
$$= A, \qquad\qquad\qquad\qquad (\text{A.1.41})$$

so on Part W

$$\Diamond\Box = 1_{\text{Part } W}. \qquad\qquad\qquad (\text{A.1.42})$$

This proves the theorem. $\Box$

This result has interesting applications. Cf. section 1.4.2, Example 3.

The following result answers a question of central interest to us here. First we shall need the following definition.

Definition A.1.6. Let L denote a bounded lattice. Then a is a *complement* of b if $a \sqcup b = 1_L$ and $a \sqcap b = 0_L$. L is said to be *uniquely complemented* if every element in L has a unique complement.

Theorem A.1.2. *Let L denote an ortholattice. Then the following are equivalent.*

1. *L is uniquely complemented;*
2. *L is Boolean;*
3. *For $a, b \in L$, $a \sqcap b = 0_L$ implies $a \sqsubseteq b'$.*

Proof. (1) implies (2). This is proved in [56], Proposition 7, p. 26.

(2) implies (3). Suppose $a \sqcap b = 0_L$ with L Boolean. Then

$$(a \sqcap b) \sqcup b' = b' \tag{A.1.43}$$
$$= (a \sqcup b') \sqcap (b \sqcup b') \quad \text{since } L \text{ is distributive} \tag{A.1.44}$$
$$= a \sqcup b', \tag{A.1.45}$$

and the result follows from equations (A.1.43) and (A.1.45).

(3) implies (1). Suppose a has complement b. That is

$$a \sqcap b = 0_L \quad \text{and} \tag{A.1.46}$$
$$a \sqcup b = 1_L \quad \text{so that} \tag{A.1.47}$$
$$a' \sqcap b' = 0_L. \tag{A.1.48}$$

Then (3) implies

$$a \sqsubseteq b' \quad \text{from equation (A.1.46) and} \tag{A.1.49}$$
$$a' \sqsubseteq b \text{ or } b' \sqsubseteq a \quad \text{from equation (A.1.48).} \tag{A.1.50}$$

Consequently

$$a = b' \text{ thus } b = a'.$$

But b was an arbitrary complement of a so the result follows. $\quad\square$

This result bears on the crucial difference between non-Boolean — presumed quantum-like — behavior, and Boolean — presumed non-quantum-like or classical — behavior. Cf. section 1.3.1.

With regard to the third statement of Theorem A.1.2 we note the following obvious fact about subsets of orthogonality spaces.

Lemma A.1.2. *For E, $F \subseteq W$ for a orthogonality space $\langle W, \perp \rangle$, $E \perp F$ iff $E \subseteq F^{\perp}$.*

The following are immediate from the theorem.

Corollary A.1.1. *An ortholattice L is not Boolean iff there exists an $a \in L$ that is not uniquely complemented.*

Corollary A.1.2. *An ortholattice L is not Boolean iff $\exists\, a, b \in L$ such that $a \sqcap b = 0_L$ and $a \not\subseteq b'$.*

Now we note the following:

Lemma A.1.3. *For W an orthogonality space, and $x, y \in W$*

$$x^{\perp\perp} \perp y^{\perp\perp} \quad \text{iff} \quad x \perp y. \tag{A.1.51}$$

Proof.

$$
\begin{aligned}
x^{\perp\perp} \perp y^{\perp\perp} \quad &\text{iff} \quad x^{\perp\perp} \subseteq y^{\perp\perp\perp} = y^{\perp} \quad \text{by M7} &(A.1.52)\\
&\text{iff} \quad x \in y^{\perp} \quad \text{since } y^{\perp} \text{ is a proposition} &(A.1.53)\\
&\text{iff} \quad x \perp y. &(A.1.54)
\end{aligned}
$$

$\square$

Then, from the lemmata and the second and third statement of the theorem:

Corollary A.1.3. *Suppose $R(W)$ is Boolean. Then for $x, y \in R(W)$*

$$x^{\perp\perp} \cap y^{\perp\perp} = \emptyset \quad \text{implies} \quad x \perp y \quad \text{or equivalently} \tag{A.1.55}$$

$$x \approx y \quad \text{implies} \quad x^{\perp\perp} \cap y^{\perp\perp} \neq \emptyset \tag{A.1.56}$$

The following corollary follows immediately from the lemmata and the second and third statements of the theorem.

Corollary A.1.4. *If W is such that $\exists \, x, y \in W$ for which*

$$x^{\perp\perp} \cap y^{\perp\perp} = \emptyset \quad and \quad x \approx y \tag{A.1.57}$$

then $R(W)$ is not Boolean.

This last corollary yields a useful test for non-Booleanness in what follows as we shall see in section 1.4.2.

Corollary A.1.5. *Suppose $R(W)$ is not Boolean. Then there exist proper propositions P and Q with $P \sqcup Q = W$ but $P \cup Q \neq W$. That is to say, there exists an $x \in P \sqcup Q$ but $x \notin P$ and $x \notin Q$.*

Proof. If $R(W)$ is not Boolean there must exist propositions E, F, say, such that $E \cap F = \emptyset$ and $E \not\subseteq F^{\perp}$, by the last corollary. Both E and F must then be proper, and $E^{\perp} \sqcup F^{\perp} = W$. Since $E \not\subseteq F^{\perp}$ there exists an $x \in E$ such that $x \notin F^{\perp}$. Since $x \in E$, $x \notin E^{\perp}$. Take $P = E^{\perp}$ and $Q = F^{\perp}$. (If there are no proper propositions, $R(W) \cong \mathbf{2}$ which is Boolean.) $\qquad\square$

Thus, in the non-Boolean $R(W)$ case, there are superpositional "states." There is a related but weaker result of the above type which should be mentioned here. It is not a corollary of Theorem A.1.2, and is easy to prove. Namely:

Proposition A.1.3. *If $R(W)$ is not Boolean there exists a proper proposition F and an element $w \in W$ such that $w \in F \sqcup F^{\perp}$ but $w \notin F$ and $w \notin F^{\perp}$.*

Proof. First note again that for elements E, F of $R(W)$, from M10,

$$E \sqcup F = (E^{\perp} \cap F^{\perp})^{\perp} = (E \cup F)^{\perp\perp} \supseteq E \cup F. \tag{A.1.58}$$

Now suppose that for all propositions E in $R(W)$ we have $E \sqcup E^{\perp} = E \cup E^{\perp}$. Since $E \sqcup E^{\perp} = W$ and $E \cap E^{\perp} = \emptyset$ this entails $E^{\perp} = E^{c}$.

Then the join in $R(W)$ is given by

$$E \sqcup F = (E^\perp \cap F^\perp)^\perp \tag{A.1.59}$$
$$= (E^c \cap F^c)^c \quad \text{by the above and Proposition A.1.2 (4)} \tag{A.1.60}$$
$$= E \cup F \tag{A.1.61}$$

so that $R(W)$ is just a Boolean lattice of subsets of W contradicting our assumption.

Consequently there exists a proposition, F, say, such that

$$W = F \sqcup F^\perp \neq F \cup F^\perp \tag{A.1.62}$$

so that the inclusion $F \cup F^\perp \subset F \sqcup F^\perp = W$ is strict and F can be neither empty nor W (since otherwise the inequality (A.1.62) would not hold). So there exists a $w \in W$ such that $w \in F \sqcup F^\perp$ but $w \notin F$ and $w \notin F^\perp$ as required. □

(We note that in view of the Goldblatt Stonean Theorem, the last four results could be formulated for a general ortholattice, since the elements of *all* ortholattices may be realized as sets of "states," namely the set of proper filters containing the element (equation (A.1.17)) but such formulations would be contorted and specious.)

For an interpretation of the crucial differences between the quantum-like non-Boolean case and the classical-like Boolean case as specified in Theorem A.1.2, please see section 1.3.1.

We have mixed the discussion above between general ortholattices and those of the type $R(W)$. In the latter type, we have lattices whose elements are subsets of a set, and so have a substructure consisting of entities which are like the states of a system in the classical meaning of the term "state," namely the elements of W. (As noted above, this is no actual restriction within the class of ortholattices in view of the Goldblatt Stonean Theorem.) To further complicate matters, in actual quantum theory, the "pure states" are identifiable with *propositions* — that is, closed subspaces of a Hilbert space — of the form $w^{\perp\perp}$, namely the smallest proposition containing the element w. In case w is a vector, this is the one-dimensional subspace, or ray, containing w. These "pure" states, in the actual quantum case, are the ones that are not statistical mixtures (i.e. convex real combinations of) other pure states, though they may be superpositions of them.

Thus, in the last proposition, the "pure state" determined by w is a Ketzi-like Limbo state, which verifies the disjunction $F \sqcup F^{\perp}$ (*being alive* quantum-OR *being dead*) but not *being alive* and not *being dead*, all such logical claims to be taken *cum multis granis salis*.

Before we leave the abstract case, it will be useful to note the following.

Lemma A.1.4. *For $P \in R(W)$,*

$$P = \bigcup_{x \in P} x^{\perp\perp}. \qquad (A.1.63)$$

Proof. If P is a proposition and $x \in P$ then $x \in x^{\perp\perp} \subseteq P$ so $P \subseteq \bigcup_{x \in P} x^{\perp\perp} \subseteq P$. $\qquad \square$

It follows that

$$P = P^{\perp\perp} = \left(\bigcup_{x \in P} x^{\perp\perp} \right)^{\perp\perp} = \bigsqcup_{x \in P} x^{\perp\perp}. \qquad (A.1.64)$$

Thus a proposition is determined by its pure states. This will be helpful to us later.

A.1.4 *Subsets of $\mathbb{R}^n \backslash \{0\}$*

We shall now specialize to the case of interest to us here, namely orthogonality spaces which are subsets $W \subset \mathbb{R}^n \backslash \{0\}$, with the inherited, or relative, orthogonality relation inherited from the ambient Euclidean space.

CAVEAT!
Within the scope of this relative case we shall reserve the use of the unadorned connectives, $\perp$, $\approx$, $(\ )^c$, etc., for those in the ambient Euclidean space and $\perp_W$, $\approx_W$, $(\ )^{c_W}$, etc., for those in the relative case.

Thus, for $\mathbf{v}, \mathbf{w} \in W$,

$$\mathbf{w} \perp_W \mathbf{v} \quad \text{iff} \quad \mathbf{w} \perp \mathbf{v}, \tag{A.1.65}$$

$$\mathbf{w} \approx_W \mathbf{v} \quad \text{iff} \quad \mathbf{w} \not\perp_W \mathbf{v} \quad \text{iff} \quad \mathbf{w} \approx \mathbf{v}, \tag{A.1.66}$$

and for $E \subseteq W$ we have

$$E^{\perp_W} := \{\mathbf{v} \in W : \mathbf{w} \in E \text{ implies } \mathbf{w} \perp_W \mathbf{v}\} \tag{A.1.67}$$

$$= \{\mathbf{v} \in W : \mathbf{w} \in E \text{ implies } \mathbf{w} \perp \mathbf{v}\} \tag{A.1.68}$$

$$= \{\mathbf{v} \in W : \mathbf{v} \in E^{\perp}\} \tag{A.1.69}$$

$$= E^{\perp} \cap W; \tag{A.1.70}$$

$$E^{\perp_W \perp_W} = (E^{\perp} \cap W)^{\perp} \cap W; \tag{A.1.71}$$

$$E^{c_W} := \{\mathbf{v} \in W : \mathbf{w} \notin E\} \tag{A.1.72}$$

$$= E^c \cap W; \tag{A.1.73}$$

$$\Diamond_W E = E^{\perp_W c_W} \tag{A.1.74}$$

$$= E^{\perp_W c} \cap W \tag{A.1.75}$$

$$= (E^{\perp} \cap W)^c \cap W. \tag{A.1.76}$$

It is worth noting again that for $\mathbf{v} \in \mathbb{R}^n$, $\mathbf{v}^{\perp\perp}$ is the one-dimensional subspace, or ray, containing $\mathbf{v}$, and that for $\mathbf{v}, \mathbf{w} \in \mathbb{R}^n$, $\mathbf{v}^{\perp\perp} \cap \mathbf{w}^{\perp\perp} = \mathbf{0}$ iff $\mathbf{v}$ and $\mathbf{w}$ are linearly independent.

Again within the scope of this relative case, we shall generally drop the reference to $\perp_W$ in denoting the ortholattice of (relative) propositions and just write $R(W)$ for this lattice.

The following are just transcriptions of Corollaries A.1.3, and A.1.4 into the relative case.

Proposition A.1.4. *If $R(W)$ is Boolean then for $\mathbf{v}, \mathbf{w} \in W$*

$$\mathbf{v}^{\perp_W \perp_W} \cap \mathbf{w}^{\perp_W \perp_W} = \emptyset \quad \text{implies} \quad \mathbf{v} \perp \mathbf{w}. \tag{A.1.77}$$

Proposition A.1.5. *If W is such that $\exists\, \mathbf{v}, \mathbf{w} \in W$ for which*

$$\mathbf{v}^{\perp_W \perp_W} \cap \mathbf{w}^{\perp_W \perp_W} = \emptyset \quad \text{and} \quad \mathbf{v} \approx \mathbf{w} \tag{A.1.78}$$

then $R(W)$ is not Boolean.

This last corollary will be our most useful criterion for determining non-Booleanness.

Relevant examples are discussed in section 1.4.2.

A.1.5 *Orthologic: Models, completeness and the Modal Embedding Theorem*

Theorem 1.2.2 (recalled)

$$\vdash_O \alpha \ \textit{iff} \ \models \alpha.$$

Proof. Suppose $\vdash_O \alpha$ and let $\mathcal{M} = \langle W, \approx, \varrho \rangle$ denote an arbitrary Kripke orthomodel. Then $\langle R(W), \varrho \rangle$ constitutes an algebraic orthomodel and Corollary 1.2.1 implies that $\varrho(\alpha) = W$ so that α is true in $\mathcal{M}$. The latter being arbitrary means α is Kripke valid.

Conversely, suppose $\models \alpha$ and let $\mathcal{A} := \langle L, v_L \rangle$ denote an arbitrary algebraic orthomodel. Then $\mathcal{M} := \langle W_L, \approx_L, \phi_L \circ v_L \rangle$ constitutes a Kripke orthomodel, where the notation is as in the Goldblatt Stonean Theorem A.1.1. Since α is assumed Kripke valid, it is true in $\mathcal{M}$, so that $\phi_L(v_L(\alpha)) = W_L$. But ϕ_L is injective, so $v_L(\alpha) = 1_L$. Thus α is true in the arbitrary algebraic orthomodel $\mathcal{A}$, so $\vdash_O \alpha$. □

Theorem 1.2.4 (The Modal Embedding Theorem, recalled)

$$\textit{For } \alpha \in \Phi, \ \vdash_O \alpha \ \textit{iff} \ \vdash_B \alpha^\circ.$$

Proof. To prove the Modal Embedding Theorem, it will first prove convenient, here and elsewhere, to introduce the notation for *truth sets*.

Definition A.1.7. For a modal formula α and a B-model $\mathcal{B} = \langle W, \approx, v \rangle$ we define

$$\|\alpha\|_{\mathcal{B}} := \{w \in W : w \Vdash_{\mathcal{B}} \alpha\}. \tag{A.1.79}$$

Then proof of the following is straightforward.

Lemma A.1.5.

$$\|\alpha \wedge \beta\|_{\mathcal{B}} = \|\alpha\|_{\mathcal{B}} \cap \|\beta\|_{\mathcal{B}} \tag{A.1.80}$$

$$\|\neg\alpha\|_{\mathcal{B}} = \|\alpha\|_{\mathcal{B}}^c \tag{A.1.81}$$

$$\|\Box\alpha\|_{\mathscr{B}} = \Box\|\alpha\|_{\mathscr{B}} \qquad \text{(A.1.82)}$$
$$\|\Diamond\alpha\|_{\mathscr{B}} = \Diamond\|\alpha\|_{\mathscr{B}} \qquad \text{(A.1.83)}$$

where the right-hand sides of the last two equations are as in equations (A.1.3) and (A.1.4).

We shall generally drop the subscript when it is not ambiguous.

To proceed with the proof of Theorem 1.2.4 we now suppose a Kripke orthomodel

$$\mathscr{M} := \langle W, \approx, \varrho \rangle \qquad \text{(A.1.84)}$$

is given (cf. Definition 1.2.2). Then a B-model

$$\mathscr{B}_{\mathscr{M}} := \langle W, \approx, \mathrm{v}_{\varrho} \rangle \qquad \text{(A.1.85)}$$

may be constructed by defining for all $w \in W$ and atomic formulas a_i:

$$\mathrm{v}_{\varrho}(a_i, w) := \begin{cases} 1 & \text{if } w \in \varrho(a_i) \\ 0 & \text{if } w \notin \varrho(a_i) \end{cases} \qquad \text{(A.1.86)}$$

and inductively extending it to all modal formulas via V1–V4 given above.

Note that we have

$$\|\alpha\|_{\mathscr{B}_{\mathscr{M}}} = \varrho(\alpha). \qquad \text{(A.1.87)}$$

Lemma A.1.6. *For atomic $a_i \in \Phi$*

1. $w \Vdash_{\mathscr{B}_{\mathscr{M}}} \Box\Diamond a_i$ *iff* $w \in \varrho(a_i)$;
2. $w \Vdash_{\mathscr{B}_{\mathscr{M}}} \Box\neg a_i$ *iff* $w \in \varrho(\sim a_i)$.

Proof.

1. $\mathrm{v}_{\varrho}(\Box\Diamond a_i, w) = 1$ iff $w \in \|\Box\Diamond a_i\| = \Box\Diamond\varrho(a_i) = \varrho(a_i)^{\perp\perp} = \varrho(a_i)$.
2. $\mathrm{v}_{\varrho}(\Box\neg a_i, w) = 1$ iff $w \in \|\Box\neg a_i\| = \Box\neg\|a_i\| = \Box\|a_i\|^c = \Box\varrho(a_i)^c = \varrho(a_i)^{\perp} = \varrho(\sim a_i)$. $\qquad\qquad\Box$

Now we note that, from the translation rules T1 and T3:

$$\|(\sim a_i)^\circ\| := \|\Box\neg\Box\Diamond a_i\| \tag{A.1.88}$$
$$= \|\Box\Diamond\neg\Diamond a_i\| \tag{A.1.89}$$
$$= \|\Box\Diamond\Box\neg a_i\| \tag{A.1.90}$$
$$= \Box\Diamond\Box\|\neg a_i\| \tag{A.1.91}$$
$$= \Box\|\neg a_i\| \qquad \text{from M8} \tag{A.1.92}$$
$$= \|\Box\neg a_i\|. \tag{A.1.93}$$

That is to say

$$v_\varrho((\sim a_i)^\circ, w) = 1 \text{ iff } v_\varrho((\Box\neg a_i)^\circ, w) = 1. \tag{A.1.94}$$

Thus, the last lemma may be restated as

Corollary A.1.6.

1. $w \Vdash_{\mathscr{B}_{\mathscr{M}}} a_i^\circ$ *iff* $w \in \varrho(a_i)$;
2. $w \Vdash_{\mathscr{B}_{\mathscr{M}}} (\neg a_i)^\circ$ *iff* $w \in \varrho(\sim a_i)$.

Proposition A.1.6. *For* $\alpha \in \Phi$

$$w \Vdash_{\mathscr{B}_{\mathscr{M}}} \alpha^\circ \text{ iff } w \in \varrho(\alpha).$$

The proof is an easy induction on the complexity (i.e. length) of α, taking the base case as in the last corollary.

Proposition A.1.7. *For* $\alpha \in \Phi$

$$\vdash_{\mathbf{B}} \alpha^\circ \text{ implies } \vdash_{\mathbf{O}} \alpha.$$

Proof. Choose a Kripke orthomodel $\mathscr{M} = \langle W, \approx, \varrho \rangle$. Then if $\vdash_{\mathbf{B}} \alpha^\circ$, α° is true in the B-model associated with $\mathscr{M}$, that is $\mathscr{B}_{\mathscr{M}} = \langle W, \approx, v_\varrho \rangle$, as in equation (A.1.85). Thus $w \Vdash_{\mathscr{B}_{\mathscr{M}}} \alpha^\circ$ for all $w \in W$. Then, by the last Proposition, $\varrho(\alpha) = W$. So α is true in the Kripke orthomodel $\mathscr{M}$ which was chosen arbitrarily: that is, $\models \alpha$. Thus $\vdash_{\mathbf{O}} \alpha$, by Theorem 1.2.2. $\qquad\square$

This proves one implication of the Modal Embedding Theorem. To prove the converse, choose a B-model $\mathscr{B} = \langle W, \approx, v \rangle$. We may use it to construct a Kripke orthomodel $\mathscr{M}^{\mathscr{B}} := \langle W, \approx, \varrho_v \rangle$, where $\varrho_v : \Phi \to R(W)$ is given on atoms a_i by

$$\varrho_v(a_i) := \|a_i^\circ\|_{\mathscr{B}} = \|\Box\Diamond a_i\|_{\mathscr{B}} = \|a_i\|_{\mathscr{B}}^{\perp\perp}. \tag{A.1.95}$$

The following lemma is again an easy induction on complexity.

Lemma A.1.7. *For any* $\alpha \in \Phi$

$$\varrho_v(\alpha) = \|\alpha^\circ\|_{\mathscr{B}}.$$

Proposition A.1.8. *For* $\alpha \in \Phi$

$$\vdash_{\mathbf{O}} \alpha \; implies \; \vdash_{\mathbf{B}} \alpha^\circ.$$

Proof. Choose any B-model $\mathscr{B} = \langle W, \approx, v \rangle$. If $\vdash_{\mathbf{O}} \alpha$, then α is true in the associated Kripke orthomodel $\mathscr{M}^{\mathscr{B}}$ by Theorem 1.2.2, which is to say $\varrho_v(\alpha) = W$. Then, by the last lemma $\|\alpha^\circ\|_{\mathscr{B}} = W$ showing that α° is true in the arbitrary B-model $\mathscr{B}$ and so B-valid: $\vdash_{\mathbf{B}} \alpha^\circ$. $\qquad\square$

The last two propositions prove Theorem 1.2.4.

Appendix B

A Mathematics Primer

B.1 Some multilinear algebra

References for this appendix include [20, 35, 57, 59, 62].

B.1.1 *Tensor products*

Let $V_1, V_2, \ldots, V_n$ denote vector spaces over a field (or modules over a ring) k which we may take to be $\mathbb{R}$, the field we will be using here. If W is another vector space (or module) over the same field or ring, denote by $f\colon V_1 \times V_2 \times \cdots \times V_n \to W$ a function that is linear in each variable separately. (The spaces need not be finite dimensional here.) Such functions are called *multilinear*. It might be thought that there is a complicated theory of multilinear functions that reduces to ordinary linear algebra in the case of a single V. However, such a theory is not necessary, since for any such collection of vector spaces $V_1, V_2, \ldots, V_n$ there exists a single vector space $T(V_1, V_2, \ldots, V_n)$ satisfying the following Universal Mapping Property (UMP):

There exists a multilinear map $\iota\colon V_1 \times V_2 \times \cdots \times V_n \to T(V_1, V_2, \ldots, V_n)$ such that for any multilinear map $f\colon V_1 \times V_2 \times \cdots \times V_n \to W$ there exists a *unique* linear map $\tilde{f}\colon T(V_1, V_2, \ldots, V_n) \to W$ such that the diagram in Figure B.1 commutes.

It is easy to prove that such a $T(V_1, V_2, \ldots, V_n)$ must be unique up to isomorphism of vector spaces. It is called the *tensor product* of the vector spaces involved and written $V_1 \otimes V_2 \otimes \cdots \otimes V_n = \bigotimes_{i=1}^{n} V_i$. Usually the base field or ring is appended to the tensor sign as in $\otimes_k$ since often algebraists have many fields and/or rings to deal with

$$V_1 \times \cdots \times V_n \xrightarrow{\quad \iota \quad} T(V_1, \ldots, V_n)$$

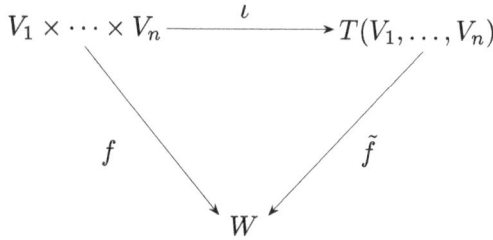

Figure B.1 The universal mapping diagram for the tensor product.

simultaneously. If the field is not in doubt, as here, it is omitted. In this way multilinear maps are turned into linear ones and there is no need for a separate theory. One may think of the tensor product as the vector space generated by basis vectors of the form $a_1 \otimes a_2 \otimes \cdots \otimes a_n, a_i \in V_i$ subject to linear additivity in each variable and the scalar multiplication property $\lambda a_1 \otimes a_2 \otimes \cdots \otimes a_n = a_1 \otimes \lambda a_2 \otimes \cdots \otimes a_n = \cdots = a_1 \otimes a_2 \otimes \cdots \otimes \lambda a_n, \lambda \in k$.

If V_i has dimension d_i then $\dim(\bigotimes_{i=1}^{n} V_i) = d_1 d_2 \ldots d_n$.

Properties of tensor products may be derived entirely through the use of the UMP stated above. For instance, suppose linear maps $f_i : V_i \to W_i, i = 1, \ldots, n$ are given. Then $f_1 \times \cdots \times f_n : V_1 \times \cdots \times V_n \to W_1 \times \cdots \times W_n$ is multilinear so that its composition with the linear ι-map of $W_1 \times \cdots \times W_n$ into $W_1 \otimes \cdots \otimes W_n$ is also multilinear so that it may be lifted to produce a linear map $V_1 \otimes \cdots \otimes V_n \to W_1 \otimes \cdots \otimes W_n$ as in the diagram above. This map is denoted $\bigotimes_{i=1}^{n} f_i = f_1 \otimes \cdots \otimes f_n$. We note also that the distributive rule analogous to the one stated for logical equivalence in Proposition 3.2.1, part 6, holds in the category of vector spaces with this equivalence replaced by isomorphism.

Finally, we note some further identities that will be useful, namely: $V \otimes W \cong W \otimes V$, $k \otimes V \cong V$, and $\text{End} V \cong V^* \otimes V$. Here V and W are k-vector spaces and for the last identity V is finite dimensional. Here, V^* denotes the *dual space* $\text{Hom}(V, k)$ where $\text{Hom}(V, W)$ denotes the vector space of linear functions from V to W, $\text{End} V := \text{Hom}(V, V)$ and $\cong$ denotes isomorphism. Note that for finite dimensional V, $V^* \cong V$, though this isomorphism will in general depend upon a chosen basis.

B.1.2 *Exterior products*

With notation as in the last section, let $V^p := \overbrace{V \times \cdots \times V}^{p}$ for $p \geqslant 2$. Then, a multilinear function $f : V^p \to W$ is said to be *alternating* if

$$f(v_1, \ldots, v_i, v_i, \ldots, v_p) = 0 \tag{B.1.1}$$

for any i. It follows from multilinearity that interchanging any pair of adjacent variables changes the sign of the value of f and from this that the same holds for the interchange of any pair of variables. Then it follows that the repetition of any pair of variables causes the value of f to vanish. There is a UMP for alternating maps similar to the one for general multilinear maps. The unique vector space that plays the rôle of the tensor product $\bigotimes^p V$ in this case is written $\bigwedge^p V$, and called the *exterior* product. It is generated by elements of the form $v_1 \wedge v_2 \wedge \cdots \wedge v_p$, $v_i \in V$. This element is multilinear in its arguments and alternating in the sense described above for f. Thus for instance, for $v, w \in V$, $v \wedge v = 0$ and if $v \wedge w = 0$ then v and w generate the same one-dimensional subspace, i.e. are *colinear*. (For, if v and w were not linearly dependent they could be included in a basis for V in which case $v \wedge w$ would be a basis element of $\bigwedge^2 V$ which could not be the zero vector. It is also easy to prove directly that if $v \wedge w = 0$ then v and w are colinear.) The map corresponding to ι in the last section sends $(v_1, \ldots, v_p)$ to $v_1 \wedge v_2 \wedge \cdots \wedge v_p$. It is not hard to show that, if the dimension of V is n, then $\dim \bigwedge^p V = \binom{n}{p}$. Note that $\dim \bigwedge^n V = 1$ and that $\bigwedge^p V = \{0\}$ if $p > n$. A useful intuition is that the exterior product $v_1 \wedge v_2 \wedge \cdots \wedge v_p$ is a vector whose length is proportional to the volume of the polytope bounded by the vectors $v_1, \ldots, v_p$.

B.1.3 *Exterior algebras*

These exterior products may be assembled into a unital associative algebra (i.e. an associative algebra containing a unit) having a certain universal mapping property with respect to linear maps $f : V \to A$ into such an algebra A, having the property that $f(v)^2 = 0$ for all $v \in V$. Namely, there exists an associative unital algebra $E(V)$ for

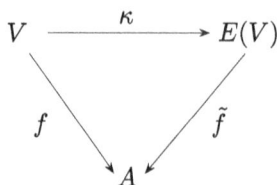

Figure B.2   The universal mapping diagram for the exterior algebra of V.

any vector space V, and a linear map $\kappa : V \to E(V)$ having the property mentioned, such that if $f : V \to A$ is any linear map into any associative unital algebra A having that "square free" property, then there exists a unique algebra morphism $\tilde{f} : E(V) \to A$ such that $\tilde{f} \circ \kappa = f$ (Figure B.2). $E(V)$ is then necessarily unique up to algebra isomorphism. One may take this algebra to be the *exterior algebra* of V defined by

$$E(V) := \bigoplus_{k \geq 0} \bigwedge^k V \tag{B.1.2}$$

which is easily seen to satisfy the UMP. In case V is finite dimensional, of dimension n, say, this direct sum terminates to give

$$E(V) = \bigwedge^0 V \oplus \bigwedge^1 V \oplus \bigwedge^2 V \oplus \cdots \oplus \bigwedge^n V \tag{B.1.3}$$

$$= \mathbb{R} \oplus V \oplus \bigwedge^2 V \oplus \cdots \oplus \bigwedge^n V. \tag{B.1.4}$$

Here we take $\bigwedge^0 V = \mathbb{R}$ and $\bigwedge^1 V = V$. Note that $\dim E(V) = 2^n$. The algebra multiplication is given by wedging together two elements of the summands — called *homogenous elements* — in the order given, and extending by linearity to the whole space, with elements in $\bigwedge^0 V = \mathbb{R}$ just acting as scalars in the usual way. The map $\kappa : V \to E(V)$ is given by $\kappa(v) = v$ considered to lie in the summand $\bigwedge^1 V = V$. This algebra has many interesting properties and symmetries which were understood by H. Grassmann in the middle of the 19th century but whose published treatment of it was famously misunderstood by his contemporaries, probably because of limitations in the notations of the time. We shall rehearse a few of these properties here. First we note that for two finite dimensional vector spaces V and W the map

$$\overset{m}{\bigwedge} V \otimes \overset{n}{\bigwedge} W \longrightarrow \overset{m+n}{\bigwedge} (V \oplus W) \tag{B.1.5}$$

given by $(v_1 \wedge \cdots \wedge v_m) \otimes (w_1 \wedge \cdots \wedge w_n) \mapsto v_1 \wedge \cdots \wedge v_m \wedge w_1 \wedge \cdots \wedge w_n$ induces an isomorphism of vector spaces

$$\overset{p}{\underset{k=0}{\bigoplus}} \left(\overset{k}{\bigwedge} V \otimes \overset{p-k}{\bigwedge} W \right) \overset{\cong}{\Rightarrow} \overset{p}{\bigwedge}(V \oplus W) \tag{B.1.6}$$

from which follows the isomorphism of vector spaces

$$\overset{m+n}{\underset{p=0}{\bigoplus}} \left[\overset{p}{\underset{k=0}{\bigoplus}} \left(\overset{k}{\bigwedge} V \otimes \overset{p-k}{\bigwedge} W \right) \right] \overset{\cong}{\Rightarrow} \overset{m+n}{\underset{p=0}{\bigoplus}} \left[\overset{p}{\bigwedge}(V \oplus W) \right]. \tag{B.1.7}$$

That is to say, we have an isomorphism of vector spaces

$$\phi : E(V) \otimes E(W) \overset{\cong}{\Rightarrow} E(V \oplus W) \tag{B.1.8}$$

which is *not* an isomorphism of algebras when the ordinary tensor product algebra multiplication is used on the left hand side of equation (B.1.8). There is, however, an algebra product structure on the left hand side that does render that isomorphism above an isomorphism of algebras: it is called the *graded* product and it is described as follows. For homogeneous elements $a, c \in E(V)$, $b, d \in E(W)$ the *graded product* on the algebra $E(V) \otimes E(W)$ is determined by the definition:

$$(a \otimes b)(c \otimes d) := (-1)^{\deg(b)\deg(c)}(ac \otimes bd), \tag{B.1.9}$$

where the degree $\deg(f)$ of a homogeneous element f is the power of the exterior product to which it belongs, also called the *grade* of f. The graded multiplication above may be canonically extended to any number of tensor products of algebras. (The notion of grading for algebraic objects was codified in the 1950s by some of the algebraists of Paris [20]).

We shall prove this result as an illustration of some of the formalism.

Theorem B.1.1. *The vector space isomorphism ϕ in equation (B.1.8) is an isomorphism of algebras with exterior product on the right and the graded product equation (B.1.9) on the left.*

Proof. It is sufficient to prove the theorem for the inverse ϕ^{-1} : $E(V \oplus W) \xrightarrow{\cong} E(V) \otimes E(W)$. Apply this isomorphism to first grade ($p = 1$) elements, namely those in $V \oplus W \cong V \otimes k \oplus k \otimes W$, to obtain $\phi^{-1}(v, w) = v \otimes 1 + 1 \otimes w$. Then note, using the graded product on $E(V) \otimes E(W)$, that

$$(\phi^{-1}(v, w))^2 = (v \otimes 1 + 1 \otimes w)(v \otimes 1 + 1 \otimes w) \qquad (B.1.10)$$

$$= (v \otimes 1)(v \otimes 1) + (1 \otimes w)(v \otimes 1)$$

$$+ (v \otimes 1)(1 \otimes w) + (1 \otimes w)(1 \otimes w) \qquad (B.1.11)$$

$$= (-1)^{\deg(w)\deg(v)}(v \otimes w) + (-1)^{\deg(1)\deg(1)}(v \otimes w) \qquad (B.1.12)$$

the outer terms being zero

$$= (-1)^{1.1}(v \otimes w) + (-1)^{0.0}(v \otimes w) \qquad (B.1.13)$$

$$= 0. \qquad (B.1.14)$$

Thus ϕ^{-1} restricted to $V \oplus W$, $\phi^{-1}|_{V \oplus W}$, satisfies the "square free" condition of the UMP for the exterior algebra, with κ as above, so there exists a *unique algebra isomorphism* $(\phi^{-1}|_{V \oplus W})^{\sim}$: $E(V \oplus W) \to E(V) \otimes E(W)$ with the graded multiplication on the latter space, which satisfies the UMP, so extends the restriction $\phi^{-1}|_{V \oplus W}$ of ϕ^{-1} to $V \oplus W$ to the whole of $E(V \oplus W)$, so must be ϕ^{-1} itself. $\qquad \square$

Let us consider the case when V is one-dimensional, with basis element e, say. Then clearly $E(V) = E(\mathbb{R}e) = \mathbb{R} \oplus V \cong \mathbb{R} \oplus \mathbb{R}e$ in our earlier notation, where, as an element in the algebra of $E(\mathbb{R}e)$, $e^2 = e \wedge e = 0$. It is immediate that $E(\mathbb{R}e)$ is commutative as an algebra. Now, for any finite dimensional vector space V with basis

$\{e_1, \ldots, e_n\}$ we have from (B.1.8)

$$E(V) \cong E(\mathbb{R}e_1 \oplus \cdots \oplus \mathbb{R}e_n) \cong E(\mathbb{R}e_1) \otimes \cdots \otimes E(\mathbb{R}e_n) \quad \text{(B.1.15)}$$

as vector spaces. As noted above, this is not an isomorphism of algebras with the ordinary tensor product multiplication on the right hand side, since this would be commutative as all of the $E(\mathbb{R}e_i)$ are, while the left hand side is not. However, as just proved, the right hand side with graded product is isomorphic with the exterior product on the left hand side. Thus, the exterior algebra may be described in terms of graded tensor products of algebras isomorphic with $E(\mathbb{R}e)$. The reader may note the similarity of such tensor products to the notion of qubit registers in the parlance of quantum computation (with $\mathbb{R}$ replaced by $\mathbb{C}$).

If $f : V \to W$ is a linear map of vector spaces, there is a unique map of algebras $E(f) : E(V) \to E(W)$ that extends $f : V \to W$. This may be proved using the UMP for exterior algebra. It is easy to see that it is given by the linear extension of the assignments $E(f)(1) = 1, E(f)(v_1 \wedge \cdots \wedge v_k) = f(v_1) \wedge \cdots \wedge f(v_k)$.

The exterior algebra allows us, among other things, to define some well-known quantities associated with square matrices in a basis independent manner.

Determinant

If $f : V \to V$ then it is easy to see that $E(f) : E(V) \to E(V)$ preserves the grades of homogeneous elements. That is, $E(f)$ maps $\bigwedge^n V$ into $\bigwedge^n V$. In particular, if $\dim V = N$, then $\bigwedge^N V$ is one-dimensional and so $E(f) : \bigwedge^N V \to \bigwedge^N V$ must be implemented by multiplication by some constant in k. This constant may be defined to be the *determinant* of f and is generally written $\det(f)$.

Derivations

If A is an algebra, a linear map $d : A \to A$ is called a *derivation* if for any $a, b \in A$

$$d(ab) = (da)b + a(db). \quad \text{(B.1.16)}$$

If $f : V \to V$ is as above, there is a unique extension $d_f : E(V) \to E(V)$ of it to that is a derivation. It is defined by

$$d_f(e_1 \wedge e_2 \wedge \cdots \wedge e_k) = f(e_1) \wedge e_2 \wedge \cdots \wedge e_k + e_1 \wedge f(e_2) \wedge \cdots$$
$$\wedge e_k + \cdots + e_1 \wedge e_2 \wedge \cdots \wedge f(e_k). \quad \text{(B.1.17)}$$

Trace

Note that d_f, like f, preserves grade so that $d_f : \bigwedge^N V \to \bigwedge^N V$ is again implemented by multiplication by a constant in k. This constant is defined to be the *trace* of f and is generally written $\mathrm{tr}(f)$.

We should note that the fact that the exterior algebra of any vector space has a graded structure does not seem obvious at all from the UMP for it. However, this graded structure turns out to be of vital importance to our model since it represents the states of all the possible cooperative subnetworks of the networks of interest to us.

For a comprehensive modern treatment of this subject see [8].

B.1.4   *The Plücker Embedding*

For a (real) vector space V, let $\mathbf{Gr}(p, V)$ denote the family of subspaces of V of dimension p, the notation $\mathbf{Gr}$ being in honor of Grassmann. The special case of $p = 1$ is called the *projective space* of V, and is denoted by $\mathbf{P}(V)$. Exterior products may be used to obtain an explicit representation of $\mathbf{Gr}(p, V)$, namely the map

$$\psi : \mathbf{Gr}(p, V) \longrightarrow \mathbf{P}\left(\bigwedge^p V\right) \qquad \text{(B.1.18)}$$

given, for a p-dimensional subspace $W \subseteq V$, with basis $\{w_1, \ldots, w_p\}$, by

$$\psi(W) := \mathbb{R} w_1 \wedge \cdots \wedge w_p \qquad \text{(B.1.19)}$$

is well-defined. For, another basis of W is related to this basis by a matrix with a non-vanishing determinant and the corresponding exterior product is the previous one, namely $w_1 \wedge \cdots \wedge w_p$, multiplied by this determinant and so specifies the same element in $\mathbf{P}(\bigwedge^p V)$. Moreover, it is not hard to see that $w \in W$ if and only if $w \wedge \psi(W) = \mathbf{0}$, showing that ψ is one-to-one, or *injective*. (For, if $\psi(W_1) = \psi(W_2)$ then $w \in W_1$ iff $w \wedge \psi(W_1) = \mathbf{0}$ iff $w \wedge \psi(W_2) = \mathbf{0}$ iff $w \in W_2$ so $W_1 = W_2$). It is called the *Plücker Embedding*.

Intuitively this last result can be interpreted as follows: if the volume of the $(p+1)$-dimensional polytope obtained by adding w as another side to the p-dimensional polytope with sides $w_1, \ldots, w_p$ is zero, then w must lie in the polytope, and conversely. This is easily seen when $p = 2$: if adding a third vector to the two-dimensional polytope with sides w_1, w_2 produces a (three-dimensional) polytope of zero volume, then w must lie in the plane determined by w_1, w_2, and conversely.

B.2   A note on convolution

An *associative algebra* B over the field k may be defined as follows. B is a vector space over k with a map $m : B \otimes B \to B$ called *multiplication* (or sometimes abusively *product*). With $m(a \otimes b) := ab$ associativity means $a(bc) = (ab)c$. This condition may be expressed in the form of a commutative diagram specifying that $m(1_B \otimes m) = m(m \otimes 1_B)$. One virtue of expressing this condition diagramatically is that it may easily be dualized by reversing the arrows.

A *coassociative colagebra* A over the field k is a vector space over k with a map $\psi : A \to A \otimes A$ called *comultiplication* (or sometimes abusively *coproduct*) with a dual version of the diagram above commuting, namely $(\psi \otimes 1_A)\psi = (1_A \otimes \psi)\psi$.

Let A be such a coalgebra and B such an algebra. Given two k-linear maps $L, M : A \to B$ we get another linear map $L * M : A \to B$ as follows.

$$L * M := m(L \otimes M)\psi. \tag{B.2.1}$$

This *-operation is called *convolution*. It is easily seen to be an associative product, due to the associativity of B and the coassociativity of A.

The simplest way to see that ordinary convolution may be expressed in this form is to consider the group algebra $k[G]$ of a finite group G written multiplicatively. This is just the space of functions on G into k, and here we ignore the algebra product on it. Each such function may be expressed in the form

$$f = \sum_{g \in G} a_g \chi_g \qquad (B.2.2)$$

where $a_g \in k$ and χ_g is the characteristic function on G which is 1 on g and zero elsewhere. We have a coassociative comultiplication on $k[G]$ induced from the associative group multiplication $G \times G \to G$, namely the map $\psi : k[G] \to k[G \times G] \cong k[G] \otimes k[G]$ given for $f \in k[G]$ by $\psi(f)(g_1, g_2) = f(g_1 g_2)$.

We can express this precisely in tensorial form for $f = \chi_{g_0}$ as follows.

$$\psi(\chi_{g_0})(g_1, g_2) = \chi_{g_0}(g_1 g_2) \qquad (B.2.3)$$

$$= \begin{cases} 1 & \text{if} \quad g_0 = g_1 g_2 \\ 0 & \text{otherwise} \end{cases} \qquad (B.2.4)$$

$$= \begin{cases} 1 & \text{if} \quad g_0 g_2^{-1} = g_1 \\ 0 & \text{otherwise} \end{cases} \qquad (B.2.5)$$

$$= \left(\sum_g \chi_{g_0 g^{-1}} \otimes \chi_g \right)(g_1, g_2). \qquad (B.2.6)$$

Now take B to be k and $m : k \otimes k \to k$ to be multiplication in k and note that any linear functional $L : k[G] \to k$ acting on some $f = \sum_{g \in G} a_g \chi_g$ takes the form $L(f) = \sum_{g \in G} a_g L(\chi_g)$. Of course, any such L may be regarded as a function on G via the one-to-one map $g \leftrightarrow \chi_g$ so that we may write $L(g) := L(\chi_g)$ and concomitantly

$L(f) = \sum_g a_g L(g)$. Then we have

$$(L * M)(f) = (m(L \otimes M)\psi)\left(\sum_h a_h \chi_h\right) \qquad \text{(B.2.7)}$$

$$= m(L \otimes M)\left(\sum_h a_h \left(\sum_g \chi_{hg^{-1}} \otimes \chi_g\right)\right)$$

from equation (B.2.6) (B.2.8)

$$= \sum_h a_h \left(\sum_g L(hg^{-1})M(g)\right) \qquad \text{(B.2.9)}$$

so that the function of $h \in G$ corresponding to the functional $L * M$ is $(L * M)(h) = (L * M)(\chi_h) = \sum_g L(hg^{-1})M(g)$. This is just the usual convolution of the functions L and M. This construction has generalizations to measures and functions on locally compact groups and in other areas of harmonic analysis.

Bibliography

[1] Abramsky, S. (1993). Computational interpretations of linear logic, *Theor. Comput. Sci.* **111**, 3–57.

[2] Adger, D. (2003). *Core Syntax: A Minimalist Approach* (Oxford University Press, Oxford).

[3] Adolphs, R. (2013). The biology of fear, *Curr. Biol.* **23**(2), R79–R93.

[4] Aggarwal, C. S. (2018). *Neural Networks and Deep Learning: A Textbook* (Springer–Verlag, Berlin, Heidelberg, New York).

[5] Alexiev, V. (1994). Applications of linear logic to computation: An overview, *Log. J. IGPL* **2**(1), 77–107.

[6] Amit, D. J. (1998). *Modeling Brain Function: The World of Attractor Neural Networks* (Cambridge University Press, Cambridge).

[7] Asperti, A., Longo, G. (1991). *Categories, Types, and Structures: An Introduction to Category Theory for the Working Computer Scientist* (The MIT Press, Cambridge and London).

[8] Barnabei, M., Brini, A., Rota, G.–C. (1985). On the exterior calculus of invariant theory, *J. Algebra* **96**, 120–160.

[9] Becker, O. (1930). Zur logik der modalitäten, *Jahrb. Philos. Phänomen. Forsch.* **11**, 497–548.

[10] Bell, J. L. (1983). Orthologic, forcing and the manifestation of attributes, *Proceedings of the Southeast Asian Conference on Logic. Studies in Logic, vol. III* (North-Holland Publishing Company, Amsterdam).

[11] Bell, J. L. (1986). A new approach to quantum logic, *Br. J. Philos. Sci.* **37**, 83–99.

[12] Berwick, R. C., Chomsky, N. (2016). *Why Only Us: Language and Evolution* (The MIT Press, Cambridge and London).

[13] Birkhoff, G., von Neumann, J. (1936). The logic of quantum mechanics, *Ann. Math.* **37**(4), 823–843.

[14] Boniolo, G., D'Agostino, M., Piazza, M., Pulcini, G. (2015). Adding logic to the toolbox of molecular biology, *Eur. J. Philos. Sci.* **5**, 399–417.

[15] Bornkessel-Schlesewsky, I., Schlesewsky, M., Small, S. L., Rauschecker, J. P. (2015). Response to Skeide and Friederici: the myth of the uniquely human 'direct' dorsal pathway, *Trends Cogn. Sci.* **(19)**9, 484–485.

[16] Bray, D. (2009). *Wetware: A Computer in Every Living Cell* (Yale University Press, New Haven and London).

[17] Busemeyer, J. R., Bruza, P. D. (2012). *Quantum Models of Cognition and Decision* (Cambridge University Press, Cambridge).

[18] Caruso, V. C., Mohl, J. T., Glynn, C., Lee, J., Willett, S. M., Zaman, A., Ebihara, A. F., Estrada, R., Freiwald, W. A., Tokdar, S. T., Groh, J. M. (2018). Single neurons may encode simultaneous stimuli by switching between activity patterns, *Nat. Commun.* **9**, 2715.

[19] Chellas, B. F. (1980). *Modal Logic: An Introduction* (Cambridge University Press, Cambridge).

[20] Chevalley, C. (1956). *Fundamental Concepts of Algebra* (Academic Press, New York).

[21] Covington, M. A., He, C., Brown, C., Naçi, L., McClain, J. T., Fjordbak, B. S., Semple, J., Brown, J. (2005). Schizophrenia and the structure of language: The Linguist's view, *Schizophr. Res.* **77**, 85–98.

[22] Craver, C. F. (2007). *Explaining the Brain* (Oxford University Press, Oxford).

[23] Curry, H. B., Feys, R. (1968). *Combinatory Logic I, Second Edition: Studies in Logic and the Foundations of Mathematics* (North Holland Publishing Company, Amsterdam).

[24] Dalla Chiara, M. L., Giuntini, R., Battiloti, G., Faggian, C. (2002). Quantum Logics, *Handbook of Philosophical Logic Vol. 6, Second Edition* (Kluwer, Dordrecht).

[25] Davis, R. L., Yi Zhong. (2017). The biology of forgetting — a perspective, *Neuron* **95**(3), 490–503.

[26] DeLisi, L. E. (2001). Speech disorder in schizophrenia: review of the literature and exploration of its relation to the uniquely human capacity for language, *Schizophr. Bull.* **27**, 48–496.

[27] Dempsey, W. P., Du, Z., Natcochiy, A., Smith, D. K., Czajkowski, A. A., Robson, D. N., Li, J. M., Applebaum, S., Truong, T. V., Kesselman, C., Fraser, S. E., Arnold, D. B. (2022). Regional synapse gain and loss accompany memory formation in larval Zebrafish, *Proc. Natl. Acad. Sci.* **119**(3), e2107661119.

[28] Dishkant, H. (1977). Imbedding of the quantum logic in the modal system of Brower (sic), *J. Symb. Log.* **42**(3), 321–328.

[29] Fain, G. L., Fain, M. J., O'Dell, T. (2014). *Molecular and Cellular Physiology of Neurons, Second Edition* (Harvard University Press, Cambridge).

[30] Finkelstein, D. R. (1963). The logic of quantum physics, *Trans. N.Y. Acad. Sci.* **25**, 621–663.

[31] Finkelstein, D. R. (1996). *Quantum Relativity* (Springer–Verlag, Berlin, Heidelberg, New York).

[32] Fitch, W. T. (2018). Bio-linguistics: Monkeys break through the syntax barrier, *Curr. Biol.* **28**, R695–R717.

[33] Friederici, A. D. (2017). *Language in our Brain* (The MIT Press, Cambridge and London).

[34] Fromm, O., Klostermann, F., Ehlen, F. (2020). A vector space model for neural network functions: Inspirations from similarities between the theory of connectivity and the logarithmic time course of word production, *Front. Syst. Neurosci.* **14**, 58.

[35] Fulton, W., Harris, J. (1991). *Representation Theory: A First Course* (Springer–Verlag, Berlin, Heidelberg, New York).

[36] Gärdenfors, P. (2004). *Conceptual Spaces: The Geometry of Thought* (The MIT Press, Cambridge and London).

[37] Gardner, R. J., Hermansen, E., Pachitariu, M., Burak, Y., Baas, N. A., Dunn, B. A., Moser, M.-B., Moser, E. I. (2022). Toroidal topology of population activity in grid cells, *Nature* **602**(7895), 123–128.

[38] Girard, J.-Y., Lafont, Y., Taylor, P. (1988). *Proofs and Types* (Cambridge University Press, Cambridge).

[39] Goaillard, J.-M., Moubarak, E., Tapia, M., Tell, F. (2020). Diversity of axonal and dendritic contributions to neuronal output, *Front. Cell. Neurosci.* **13**, 570.

[40] Goldblatt, R. I. (1973). The Stone space of an ortholattice, *Bull. London Math. Soc.* **7**, 45–48.

[41] Goldblatt, R. I. (1974). Semantic analysis of orthologic. *J. Philos. Log.* **3**, 19–35.

[42] Gollo, L. L., Mirasso, C., Sporns, O., Breakspear, M. (2014). Mechanisms of zero-Lag synchronization in cortical motifs, *PLoS Comput. Biol.* **10**(4), e1003548.

[43] Gonzalez, W. G., Zhang, H., Harutyunyan, A., Lois, C. (2019). Persistence of neuronal representations through time and damage in the hippocampus. *Science* **365**(6455), 821–825.

[44] Hartle, J. B. (1968). Quantum mechanics of individual systems, *Am. J. Phys.* **36**(8), 704–712.

[45] Hartle, J. B. (2021). What do we learn by deriving Born's rule? arXiv:2107.02297v1 [quant-ph]

[46] Hopfield, J. J. (1982). Neural networks and physical systems with emergent collective computational abilities, *Proc. Natl. Acad. Sci.* **79**, 2554–2558.

[47] Hughes, G. E., Cresswell, M. J. (1968). *An Introduction to Modal Logic* (Methuen and Co., London, England).

[48] Hughes, G. E., Cresswell, M. J. (1984). *A Companion to Modal Logic* (Methuen and Co., London, England).

[49] Izhikevich, E. M. (2007). *Dynamical Systems in Neuroscience: The Geometry of Excitability and Bursting* (The MIT Press, Cambridge and London).

[50] Jacobson, N. (1962). *Lie Algebras* (John Wiley & Sons, New Jersey).

[51] Jennings, R. E. (1994). *The Genealogy of Disjunction* (Oxford University Press, Oxford).

[52] Jiruska, P., Csicsvari, J., Powell, A. D., Fox, J. E., Chang, W. C., Vreugdenhil, M., Li, X., Palus, M., Bujan, A. F., Richard, W., Dearden, R. W., Jefferys, J. G. R. (2010). High-frequency network activity, global increase in neuronal activity, and synchrony expansion precede epileptic seizures in vitro, *J. Neurosci.* **30**(16), 5690–5701.

[53] Johansen, J. P., Diaz-Mataix, L., H. Hamanaka, H., Ozawa, T., Ycu, E., Koivumaa, J., Kumar, A., Hou, M., Deisseroth, K., Boyden, E. S., LeDoux, J. E. (2014). Hebbian and modularity mechanisms interact to trigger associative memory formation, *Proc. Natl. Acad. Sci.* **111**(51), E5584–E5592.

[54] Jonides, J., Lewis, R. L, Nee, D. E., Lustig, C. A., Berman, M. G., Moore, K. S. (2008). The mind and brain of short-term memory, *Annu. Rev. Psychol.* **59**, 93–224.

[55] Jordan, P., Wigner, E. (1928). Über das paulische äquivalenzverbot, *Z. Phys.* **47**, 631.

[56] Kalmbach, G. (1983). *Orthomodular Lattices* (Academic Press, New York).

[57] Knapp, A. W. (1988). *Lie Groups, Lie Algebras, and Cohomology* (Princeton University Press, Princeton).

[58] Kutter, E. F., Boström, J., Elger, C., Nieder, A., Mormann, F. (2022). Neuronal codes for arithmetic rule processing in the human brain, *Curr. Biol.* **32**(6), 1275–1284.

[59] Lang, S. (1993). *Algebra, Third Edition* (Addison–Wesley, Reading).

[60] Li, M., Liu, J., Tsien, J. Z. (2016). Theory of connectivity: Nature and nurture of cell assemblies and cognitive computation, *Front. Neural Circuits* **10**, 34.

[61] Lin, L., Osan, R., Shoham, S., Jin, W., Zuo, W., Tsien, J. Z. (2005). Identification of network-level coding units for real-time representation of episodic experiences in the hippocampus, *Proc. Natl. Acad. Sci. USA* **102**, 6125–6130.

[62] Mac Lane, S. (1963). *Homology* (Springer–Verlag, Berlin, Heidelberg, New York).

[63] Jiang, M., Zhu, J., Liu, Y., Yang, M., Tian, C., Jiang, S., Wang, Y., Guo, H., Wang, K., Shu, Y. (2012). Enhancement of asynchronous release from fast-spiking interneuron in human and rat epileptic neocortex, *PLoS Biol.* **10**(5), e1001324.

[64] Maruhn, J. A., Reinhard, P. G., Suraud, E. (2010). *Simple Models of Many-Fermion Systems* (Springer–Verlag, Berlin, Heidelberg, New York).

[65] Matchin, W. (2018). A neuronal retuning hypothesis of sentence specificity in Broca's area, *Psychon. Bull. Rev.* **25**, 1682–1694.

[66] Matchin, W., Hickok, G. (2020). The cortical organization of syntax, *Cereb. Cortex* **30**, 1481–1498.

[67] Morrow, J., Mosher, C., Gothard, K. (2019). Multisensory neurons in the primate amygdala. *J. Neurosci.* **39**(19), 3663–3675.

[68] Mumford, D., Oda, T. (2015). *Algebruic Geometry II* (Hindustan Book Agency, New Delhi).

[69] Nielsen, M., Chuang, I. (2000). *Quantum Computation and Quantum Information* (Cambridge University Press, Cambridge).

[70] Ouattara, K., Lemasson, A., Zuberbühler, K. (2009). Campbell's monkeys concatenate vocalizations into context-specific call sequences, *Proc. Natl. Acad. Sci.* **106**(51), 22026–22031.

[71] Peterson, L. R., Peterson, M. J. (1959). Short-term retention of individual verbal items, *J. Exp. Psychol.* **58**(3), 193–198.

[72] Piccinini, G. (2020). *Neurocognitive Mechanisms: Explaining Biological Cognition* (Oxford University Press, Oxford).

[73] Poo, M., Pignatelli, M., Ryan, T. J., Tonegawa, S., Bonhoeffer, T., Martin, K. C., Rudenko, A., Tsai, L.-H., Tsien, R. W., Fishell, G., Mullins, C., Gonçalves, J. T., Shtraman, M., Johnston, S. T., Gage, F. H., Dan, Y., Long, J., Buzsáki, G., Stevens, C. (2016). What is memory? The present state of the engram, *BMC Biol.* **14**, 40.

[74] Price, T., Wadewitz, P., Cheney, D., Steyfarth, R., Hammerschmidt, K., Fischer, J. (2015). Vervets revisited: A quantitative analysis of alarm call structure and context specificity, *Sci. Rep.* **5**, 13220.

[75] Quiroga, R. Q. (2013). Gnostic cells in the 21st century, *Acta Neurobiol. Exp.* **73**(4), 463–471.

[76] Rawling, J. P., Selesnick, S. A. (2000). Orthologic and quantum logic: models and computational elements, *J. ACM* **47**(4), 721–751.

[77] Reimann, M. W., Nolte, M., Scolamiero, M., Turner, K., Perin, R., Chindemi, G., Dłotko, P., Levi, R., Hess, K., Markram, H. (2017). Cliques of neurons bound into cavities provide a missing link between structure and function, *Front. Comput. Neurosci.* **11**, 48.

[78] Rilling, J. K. (2014). Comparative primate neurobiology and the evolution of brain language systems, *Curr. Opin. Neurobiol.* **28**, 10–14.

[79] Sierpowska, J., Bryant, K. L., Janssen, N., Freches, G. B., Römkens, M., Mangnus, M., Mars, R. B., Piai, V. (2022). Comparing human and chimpanzee temporal lobe neuroanatomy reveals modifications to human language hubs beyond the frontotemporal arcuate fasciculus, *Proc. Natl. Acad. Sci. U.S.A.* **119**(28), e2118295119.

[80] Schwinger, J. (2001). *Quantum Mechanics: Symbolism of Atomic Measurements (edited by Englert, B.-G.)* (Springer–Verlag, Berlin, Heidelberg, New York).

[81] Selesnick, S. A. (2003). Foundation for quantum computing, *Int. J. Theor. Phys.* **42**(3), 383–426.

[82] Selesnick, S. A. (2003). *Quanta, Logic and Spacetime, Second Edition* (World Scientific Publishing, Singapore, London and Hong Kong).

[83] Selesnick, S. A., Owen, G. S. (2012). Quantum-like logics and schizophrenia, *J. Appl. Log.* **10**(1), 115–126.

[84] Selesnick, S. A., Piccinini, G. (2018). Quantum-like Behavior without Quantum Physics II. A quantum-like model of neural network dynamics, *J. Biol. Phys.* **44**, 501–538.

[85] Selesnick, S., Piccinini, G. (2019). Quantum-like Behavior without Quantum Physics III. Logic and memory, *J. Biol. Phys.* **45**, 335–366.

[86] Sestini, F., Crafa, S. (2018). Proof search in context-sensitive logic for molecular biology, *J. Log. Comput.* **28**(7), 1565–1600.

[87] Smalheiser, N. R. (2007). Exosomal transfer of proteins and RNAs at synapses in the nervous system, *Biol. Direct* **2**, 35.

[88] Smith, S. M. (1988). Environmental context-dependent memory, in *Memory in Context* (John Wiley & Sons, New Jersey).

[89] Smythies, J. (2015). Off the beaten track: the molecular structure of long-term memory: Three novel hypotheses — electrical, chemical and anatomical (allosteric), *Front. Integr. Neurosci.* **9**, 4.

[90] Sporns, O. (2011). *Networks of the Brain* (The MIT Press, Cambridge and London).

[91] Sportiche, D., Koopman, H., Stabler, E. (2014). *An Introduction to Syntactic Analysis and Theory* (John Wiley & Sons, New Jersey).

[92] Sterling, P., Laughlin, S. (2015). *Principles of Neural Design* (The MIT Press, Cambridge and London).

[93] Szkudlarek, E., Zhang, H., DeWind, N. K., Brannon, E. M. (2022). Young children intuitively divide before they recognize the division symbol, *Front. Hum. Neurosci.* **16**, 752190.

[94] Troelstra, A. S. (1991). *Lectures on Linear Logic* (CSLI, Stanford).

[95] Troelstra, A. S., Schwichtenberg, H. (2000). *Basic Proof Theory, Second Edition* (Cambridge University Press, Cambridge).

[96] Tsien, J. Z. (2015). A postulate on the brain's basic wiring logic, *Trends Neurosci.* **38**(11), 669–671.

[97] Tsien, J. Z. (2016). Principles of intelligence: On evolutionary logic of the brain, *Front. Syst. Neurosci.* **9**, 186.

[98] Uhlhaas, P. J., Pipa, G., Lima, B., Melloni, L., Neuenschwander, S., Nikolić, D., Singer, W. (2009). Neural synchrony in cortical networks: history, concept and current status, *Front. Integr. Neurosci.* **3**, 17.

[99] Vecoven, N., Ernst, D., Wehenkel, A., Drion, G. (2002). Introducing neuromodulation in deep neural networks to learn adaptive behaviours, *PLoS ONE* **15**(1), e0227922.

[100] Widdows, D. (2004). *Geometry and Meaning* (CSLI, Stanford).

[101] Xie, K., Fox, G. E., Liu, J., Lyu, C., Lee, J. C., Kuang, H., Jacobs, S., Li, M., Liu, T., Song, S., Tsien, J. Z. (2016). Brain computation is organized via power-of-two-based permutation logic, *Front. Syst. Neurosci.* **10**, 95.

[102] Yang, Y., Wang, J.-Z. (2017). From structure to behavior in basolateral amygdala-hippocampus circuits, *Front. Neural Circuits* **11**, 86.

[103] Ying, M., Yu, N., Feng, Y. (2014). Alternation in quantum programming: From superposition of data to superposition of programs, arXiv:1402.5172v1.

Index